AF361018

Symmetry and Complexity

Symmetry and Complexity

Editor

Carlo Cattani

MDPI • Basel • Beijing • Wuhan • Barcelona • Belgrade • Manchester • Tokyo • Cluj • Tianjin

Editor
Carlo Cattani
Engineering School (DEIM),
University of Tuscia,
Largo dell'Università
Italy

Editorial Office
MDPI
St. Alban-Anlage 66
4052 Basel, Switzerland

This is a reprint of articles from the Special Issue published online in the open access journal *Symmetry* (ISSN 2073-8994) (available at: https://www.mdpi.com/journal/symmetry/special_issues/symmetry_complexity).

For citation purposes, cite each article independently as indicated on the article page online and as indicated below:

LastName, A.A.; LastName, B.B.; LastName, C.C. Article Title. *Journal Name* **Year**, *Article Number*, Page Range.

ISBN 978-3-03936-846-4 (Hbk)
ISBN 978-3-03936-847-1 (PDF)

Contents

About the Editor

Carlo Cattani is a professor of Mathematical Physics and Applied Mathematics at the Enginering School (DEIM) of Tuscia University (VT)—Italy, since 2015. Previously, he was a professor at the Dept. of Mathematics, University of Rome "La Sapienza" (1980–2004) and the Department of Mathematics, University of Salerno (2004–2015). His scientific research interests focus on wavelets, dynamical systems, fractals, fractional and stochastic equations, computational and numerical methods, nonlinear analysis, the complexity of living systems, pattern analysis, data mining, and artificial intelligence.

He has been awarded as honorary professor at the Azerbaijan University Baku (2019) and (in 2009) at the BSP University, Ufa (Russia). Since 2018, he has also been an adjunct professor at the Ton Duc Thang University—HCMC Vietnam.

He has been a visiting professor at the Dep. De Matematica Aplicada, EUTII, Politecnico di Valencia (2002), East China University (Shanghai, 2007, 2009), BSP University (Ufa, 2008, 2010), and a research fellow at the Physics Institute of Stockholm University (1987–1988).

Preface to "Symmetry and Complexity"

Symmetry and complexity are the two fundamental features of almost all phenomena in nature and science. Any complex physical model is characterized by the existence of some symmetry groups at different scales. On the other hand, breaking the symmetry of a scientific model has always been considered as the most challenging direction for new discoveries. Modeling complexity has recently become an increasingly popular subject, with an impressive growth when it comes to applications. The main goal of modeling complexity is the search for hidden or broken symmetries.

Usually, complexity is modeled by dealing with big data or dynamical systems, depending on a large number of parameters. Nonlinear dynamical systems and chaotic dynamical systems are also used for modeling complexity. Complex models are often represented by un-smooth objects, non-differentiable objects, fractals, pseudo-random phenomena, and stochastic processes.

The discovery of complexity and symmetry in mathematics, physics, engineering, economics, biology and medicine have opened new challenging fields of research. Therefore, new mathematical tools were developed in order to obtain quantitative information from models, newly reformulated in terms of nonlinear differential equations.

This Special Issue focuses on the most recent advances in calculus, applied to dynamical problems, linear and nonlinear (fractional, stochastic) ordinary and partial differential equations, integral differential equations and stochastic integral problems, arising in all fields of science, engineering applications, and other applied fields dealing with complexity.

Carlo Cattani
Editor

Article

A Framework for Circular Multilevel Systems in the Frequency Domain

Guomin Sun [1,*,†], **Jinsong Leng** [1,†] **and Carlo Cattani** [2,*,†]

[1]   School of Mathematical Sciences, University of Electronic Science and Technology of China, Chengdu 610054, China; lengjs@uestc.edu.cn
[2]   Engineering School, DEIM, University of Tuscia, 01100 Viterbo, Italy
*   Correspondence: guominsun0120@gmail.com (G.S.); cattani@unitus.it (C.C.)
†   These authors contributed equally to this work.

Received: 9 February 2018; Accepted: 31 March 2018; Published: 8 April 2018

Abstract: In this paper, we will construct a new multilevel system in the Fourier domain using the harmonic wavelet. The main advantages of harmonic wavelet are that its frequency spectrum is confined exactly to an octave band, and its simple definition just as Haar wavelet. The constructed multilevel system has the circular shape, which forms a partition of the frequency domain by shifting and scaling the basic wavelet functions. To possess the circular shape, a new type of sampling grid, the circular-polar grid (*CPG*), is defined and also the corresponding modified Fourier transform. The *CPG* consists of equal space along rays, where different rays are equally angled. The main difference between the classic polar grid and *CPG* is the even sampling on polar coordinates. Another obvious difference is that the modified Fourier transform has a circular shape in the frequency domain while the polar transform has a square shape. The proposed sampling grid and the new defined Fourier transform constitute a completely Fourier transform system, more importantly, the harmonic wavelet based multilevel system defined on the proposed sampling grid is more suitable for the distribution of general images in the Fourier domain.

Keywords: harmonic wavelet; filtering; multilevel system

1. Introduction

Wavelet multiresolution representations are one of the effective techniques for analyzing signals and images. The wavelet multiresolution analysis (MRA) technology has been widely used in signal and image processing. It was first given by Mallat [1], and the authors study the difference of information between approximation of a signal at the resolution 2^{j+1} and 2^j, by decomposing this signal on a wavelet basis of $L^2(R)$. The 2D general MRA technique possesses a square shape in the frequency domain [2–4]. To design the filter of circular-shape in the Fourier domain, the classical polar Fourier transformation is considered. However, the classical polar Fourier transform retains the same shape as in the space domain, so new approaches are investigated. One way is to redefine the sampling grid in the Fourier domain. In [5], the authors introduce a pseudo-polar Fourier transform that samples the Fourier transform on the pseudo-polar grid, also known as the concentric squares grid. We will give more details in Section 3. In addition, [6] samples on points that are equally spaced on an arbitrary arc of the unit circle, which brings about the Fractional Fourier transform; and, in [7], the sampling is on spirals of the form AW^k, with $A, W \in C$. Using this type of sampling, the authors develops a computation algorithm for numerically evaluating the $z - transform$. Our goal is to obtain the sampling grid in a circular shape; therefore, we hope to design a new type of sampling that ensures the sampling points concentrated in a circular region. Then, the sampling grid has a circular shape in the Fourier domain. Inspired by the pseudo-polar Fourier transform in [5], we will also redefine the Fourier transform on circular sampling grid.

In recent years, many kinds of directional wavelets filters have been designed, in order to further efficiently capture the details of signals. The most widely used directional multilevel system includes curvelets [8], contourlets [9] and shearlets [10,11]. What these wavelets have in common is that they have compact support multiscale structure in the space domain. In the Fourier domain, the support of a multilevel system constitutes a high redundant partition. To reduce the redundancy, we consider designing the multilevel system in the frequency domain directly. We also must ensure that the multilevel system constructs the basis of $L^2(R)$ in the space domain.

To design the multilevel system in the Fourier domain, wavelets with compact support in frequency are needed. According to the definition of the harmonic wavelet [12–15], it is suitable to construct a directional multilevel structure with harmonic wavelets whose Fourier transforms are compact and are constructed from simple functions like Haar wavelets [16] in the space domain. We will review the basic definition and property of harmonic wavelet in Section 2.

In this work, by defining the circular-shape Fourier transform (CFT), we will construct the circular-shape directional multilevel system (CMS) in the Fourier domain due to the compact support of harmonic wavelets [12–15,17]. The specific structure is totally different from the general Cartesian system. By introducing the CFT, we plan to give a parallel analogy with the general classical Descartes Fourier transform, and the corresponding circular-shape directional multilevel system is constructed naturally, which is suitable for the circular shape of images in the Fourier domain. More details will be given in Section 4.

This paper is organized as follows: Section 2 reviews the basic definition and property of harmonic wavelets. The design of CFT is given in Section 3. Then, in Section 4, the multilevel system in the frequency domain based harmonic wavelet is constructed. The quantitative test measures and test results are displayed in Sections 5 and 6.

2. Preliminary

The Basic Definition

Harmonic wavelets are complex wavelets defined in the Fourier domain. It is consists of an even function $H_e(\omega)$ (see Figure 1a) as the real part and an odd function $H_o(\omega)$ (see Figure 1b) as the imaginary part, which are defined by

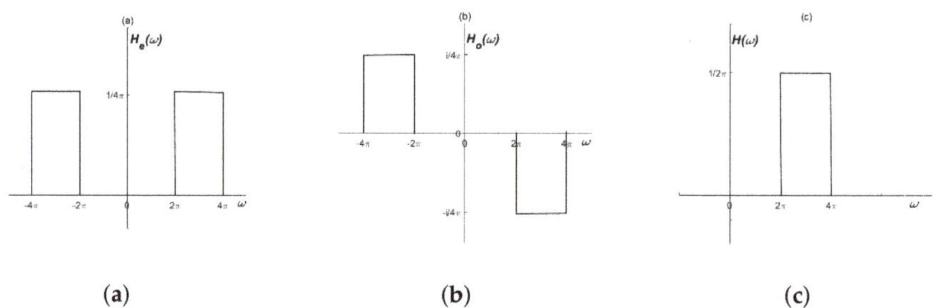

(a) **(b)** **(c)**

Figure 1. The harmonic wavelet function. (**a**) the even part $H_e(\omega)$; (**b**) the odd part $H_o(\omega)$; (**c**) the harmonic wavelet function $H(\omega)$.

$$H_e(\omega) = \begin{cases} 1/4\pi, & \omega \in [-4\pi, -2\pi) \cup [2\pi, 4\pi), \\ 0, & \text{otherwise.} \end{cases} \qquad H_o(\omega) = \begin{cases} i/4\pi, & \omega \in [-4\pi, -2\pi), \\ -i/4\pi, & \omega \in [2\pi, 4\pi), \\ 0, & \text{otherwise.} \end{cases} \qquad (1)$$

Combining H_e and H_o, we get the harmonic function

$$H(\omega) = H_e(\omega) + iH_o(\omega). \tag{2}$$

From Label (1), we have

$$H(\omega) = \begin{cases} 1/2\pi, & \omega \in [2\pi, 4\pi), \\ 0, & \text{otherwise.} \end{cases} \tag{3}$$

This is shown in Figure 1c.

The corresponding scaling function S is given in the same way, and the even and odd functions are defined as

$$S_e(\omega) = \begin{cases} 1/4\pi, & \omega \in [-2\pi, 2\pi), \\ 0, & \text{otherwise.} \end{cases} \qquad S_o(\omega) = \begin{cases} i/4\pi, & \omega \in [-2\pi, 0), \\ -i/4\pi, & \omega \in [0, 2\pi), \\ 0, & \text{otherwise.} \end{cases} \tag{4}$$

so that, from Label (4),

$$S(\omega) = S_e(\omega) + iS_o(\omega). \tag{5}$$

Therefore, we have

$$S(\omega) = \begin{cases} 1/2\pi, & \omega \in [0, 2\pi), \\ 0, & \text{otherwise,} \end{cases} \tag{6}$$

shown in Figure 2.

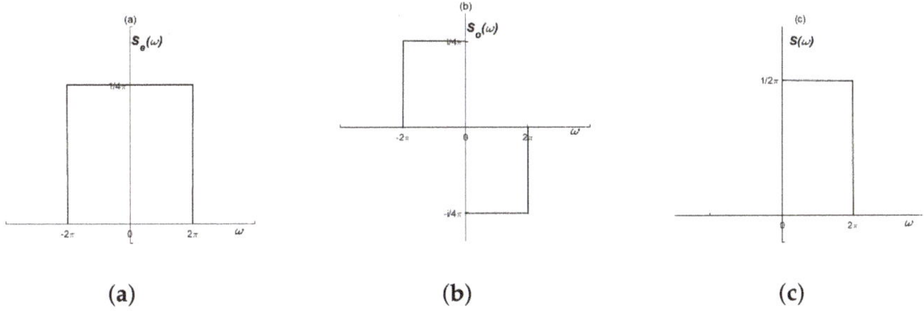

Figure 2. The harmonic scaling function. (**a**) the even part $S_e(\omega)$; (**b**) the odd part $S_o(\omega)$; (**c**) the scaling function $S(\omega)$.

Then, the shifting and scaling of basic functions are denoted as $S_{j,\ell}(\omega)$ and $H_{j,\ell}(\omega)$, which are given as

$$\begin{aligned} S_{j,\ell}(\omega) &= 1/2^j S(\omega/2^j - \ell), \\ H_{j,\ell}(\omega) &= 1/2^j H(\omega/2^j - \ell), \end{aligned} \tag{7}$$

where $j \in Z$ is the scaling parameter, and $\ell \in R$ is the shifting parameter. According to Label (7), the harmonic wavelet system constructs a basis of $L^2(R)$ in the frequency domain; then, for $f \in L^2(R)$, we have

$$f(\omega) = \sum_{j=-\infty}^{+\infty} \sum_{\ell=-\infty}^{+\infty} a_{j,\ell} H(2^j \omega - \ell), \tag{8}$$

$$f(\omega) = \sum_{\ell=-\infty}^{+\infty} a_\ell S(\omega - \ell) + \sum_{j=0}^{+\infty} \sum_{\ell=-\infty}^{+\infty} a_{j,\ell} H(2^j \omega - \ell), \tag{9}$$

where

$$a_\ell = \int_{-\infty}^{+\infty} f(\omega)S(x-\ell)dx, \quad a_{j,\ell} = \int_{-\infty}^{+\infty} f(\omega)H(2^j x - \ell)dx. \tag{10}$$

3. The Circular-Shape Fourier Transform (CFT)

This section describes the circular-shape Fourier transform (CFT). We begin with the sampling grid in the Fourier domain, including the Cartesian coordinates (see Figure 3a) for classical Fourier transform and the pseudo-polar grid in [5] (see Figure 3b). This grid samples points of equally spaced long rays but not equally angles. In order to have a circular structure, the sampling grid in concentric circles (see Figure 3c) is designed, which has equally arc and angle in each circle. This type of sampling is consistent with the distribution of images in the frequency domain.

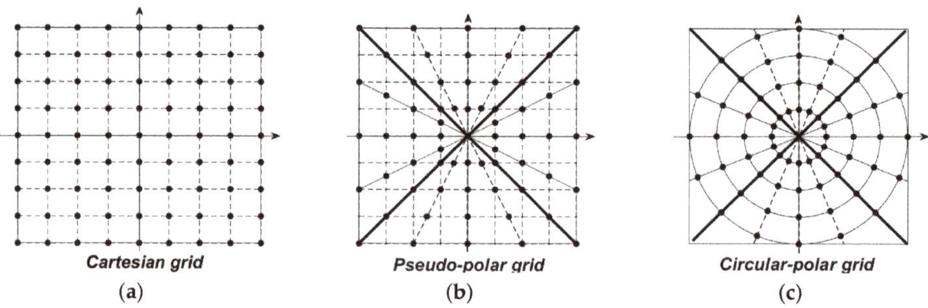

Figure 3. Three different grids.

3.1. The Pseudo-Polar Grid

The pseudo-polar grid Ω_R is given as

$$\Omega_R = \Omega_R^1 \cup \Omega_R^2, \tag{11}$$

where

$$\Omega_R^1 = \{(\frac{-4\ell k}{RN}, \frac{2k}{R}) : |\ell| \le N/2, |k| \le RN/2\},$$
$$\Omega_R^2 = \{(\frac{2k}{R}, -\frac{4\ell k}{RN}) : |\ell| \le N/2, |k| \le RN/2\}, \tag{12}$$

with $R = 2$ the oversampling parameter. The nod of Ω_R^1 is on the *solid line* in Figure 3b and the nod of Ω_R^2 is on *dotted line*.

For an $N \times N$ image u, the general discrete Fourier transform $\hat{u}$ is evaluated on the $N \times N$ Cartesian grid in the form

$$\hat{u}(\omega_x, \omega_y) = \sum_{x,y=-N/2}^{N/2-1} u(x,y)e^{-\frac{2\pi i}{N}(x\omega_x + y\omega_y)}, \tag{13}$$

where $\{(\omega_x, \omega_y) : \omega_x, \omega_y = -N/2, ..., N/2.\}$, and

$$\sum_{x,y=-N/2}^{N/2-1} |u(x,y)|^2 = \frac{1}{N^2} \sum_{\omega_x,\omega_y=-N/2}^{N/2-1} |\hat{u}(\omega_x, \omega_y)|^2. \tag{14}$$

Analogously, the pseudo-polar Fourier transform is the same as (13) (see [5]), but $\{(\omega_x, \omega_y) \in \Omega_R.\}$. According to the Plancherel theorem, (14) can be modified by introducing the weighting function w

$$\sum_{x,y=-N/2}^{N/2-1} |u(x,y)|^2 = \sum_{(\omega_x,\omega_y)\in\Omega_R} w(\omega_x,\omega_y)|\hat{u}(\omega_x,\omega_y)|^2. \tag{15}$$

3.2. The Circular-Polar Grid (CPG)

In this section, the circular-polar grid (CPG) is designed (see Figure 3c), which is defined as

$$C_R = C_R^0 \cup C_R^\sharp; \tag{16}$$

where $C_R^0 = \{(0,0)\}$ and $C_R^\sharp = C_R^1 \cup C_R^2$

$$
\begin{aligned}
C_R^1 &= \{(r\cos(\frac{\ell\pi}{m_0}), r\sin(\frac{\ell\pi}{m_0})) : 1 \le |r| \le R, |\ell| \le \frac{m_0}{2}\}, \\
C_R^2 &= \{(r\sin(\frac{\ell\pi}{m_0}), r\cos(\frac{\ell\pi}{m_0})) : 1 \le |r| \le R, |\ell| \le \frac{m_0}{2}\},
\end{aligned}
\tag{17}
$$

where m_0 is the sampling number in each circle.

As can be seen from Figure 3c, the nod of C_R^1 is on a *solid line* and the nod on a *dotted line* belongs to C_R^2. In addition, r in (17) serves as the radius and ℓ serves as the parameter of angle. $m_0 = 16$ is the sampling number. In the *CPG* coordinates, the nod has the following characteristics, for

$$C_R^1(\omega_x, \omega_y) = (r^1, \theta^1), \quad C_R^2(\omega_x, \omega_y) = (r^2, \theta^2), \tag{18}$$

where

$$
\begin{aligned}
r^1 &= k_1, \quad r^2 = k_2, \\
\theta^1 &= \ell_1\pi/m_0; \quad \theta^2 = \ell_2\pi/m_0.
\end{aligned}
\tag{19}
$$

$k_i = 0, ...R; i = 1,2$ and $\ell_i = -m_0/2, ...m_0/2; i = 1,2$. For each fixed angle θ, the samples of the *CPG* are equally spaced in the radial direction, and, for each fixed radius r, the grid possesses the same angle. Formally,

$$
\begin{aligned}
\Delta r^1 &\triangleq (k_1 + 1) - k_1 = 1; \quad \Delta r^2 \triangleq (k_2 + 1) - k_2 = 1, \\
\Delta\theta^1 &\triangleq (\ell_1 + 1)\pi/m_0 - \ell_1\pi/m_0 = \pi/m_0, \\
\Delta\theta^2 &\triangleq (\ell_2 + 1)\pi/m_0 - \ell_2\pi/m_0 = \pi/m_0,
\end{aligned}
\tag{20}
$$

where r^1, r^2 and θ^1, θ^2 are given by (19).

For an $N \times N$ image u, the *CFT* of $\hat{u}$ on *CPG* holds

$$\sum_{x,y=-N/2}^{N/2-1} |u(x,y)|^2 = \sum_{(\omega_x,\omega_y)\in C_R} w_c(\omega_x,\omega_y)|\hat{u}(\omega_x,\omega_y)|^2, \tag{21}$$

and

$$\hat{u}_{C_R}(\omega_x, \omega_y) = \sum_{x,y=-N/2}^{N/2-1} u(x,y)e^{-\frac{2\pi i}{Rm_0+1}(x\omega_x + y\omega_y)}. \tag{22}$$

Using operator notation, we denote the refined *CFT* of an image u as $\mathcal{F}_p$, where

$$(\mathcal{F}_p u)(r, \ell) \triangleq \hat{u}_{C_R}(r, \ell), \tag{23}$$

with $r = -R, ..., R$, $\ell = -m_0/2, ..., m_0/2$. Now, our goal is to choose weight w_c, such that w_c satisfies (21), and we have

$$
\sum_{(\omega_x,\omega_y)\in C_R} w_c(\omega_x,\omega_y)|\hat{u}(\omega_x,\omega_y)|^2
$$

$$
= \sum_{(\omega_x,\omega_y)\in C_R} w_c(\omega_x,\omega_y)| \sum_{x,y=-N/2}^{N/2-1} u(x,y)E(x,y)|^2
$$

$$
= \sum_{(\omega_x,\omega_y)\in C_R} w_c(\omega_x,\omega_y)[\sum_{x,y=-N/2}^{N/2-1} \sum_{x',y'=-N/2}^{N/2-1} u(x,y)E(x,y)\overline{u(x',y')E(x',y')}] \tag{24}
$$

$$
= \sum_{(\omega_x,\omega_y)\in C_R} w_c(\omega_x,\omega_y) \sum_{x,y=-N/2}^{N/2-1} |u(x,y)|^2
$$

$$
+ \sum_{(x,y)\neq(x',y')} u(x,y)\overline{u(x',y')}[\sum_{(\omega_x,\omega_y)\in C_R} w_c(\omega_x,\omega_y)E(x,y)\overline{E(x',y')}],
$$

where $E(x,y) \triangleq e^{-\frac{2\pi i}{Rm_0}(x\omega_x+y\omega_y)}$. Compared with the left of equation (21), the weights w_c holds

$$
\sum_{(\omega_x,\omega_y)\in C_R} w_c(\omega_x,\omega_y)e^{-\frac{2\pi i}{Rm+1}(x\omega_x+y\omega_y)} = \delta(x,y); \tag{25}
$$

with $-N/2 \leq x,y \leq N/2-1$.

3.3. The Choice of Weights w_c

In the following, we present the basic condition of weights w_c, according to (25), which satisfies that

$$
0 = \sum_{(\omega_x,\omega_y)\in C_R^\sharp} w_c(\omega_x,\omega_y)[\cos(\frac{2\pi}{Rm_0+1}x\omega_x)\cos(\frac{2\pi}{Rm_0+1}y\omega_y)
$$

$$
- \sin(\frac{2\pi}{Rm_0+1}x\omega_x)\sin(\frac{2\pi}{Rm_0+1}y\omega_y)];
$$

$$
0 = \sum_{(\omega_x,\omega_y)\in C_R^\sharp} w_c(\omega_x,\omega_y)[\sin(\frac{2\pi}{Rm_0+1}x\omega_x)\cos(\frac{2\pi}{Rm_0+1}y\omega_y) \tag{26}
$$

$$
+ \cos(\frac{2\pi}{Rm_0+1}x\omega_x)\sin(\frac{2\pi}{Rm_0+1}y\omega_y)].
$$

According to the symmetry of the *CFT*, the weighting function w_c is assumed to satisfy

$$
\begin{aligned}
w_c(\omega_x,\omega_y) &= w_c(\omega_y,\omega_x), (\omega_x,\omega_y) \in C_R, \\
w_c(\omega_x,\omega_y) &= w_c(\omega_y,-\omega_x), (\omega_x,\omega_y) \in C_R, \\
w_c(\omega_x,\omega_y) &= w_c(-\omega_y,-\omega_x), (\omega_x,\omega_y) \in C_R,
\end{aligned} \tag{27}
$$

where four equations of (27) describe the $(\omega_y = \omega_x)$-symmetry, $(\omega_y = -\omega_x)$-symmetry and the *origin*-symmetry.

In addition,

$$
\sum_{(\omega_x,\omega_y)\in C_R} w_c(\omega_x,\omega_y) = 1. \tag{28}
$$

To avoid high complexity, we choose the weight $w(\omega_x,\omega_y)$ in the form:

$$
w(\omega_x,\omega_y) = \frac{w_0(\omega_x,\omega_y)}{\sum_{\omega_x,\omega_y} w_0(\omega_x,\omega_y)}, \tag{29}
$$

where

$$w_0(\omega_x, \omega_y) = \frac{1}{m_0 r}, \{(\omega_x, \omega_y) : \omega_x^2 + \omega_y^2 = r^2\}, \tag{30}$$

with $r \in [1, R]$, and $w(0,0) = 1$.

4. The Construction of Multilevel System in Frequency Domain

In this section, we construct a new type multilevel system on CPG in the frequency domain.

4.1. 2D Basic Harmonic Function

First, we define the 2D basic harmonic wavelet functions in the Fourier domain. For deriving convenience, the wavelet function H and scaling function S can be normalized in the form given by **Definition 1**.

Definition 1. *The 2D harmonic basic functions are defined as*

$$\begin{aligned}
H(r,\theta) &:\triangleq H(2\pi|r|\cos(|\theta|))S(2\pi|r|\sin(|\theta|)), \\
S(r,\theta) &:\triangleq S(2\pi|r|\cos(|\theta|))S(2\pi|r|\sin(|\theta|)),
\end{aligned} \tag{31}$$

with $r \in [-R, R]$, $\theta \in [0, \pi]$.

Then, the support of $H(r,\theta)$ and $S(r,\theta)$ are investigated, according to (3) and (6),

$$H(2\pi|r|\cos(|\theta|)) \neq 0, \quad S(2\pi|r|\sin(|\theta|)) \neq 0 \tag{32}$$

hold simultaneously; therefore,

$$1 \leq |r|\cos(|\theta|) \leq 2, \quad 0 \leq |r|\sin(|\theta|) \leq 1. \tag{33}$$

Thus, the support of $H(r,\theta)$ is given as

$$\sqrt{2} \leq |r| \leq 2, \quad |\theta| \leq \pi/4. \tag{34}$$

Similarly,

$$\text{supp}S(r,\theta) = \{(r,\theta) : 0 \leq |r| \leq 1, |\theta| \leq \pi/4\}. \tag{35}$$

Next, the 2D scaling and shifting of $H(r,\theta)$ and $S(r,\theta)$ are defined.

Definition 2. *The 2D scaling and shifting of harmonic basic functions in the frequency domain are defined as*

$$\begin{aligned}
H_{j,\ell}(r,\theta) &:\triangleq H(2\pi 2^{-j}|r|\cos(2^j|\theta - \ell|))S(2\pi 2^{-j}|r|\sin(2^j|\theta - \ell|)), \\
S_{j,\ell}(r,\theta) &:\triangleq S(2\pi 2^{-j}|r|\cos(2^j|\theta - \ell|))S(2\pi 2^{-j}|r|\sin(2^j|\theta - \ell|)), \\
H_{j,\ell}^*(r,\theta) &:\triangleq H(2\pi 2^{-j}|r|\sin(2^j|\theta - \ell|))S(2\pi 2^{-j}|r|\cos(2^j|\theta - \ell|)), \\
S_{j,\ell}^*(r,\theta) &:\triangleq S(2\pi 2^{-j}|r|\sin(2^j|\theta - \ell| - \frac{\pi}{2}))S(2\pi 2^{-j}|r|\sin(2^j|\theta - \ell|)),
\end{aligned} \tag{36}$$

with $j, \ell \in R$.

4.2. The Polar Harmonic Multilevel System in the Frequency Domain (PHMS) on CPG

In this section, we give the definition of the polar harmonic multilevel system (PHMS) defined on CPG.

Definition 3. *The 2D PHMS on CPG is defined as*

$$PHMS :\triangleq H_{j,\ell_j}(r,\theta) \cup S_{j,\ell_j}(r,\theta) \cup H^*_{j,\ell_j}(r,\theta) \cup S^*_{j,\ell_j}(r,\theta),\tag{37}$$

where $H_{j,\ell}$, $H^*_{j,\ell}$, $S_{j,\ell}$ and $S^*_{j,\ell}$ are given in (36). The level parameter $j \leq [\log_2 R]$, the shifting parameter is related to j, and we defined the $\ell_j = \ell 2^{-j}\frac{\pi}{4}$ with $|\ell| \leq 2^j$, $\ell \in Z$.

From $|\ell| \leq 2^j$, we have $2(2^{j+1} + 1)$ subbands in each level j, in order to reduce the overlap, we choose $2(2^{j+1})$ subbands; then, the *PHMS* constructs a partition of the Fourier domain. We displayed the *PHMS* structure in Figure 4.

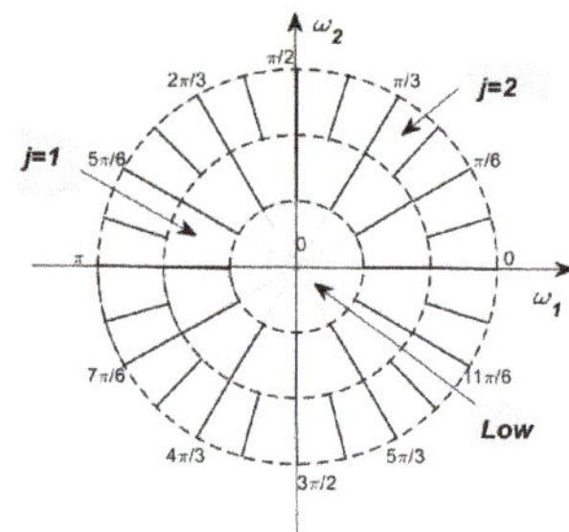

Figure 4. The polar harmonic multilevel system (PHMS) in the Fourier domain $j \leq 2$, $\ell_j \in Z$ and $-2^j \leq \ell_j < 2^j$.

For a signal or image u, the corresponding *PHMS* transform $\mathcal{P}(u)$ in the frequency domain can be defined as

$$\mathcal{P}(u) \triangleq <\hat{u}_{C_R}, PHMS> = \sum_{j=0}^{J} \sum_{\ell_j=-\ell(2^{-j}\frac{\pi}{4})}^{\ell(2^{-j}\frac{\pi}{4})} \sum_{\ell=-2^j}^{2^j} (H_{j,\ell_j}\cdot * \hat{u}_{C_R} + S_{j,\ell_j}\cdot * \hat{u}_{C_R}$$
$$+ H^*_{j,\ell_j}\cdot * \hat{u}_{C_R} + S^*_{j,\ell_j}\cdot * \hat{u}_{C_R}),\tag{38}$$

with $\ell_j = \ell 2^{-j}\frac{\pi}{4}$, $|\ell| \leq 2^j$, $\ell \in Z$, where $\hat{u}_{C_R}$ is the *CPFT* of u, defined in (22). In addition, $'.*'$ is the dot product, and the matrix $M_1.* M_2$ is defined as

$$(M_1.* M_2)_{i,j} = (M_1)_{i,j}(M_2)_{i,j}.\tag{39}$$

Theorem 1. *The discrete polar harmonic multilevel system* PHMS *defined on CPG forms a framelet of $L^2(R^2)$.*

According to the framelet defined in [18], for the signal U in (38),

$$\|U\|^2 \leq \| <U, PHMS> \|^2 \leq c\|U\|^2,\tag{40}$$

where $c < +\infty$ is the constant; therefore, *PHMS* forms a framelet of $L^2(R^2)$.

Then two denoising reconstruction tests of *PHMS* are shown in Figure 5.

Figure 5. Recovery results by *PHMS* with scale $j = 4$, the random noise level $\sigma = 20$.

5. Quantitative Test Measures

In the following, several performance measures are introduced to test the quality of the *PHMS*. The quality measure is the Monte Carlo estimate for the different operator norm by generating a sequence of five random images $u_i, i = 1, ..., 5$ on *CPG* for $R = 256, \ell = 8$ with standard normally distributed entries.

1. *Isometry of CFT*:

   (a) Closeness to tight: $M_{clo} = \max_{i=1,...,5} \dfrac{\|\mathcal{F}_p^* \mathcal{F}_p u_i - u_i\|_2}{\|u_i\|_2}$,

   (b) Quality of preconditioning. $M_{qua} = \dfrac{\lambda_{\max}(\mathcal{F}_p^* \mathcal{F}_p)}{\lambda_{\min}(\mathcal{F}_p^* w \mathcal{F}_p)}$.

2. *Tight Frame Property*: The operator norm $\|\mathcal{P}^*\mathcal{P} - I\|_{op}$, which is defined as $M_{tig} = \max_{i=1,...,5} \dfrac{\|\mathcal{P}^*\mathcal{P} u_i - u_i\|_2}{\|u_i\|_2}$.

3. *Robustness*:

   (a) Thresholding: Let u be the regular sampling of a Gaussian function with mean 0 and variance 512 on $[257]^2$ generating an 512×512 image. Two types of robustness are considered, for $k = 1, 2$, and $M_{p_k} = \dfrac{\|\mathcal{P}^* T_{p_k} \mathcal{P} u - u\|_2}{\|u\|_2}$.

   T_{p_1}: T_{p_1} discards $100(1 - 2^{-p_1})$ percent of coefficient, with $p_1 = [2 : 2 : 20]$.

   T_{p_2}: T_{p_2} keeps the absolute value of coefficients bigger than $m/2^{p_2}$ with m is the maximal absolute value of all coefficients, where $p_2 = [0.5 : 0.5 : 5]$.

   (b) Quantization: The quality measure is given as $M_p = \dfrac{\|\mathcal{P}^* Q_q \mathcal{P} u - u\|_2}{\|u\|_2}$, where $Q_q(c) = round(c/(m/2^q)) \cdot (m/2^q)$, and $q \in [5 : -0.5 : 0.5]$.

6. Test Results

In this section, the test results of *PHMS* on *CPG* for quantitative measure in (1)–(3) are shown. First the performance with respect to quantitative measures in (1), (2) are presented in Table 1.

Table 1. Results of Labels (1)–(2).

M_{clo}	M_{qua}	M_{tig}
0.00946092	1.9342627	0.1036731

From Table 1, the error of tightness M_{tig} for the $PHMS$ transform is about 0.1, which confirms that the multilevel system is indeed not a tight frame. The main reason is the redundancy of the multilevel structure due to the radius in (36). The quantity $M_{clo} \approx 0.0095$ and $M_{qua} \approx 1.934$ suggests that the circular-polar Fourier transform provides good properties in terms of isometry, which allows us to employ the conjugate gradient method to compute the inverse of $\mathcal{F}_p$. The weight w should also be chosen carefully.

Second, the robustness measurements in (3) displayed in Table 2:

Table 2. Results of Label (3).

M_{p_1}	3.6×10^{-6}	1.7×10^{-5}	5.9×10^{-3}	2.1×10^{-2}	0.9×10^{-2}
M_{p_2}	0.009	0.063	0.103	0.172	0.197
M_p	0.051	0.065	0.083	0.126	0.143

Table 2 shows the robustness of $PHMS$. Even discarding $100(1 - 2^{-10}) \approx 99.9\%$ of the coefficients, the image still can be recovered with error $M_{P_1} = 0.9 \times 10^{-2}$. The second row suggests that just the coefficient greater than thresholding value $m(1 - 1/2^{0.001}) \approx 0.1\%$ can give a good reconstruction with $M_{p_2} = 0.009$. In the third row, the quantization of robustness M_p is displayed.

7. Conclusions

In this work, we developed and implemented a polar harmonic multilevel system on the circular-polar grid based on the multiscale theory, and testified to the performance of the $PHMS$ in four different quantitative measures, which suggests that the $PHMS$ transform is suitable for the circular-shape Fourier transform. Another advantage of the $PHMS$ is that the circular low frequency region in the frequency domain is consistent with the image spectrum distribution, which can process the multilevel structure more effectively.

Acknowledgments: The first author Guomin Sun is very grateful to the China Scholarship Council for funding the first author visiting the University of Tuscia. This work was supported by the National Natural Science Foundation of China under Grant 11271001, Grant 61370147, and Grant 61573085.

Author Contributions: Guomin Sun and Carlo Cattani designed the circular multilevel system and performed the experiments, Jinsong Leng analyzed the results. Guomin Sun and Carlo Cattani wrote the paper. All authors have read and approved the final manuscript.

Conflicts of Interest: The authors declare no conflict of interest.

Abbreviations

The following abbreviations are used in this manuscript:

CPG	Circular-polar grid
MRA	Multiresolution analysis
CFT	Circular-shape Fourier transform
CMS	Circular-shape directional multilevel system
PHMS	Polar harmonic multilevel system

References

1. Mallat, S.G. A Theory for Multiresolution Signal Decomposition: The Wavelet Representation. *IEEE Trans. Pattern Anal. Mach. Intell.* **1989**, *11*, 674–693.
2. Leng, J.; Huang, T.; Jing, Y.; Jiang, W. A Study on Conjugate Quadrature Filters. *EURASIP J. Adv. Signal Process.* **2011**, *2011*, 2317–2329.
3. Leng, J.; Cheng, Z.; Huang, T.; Lai, C. Construction and properties of multiwavelet packets with arbitrary scale and the related algorithms of decomposition and reconstruction. *Comput. Math. Appl.* **2006**, *51*, 1663–1676.
4. Leng, J; Huang, T; Fu, Y. Construction of bivariate nonseparable compactly supported biorthogonal wavelets. In proceedings of the International Conference on Machine Learning and Cybernetics IEEE, Kunming, China, 12–15 July 2008.
5. Averbuch, A.; Coifman, R.R.; Donoho, D.L.; Israeli, M.; Shkolnisky, Y. A Framework for Discrete Integral Transformations I-The Pseudopolar Fourier Transform. *Siam J. Sci. Comput.* **2007**, *30*, 764–784.
6. Bailey, D.H.; Swarztrauber, P.N. The Fractional Fourier Transform and Applications. *SIAM Rev.* **1991**, *33*, 389–404.
7. Rabiner, L.R. Chirp Z-transform algorithm. *IEEE Trans. Audio Electroacoust.* **1969**, *17*, 86–92.
8. Candes, E.J.; Donoho, D.L. New tight frames of curvelets and optimal representations of objects with smooth singularities. *Commun. Pure Appl. Math.* **2004**, *57*, 219–266.
9. Do, M.N.; Vetterli, M. The contourlet transform: an efficient directional multiresolution image representation. *IEEE Trans. Image Process.* **2005**, *14*, 2091–2106.
10. Kutyniok, G.; Shahram, M.; Zhuang, X. ShearLab: A Rational Design of a Digital Parabolic Scaling Algorithm. *Siam J. Imaging Sci.* **2011**, *5*, 1291–1332.
11. Kutyniok, G. Shearlets. In *Applied and Numerical Harmonic Analysis*; Springer: Birkhauser Basel, Berlin, Germany, 2012.
12. Newland, D.E. Harmonic Wavelet Analysis. *Proc. R. Soc. Math. Phys. Eng. Sci.* **1993**, *443*, 203–225.
13. Cattani, C. Harmonic wavelets towards solution of nonlinear PDE. *Comput. Math. Appl.* **2005**, *50*, 1191–1210.
14. Cattani, C. Fractals Based on Harmonic Wavelets. In Proceedings of the International Conference on Computational Science and Its Applications, Seoul, South Korea, 29 June–2 July 2009; Springer: Berlin, Heidelberg, 2009; pp. 729–744.
15. Cattani, C. Harmonic wavelet approximation of random, fractal and high frequency signals. *Telecommun. Syst.* **2010**, *43*, 207–217.
16. Haar, A. Zur Theorie der orthogonalen Funktionensysteme. *Math. Ann.* **1911**, *71*, 38–53.
17. Duan, X.; Leng, J.; Cattani, C.; Li, C. A Shannon-Runge-Kutta-Gill Method for Convection-Diffusion Equations. *Math. Probl. Eng.* **2013**, *46*, 532–546.
18. Cai, J.; Ji, H.; Liu, C.; Shen, Z. Framelet. *IEEE Trans. Image Process.* **2012**, *21*, 562–572.

Article

Three Classes of Fractional Oscillators

Ming Li [1,2]

[1] Shanghai Key Laboratory of Multidimensional Information Processing, No. 500, Dong-Chuan Road, East China Normal University, Shanghai 200241, China; ming_lihk@yahoo.com or mli@ee.ecnu.edu.cn
[2] School of Information Science and Technology, East China Normal University, Shanghai 200241, China

Received: 2 November 2017; Accepted: 18 December 2017; Published: 30 January 2018

Abstract: This article addresses three classes of fractional oscillators named Class I, II and III. It is known that the solutions to fractional oscillators of Class I type are represented by the Mittag-Leffler functions. However, closed form solutions to fractional oscillators in Classes II and III are unknown. In this article, we present a theory of equivalent systems with respect to three classes of fractional oscillators. In methodology, we first transform fractional oscillators with constant coefficients to be linear 2-order oscillators with variable coefficients (variable mass and damping). Then, we derive the closed form solutions to three classes of fractional oscillators using elementary functions. The present theory of equivalent oscillators consists of the main highlights as follows. (1) Proposing three equivalent 2-order oscillation equations corresponding to three classes of fractional oscillators; (2) Presenting the closed form expressions of equivalent mass, equivalent damping, equivalent natural frequencies, equivalent damping ratio for each class of fractional oscillators; (3) Putting forward the closed form formulas of responses (free, impulse, unit step, frequency, sinusoidal) to each class of fractional oscillators; (4) Revealing the power laws of equivalent mass and equivalent damping for each class of fractional oscillators in terms of oscillation frequency; (5) Giving analytic expressions of the logarithmic decrements of three classes of fractional oscillators; (6) Representing the closed form representations of some of the generalized Mittag-Leffler functions with elementary functions. The present results suggest a novel theory of fractional oscillators. This may facilitate the application of the theory of fractional oscillators to practice.

Keywords: fractional differential equations; fractional oscillations (vibrations); fractional dynamical systems; nonlinear dynamical systems

---

1. Introduction

Any systems that consist of three elements, namely, inertia, restoration, and damping, may oscillate. Therefore, oscillations are common phenomena encountered in various fields, ranging from physics to mechanical engineering, see, e.g., [1–17].

Fractional oscillators and their processes attract the interests of researchers, see, e.g., [18–53]. There are problems worth studying with respect to fractional oscillators. On the one hand, the analytical expressions in the closed forms of responses to certain fractional oscillators, e.g., those described by (42) and (43) in Section 2, remain unknown. In addition, closed form representations of some physical quantities in fractional oscillators, such as mass, damping, natural frequencies, in the intrinsic sense, are lacking. On the other hand, technology and analysis methods, based on 2-order linear oscillations, almost dominate the preference of engineers although nonlinear oscillations have been paid attention to. Therefore, from a view of engineering, it is meaningful to establish a theory to deal with fractional oscillators with equivalent linear oscillation systems of order 2. This article contributes my results in this aspect.

This research studies three classes of fractional oscillators.

Class I: The first class contains oscillators with fractional inertia force $m\frac{d^{\alpha}x(t)}{dt^{\alpha}}(1 < \alpha \le 2)$ only. Its oscillation equation is in the form of (31), see, e.g., Duan ([24], Equation (3)), Mainardi ([25], Equation (27)), Zurigat ([26], Equation (16)), Blaszczyk and Ciesielski ([27], Equation (1)), Blaszczyk et al. ([28], Equation (10)), Al-rabtah et al. ([29], Equation (3.1)), Drozdov ([30], Equation (9)), Stanislavsky [31], Achar et al. ([32], Equation (1), [33], Equation (9), [34], Equation (2)), Tofighi ([35], Equation (2)), Ryabov and Puzenko ([36], Equation (1)), Ahmad and Elwakil ([37], Equation (1)), Uchaikin ([38], Chapter 7), Duan et al. ([39], Equation (4.2)).

Class II: The second consists of oscillators only with fractional damping term $c\frac{d^{\beta}x(t)}{dt^{\beta}}(0 < \beta \le 1)$, see, e.g., Lin et al. ([40], Equation (2)), Duan ([41], Equation (31)), Alkhaldi et al. ([42], Equation (1a)), Dai et al. ([43], Equation (1)], Ren et al. ([44], Equation (1)), Xu et al. ([45], Equation (1)), He et al. ([46], Equation (4)), Leung et al. ([47], Equation (2)), Chen et al. ([48], Equation (1)), Deü and Matignon ([49], Equation (1)), Drăgănescu et al. ([50], Equation (4)), Rossikhin and Shitikova ([51], Equation (3)), Xie and Lin ([52], Equation (1)), Chung and Jung [53]. That takes the form of (42) in the next Section.

Class III: The third includes the oscillators with both fractional inertia force $m\frac{d^{\alpha}x(t)}{dt^{\alpha}}(1 < \alpha \le 2)$ and fractional friction $c\frac{d^{\beta}x(t)}{dt^{\beta}}(0 < \beta \le 1)$, see, e.g., Liu et al. ([54], Equation (1)), Gomez-Aguilar ([53], Equation (10)), Leung et al. ([50], Equation (3)). This class of oscillators is expressed by (43).

By fractional oscillating in this research, we mean that either the inertia term (31) or the damping (42) or both (43) are described by fractional derivative. Thus, this article studies all described above from Class I to III except those fractional nonlinear ones, such as fractional van der Pol oscillators (Leung et al. [47,55], Xie and Lin [52], Kavyanpoor and Shokrollahi [56], Xiao et al. [57]), fractional Duffing ones (Xu et al. [45], Liu et al. [54], Chen et al. [58], Wen et al. [59], Liao [60]). Besides, the meaning of fractional oscillation in this research neither implies those with fractional displacement such as Abu-Gurra et al. [61] discussed nor those in the sense of subharmonic oscillations as stated by Den Hartog ([3], Sections 8–10, Chapter 4), Ikeda [62], Fudan Univ. ([63], pp. 96–97), Andronov et al. ([64], Section 5.1).

Fractional differential equations represented by (31), (42), and (43) are designated as fractional oscillators in Class I, II, and III, respectively, in what follows. Note that closed form analytic expressions for the responses (free, impulse, step, frequency, and sinusoidal) to fractional oscillators in Class II and III are rarely reported. For oscillators in Class I, analytic expressions for the responses (free, impulse, step) are only represented by a type of special functions called the Mittag-Leffler functions but lack in representing the intrinsic properties, such as damping. This article aims at presenting a unified approach to deal with three classes of fractional oscillators.

The present highlights are as follows.

- Establishing three equivalent 2-order differential equations respectively corresponding to three classes of fractional oscillators.
- Presenting the analytical representations, in the closed form, of equivalent masses, equivalent dampings, equivalent damping ratios, equivalent natural frequencies, and equivalent frequency ratios, for each class of fractional oscillators.
- Proposing the analytic expressions, in the closed form by using elementary functions, of the free, impulse, step, frequency, and sinusoidal responses to three classes of fractional oscillators.
- Revealing the power laws of equivalent mass and equivalent damping for each class of fractional oscillators.
- Representing some of the generalized Mittag-Leffler functions by using elementary functions.

Note that this article studies fractional oscillators by the way of dealing with fractional inertia force and or fractional friction equivalently using inertia force and or fractional friction of integer order. In doing so, methodologically speaking, the key point is about three equivalent oscillation models, which transform fractional inertia force and or fractional friction equivalently into inertia force and or fractional friction of integer order, which we establish with Theorems 1–7. Though they may yet imply a novel way to study fractional derivatives from the point of view of mathematics, my focus in this research is on treating fractional oscillators from a view of physical or engineering oscillations (vibrations).

The rest of the article is organized as follows. Section 2 is about preliminaries. The problem statement and research thoughts are described in Section 3. We establish three equivalent 2-order oscillation equations respectively corresponding to three classes of fractional oscillators in Section 4. The analytical representations of equivalent masses, equivalent dampings, equivalent damping ratios, equivalent natural frequencies for three classes of fractional oscillators are proposed in Section 5. We present the analytic expressions of the free responses to three classes of fractional oscillators in Section 6, the impulse responses to three classes of fractional oscillators in Section 7, the step responses in Section 8, the frequency responses in Section 9, and the sinusoidal ones in Section 10. Discussions are in Section 11, which is followed by conclusions.

2. Preliminaries

This Section consists of two parts. One is to describe the basic of linear oscillations and fractional ones related to the next sections. The other the solutions to fractional oscillators in Class I based on the generalized Mittag-Leffler functions.

2.1. Brief of Linear Oscillations of Order 2

2.1.1. Simple Oscillation Model

The simplest model of an oscillator of order 2 is with single degree of freedom (SDOF). It consists of a constant mass m and a massless damper with a linear viscous damping constant c. The stiffness of spring is denoted by spring constant k. That SDOF mass-spring system is described by

$$\begin{cases} m\frac{d^2q(t)}{dt^2} + c\frac{dq(t)}{dt} + kq(t) = e(t) \\ q(0) = q_0, q'(0) = v_0, \end{cases} \tag{1}$$

where $e(t)$ is the forcing function. The solution $q(t)$ may be the displacement in mechanical engineering [1–7] or current in electronics engineering [8].

In physics and engineering, for facilitating the analysis, one usually rewrites (1) by

$$\begin{cases} \frac{d^2q(t)}{dt^2} + \frac{c}{m}\frac{dq(t)}{dt} + \frac{k}{m}q(t) = \frac{e(t)}{m} \\ q(0) = q_0, q'(0) = v_0, \end{cases} \tag{2}$$

and further rewrites it by

$$\begin{cases} \frac{d^2q(t)}{dt^2} + 2\varsigma\omega_n\frac{dq(t)}{dt} + \omega_n^2 q(t) = \frac{e(t)}{m} \\ q(0) = q_0, q'(0) = v_0, \end{cases} \tag{3}$$

where ω_n is called the natural angular frequency (natural frequency for short) with damping free given by

$$\omega_n = \sqrt{\frac{k}{m}}, \tag{4}$$

and the parameter ς is the damping ratio expressed by

$$\varsigma = \frac{c}{2\sqrt{mk}}. \tag{5}$$

The characteristic equation of (3) is in the form

$$p^2 + 2\varsigma\omega_n p + \omega_n^2 = 0, \tag{6}$$

which is usually called the frequency equation in engineering [1–7]. The solution to the above is given by

$$p_{1,2} = -\varsigma\omega_n \pm i\omega_n\sqrt{1 - \varsigma^2}, \tag{7}$$

where $i = \sqrt{-1}$. Taking into account damping, one uses the term damped natural frequency denoted by ω_d. It is given by

$$\omega_d = \omega_n\sqrt{1 - \varsigma^2}. \tag{8}$$

Note 2.1: All parameters above, namely, m, c, k, ς, ω_n, and ω_d, are constants.

2.1.2. Responses

The free response, meaning that the response with $e(t) = 0$, is driven by initial conditions only. It is given by

$$q(t) = e^{-\varsigma\omega_n t}\left(q_0 \cos\omega_d t + \frac{v_0 + \varsigma\omega_n q_0}{\omega_d}\sin\omega_d t\right), t \geq 0. \tag{9}$$

If $e(t) = \delta(t)$, where $\delta(t)$ is the Dirac-delta function, the response with zero initial conditions is called the impulse response. In the theory of linear systems (Gabel and Roberts [65], Zheng et al. [66]), the symbol $h(t)$ is used for the impulse response. Thus, consider the equation

$$\frac{d^2h(t)}{dt^2} + 2\varsigma\omega_n\frac{dh(t)}{dt} + \omega_n^2 h(t) = \frac{\delta(t)}{m}. \tag{10}$$

One has

$$h(t) = \frac{e^{-\varsigma\omega_n t}}{m\omega_d}\sin\omega_d t, t \geq 0. \tag{11}$$

Let $u(t)$ be the Heaviside unit step (unit step for short) function. Then, the response to (3) with zero initial conditions is called the unit step response. As usual, it is denoted by $g(t)$ in practice. Thus, consider

$$\frac{d^2g(t)}{dt^2} + 2\varsigma\omega_n\frac{dg(t)}{dt} + \omega_n^2 g(t) = \frac{u(t)}{m}. \tag{12}$$

One has

$$g(t) = \int_0^t h(\tau)d\tau = \frac{1}{k}\left[1 - \frac{e^{-\varsigma\omega_n t}}{\sqrt{1 - \varsigma^2}}\cos(\omega_d t - \phi)\right], \tag{13}$$

where

$$\phi = \tan^{-1}\frac{\varsigma}{\sqrt{1 - \varsigma^2}}. \tag{14}$$

Denote by $H(\omega)$ the Fourier transform of $h(t)$. Then, $H(\omega)$ is usually called the frequency response to the oscillator described by (3). It is in the form

$$H(\omega) = \frac{1}{m(\omega_n^2 - \omega^2 + i2\varsigma\omega_n\omega)} = \frac{1}{m\omega_n^2\left(1 - \frac{\omega^2}{\omega_n^2} + i2\varsigma\frac{\omega}{\omega_n}\right)}. \tag{15}$$

With the parameter γ defined by

$$\gamma = \frac{\omega}{\omega_n}, \tag{16}$$

which is called frequency ratio, $H(\omega)$ may be rewritten by

$$H(\omega) = \frac{1}{m\omega_n^2(1 - \gamma^2 + i2\varsigma\gamma)}. \tag{17}$$

The amplitude of $H(\omega)$ is called the amplitude frequency response. It is in the form

$$|H(\omega)| = \frac{1}{m\omega_n^2\sqrt{(1 - \gamma^2)^2 + (2\varsigma\gamma)^2}}. \tag{18}$$

Its phase is termed the phase frequency response given by

$$\varphi(\omega) = \tan^{-1}\frac{2\varsigma\gamma}{1 - \gamma^2}. \tag{19}$$

When the oscillator is excited by a sinusoidal function, the solution to (3) is termed the sinusoidal or simple harmonic response. Suppose the sinusoidal excitation function is $A\cos\omega t$, where A is a constant. Then, the solution to

$$\begin{cases} \frac{d^2q(t)}{dt^2} + 2\varsigma\omega_n\frac{dq(t)}{dt} + \omega_n^2 q(t) = \frac{A\cos\omega t}{m} \\ q(0) = q_0, q'(0) = v_0, \end{cases} \tag{20}$$

is the sinusoidal response in the form

$$q(t) = \frac{\frac{A}{m\omega_d}}{\left(\omega_n^2 - \omega^2\right)^2 + (2\varsigma\omega_n\omega)^2} \left\{ \begin{array}{l} \left(\omega_n^2 - \omega^2\right)\cos\omega t + 2\varsigma\omega_n\omega\sin\omega t \\ +e^{-\varsigma\omega_n t}\left[\left(\omega_n^2 - \omega^2\right)\cos\omega_d t - \frac{\varsigma}{\sqrt{1-\varsigma^2}}\left(\omega_n^2 + \omega^2\right)\sin\omega_d t\right] \end{array} \right\}. \tag{21}$$

The responses mentioned above are essential to linear oscillators. We shall give our results for three classes of fractional oscillators with respect to those responses in this research.

2.1.3. Spectra of Three Excitations

The spectrum of $\delta(t)$ below means that $\delta(t)$ contains the equal frequency components for $\omega \in (0, \infty)$.

$$\int_{-\infty}^{\infty} \delta(t)e^{-i\omega t}dt = 1. \tag{22}$$

The spectrum of $u(t)$ is in the form

$$\int_{-\infty}^{\infty} u(t)e^{-i\omega t}dt = \pi\delta(\omega) + \frac{1}{i\omega}. \tag{23}$$

The Fourier transform of $\cos\omega_1 t$ is given by

$$\int_{-\infty}^{\infty} \cos\omega_1 t e^{-i\omega t}dt = \pi[\delta(\omega + \omega_1) + \delta(\omega - \omega_1)]. \tag{24}$$

Three functions or signals above, namely, $\delta(t)$, $u(t)$, and sinusoidal functions, are essential to the excitation forms in oscillations. However, their spectra do not exist in the domain of ordinary functions but they exist in the domain of generalized functions. Due to the importance of generalized functions in oscillations, for example, $\delta(t)$ and $u(t)$, either theory or technology of oscillations nowadays is in the domain of generalized functions. In the domain of generalized functions, any function is differentiable of any times. The Fourier transform of any function exists (Gelfand and Vilenkin [67], Griffel [68]).

2.1.4. Generalization of Linear Oscillators

Let us be beyond the scope of the conventionally physical quantities, such as displacement, velocity, acceleration in mechanics, or current, voltage in electronics. Then, we consider the response of the quantity $q^{(n)}(t)$, where n is a positive integer. Precisely, we consider the following oscillation equation

$$
\begin{cases}
\dfrac{d^2}{dt^2}\left[\dfrac{d^n q(t)}{dt^n}\right] + 2\varsigma\omega_n \dfrac{d}{dt}\left[\dfrac{d^n q(t)}{dt^n}\right] + \omega_n^2 \dfrac{d^n q(t)}{dt^n} = \dfrac{e(t)}{m} \\
q^{(n)}(0) = q_0, q^{(n+1)}(0) = v_0.
\end{cases}
\tag{25}
$$

The above may be taken as a generalization of the conventional oscillator described by (3). Another expression of the above may be given by

$$
\begin{cases}
\dfrac{d^n}{dt^n}\left[\dfrac{d^2 q(t)}{dt^2}\right] + 2\varsigma\omega_n \dfrac{d^n}{dt^n}\left[\dfrac{dq(t)}{dt}\right] + \omega_n^2 \dfrac{d^n q(t)}{dt^n} = \dfrac{e(t)}{m} \\
q^{(n)}(0) = q_0, q^{(n+1)}(0) = v_0.
\end{cases}
\tag{26}
$$

Alternatively, we have a linear oscillation system described by

$$
\begin{cases}
\dfrac{d^{n+2} q(t)}{dt^{n+2}} + 2\varsigma\omega_n \dfrac{d^{n+1} q(t)}{dt^{n+1}} + \omega_n^2 \dfrac{d^n q(t)}{dt^n} = \dfrac{e(t)}{m} \\
q^{(n)}(0) = q_0, q^{(n+1)}(0) = v_0.
\end{cases}
\tag{27}
$$

Physically, the above item with $q^{(n+2)}(t)$ corresponds to inertia, the one with $q^{(n)}(t)$ to restoration, and the one with $q^{(n+1)}(t)$ damping.

Note that (27) remains a linear oscillator after all. Nevertheless, when generalizing n to be fractions, for instance, considering $-1 < \varepsilon_1 \leq 0$ and $-1 < \varepsilon_2 \leq 0$, we may generalize (27) to be

$$
\begin{cases}
m\dfrac{d^{\varepsilon_1+2} q(t)}{dt^{\varepsilon_1+2}} + c\dfrac{d^{\varepsilon_2+1} q(t)}{dt^{\varepsilon_2+1}} + kq(t) = e(t) \\
q(0) = q_0, q'(0) = v_0.
\end{cases}
\tag{28}
$$

Then, we go into the scope of fractional oscillations.

2.2. Three Classes of Fractional Oscillators

Denote by $\dfrac{d^v}{dt^v} = {}_{-\infty}D_t^v$ the Weyl fractional derivative of order $v > 0$. Then (Uchaikin [38], Miller and Ross [69], Klafter et al. [70]),

$$
{}_{-\infty}D_t^v f(t) = \frac{1}{\Gamma(-v)}\int_{-\infty}^{t}\frac{f(u)du}{(t-u)^{1+v}},
\tag{29}
$$

where $\Gamma(v)$ is the Gamma function. The Weyl fractional derivative is used in this research because it is suitable for the Fourier transform in the domain of fractional calculus (Lavoie et al. ([71], p. 247)).

The Fourier transform of $\dfrac{d^v f(t)}{dt^v}$, following Uchaikin ([72], Section 4.5.3), is given by

$$
\int_{-\infty}^{\infty}\frac{d^v f(t)}{dt^v}e^{-i\omega t}dt = (i\omega)^v F(\omega),
\tag{30}
$$

where $F(\omega)$ is the Fourier transform of $f(t)$.

This article relates to three classes of fractional oscillators as follow. We denote the following oscillation equation as a fractional oscillator in Class I.

$$
\begin{cases}
m\dfrac{d^\alpha y_1(t)}{dt^\alpha} + ky_1(t) = e(t) \\
y_1(0) = y_{10}, y'_1(0) = y'_{10}
\end{cases}
,1 < \alpha \leq 2.
\tag{31}
$$

The free response to (31) is in the form (Mainardi [25], Achar et al. [33], Uchaikin ([38], Chapter 7))

$$y_1(t) = y_{10}E_{\alpha,1}\left[-(\omega_n t)^\alpha\right] + y'_{10}tE_{\alpha,2}\left[-(\omega_n t)^\alpha\right], 1 < \alpha \leq 2, t \geq 0, \tag{32}$$

where $E_{a,b}(z)$ is the generalized Mittag-Leffler function given by

$$E_{a,b}(z) = \sum_{k=0}^{\infty} \frac{z^k}{\Gamma(ak+b)}, a, b \in C, \mathrm{Re}(a) > 0, \mathrm{Re}(b) > 0. \tag{33}$$

The Mittag-Leffler function denoted by $E_a(t)$ is in the form

$$E_a(z) = \sum_{k=0}^{\infty} \frac{z^k}{\Gamma(ak+1)}, a \in C, \mathrm{Re}(a) > 0, \tag{34}$$

referring Mathai and Haubold [73], or Gorenflo et al. [74], or Erdelyi et al. [75] for the Mittag-Leffler functions.

Denote by $h_{y1}(t)$ the impulse response to a fractional oscillator in Class I. Then (Uchaikin ([38], Chapter 7)),

$$h_{y1}(t) = t^{\alpha-1}E_{\alpha,\alpha}\left[-(\omega_n t)^\alpha\right], 1 < \alpha \leq 2, t \geq 0. \tag{35}$$

Let $g_{y1}(t)$ be the step response to a fractional oscillator of Class I type. Then,

$$g_{y1}(t) = t^\alpha E_{\alpha,\alpha+1}\left[-(\omega_n t)^\alpha\right], 1 < \alpha \leq 2, t \geq 0. \tag{36}$$

For a fractional oscillator in Class I, its sinusoidal response driven by $\sin\omega t$ is expressed by

$$y_1(t) = A_1 \sin(\omega t - \theta_1) + A_2 e^{-\beta t} \cos\left[\omega_n t \sin\frac{\pi}{\alpha} - \theta_2\right] + \int_0^{\infty} e^{-st} K_\alpha(s)ds, \tag{37}$$

where

$$A_1 = \frac{1}{\sqrt{\omega_n^{2\alpha}+\omega^{2\alpha}+2\omega_n^\alpha\omega^\alpha\cos\frac{\alpha\pi}{2}}},$$

$$A_2 = \frac{2\omega}{\alpha\omega_n^{\alpha-1}\sqrt{\omega_n^4+\omega^4+2\omega_n^2\omega^2\cos\frac{2\pi}{\alpha}}}, \tag{38}$$

$$\beta = -\omega_n \cos\frac{\pi}{\alpha}, \tag{39}$$

$$\theta_1 = \tan^{-1}\frac{\omega^\alpha\sin\frac{\alpha\pi}{2}}{\omega_n^\alpha+\omega^\alpha\cos\frac{\alpha\pi}{2}},$$

$$\theta_2 = \tan^{-1}\left[\frac{\omega_n^2\sin\frac{(1+\alpha)\pi}{\alpha}-\omega^2\sin\frac{(1-\alpha)\pi}{\alpha}}{\omega_n^2\cos\frac{(1+\alpha)\pi}{\alpha}+\omega^2\cos\frac{(1-\alpha)\pi}{\alpha}}\right], \tag{40}$$

$$K_\alpha(s) = \frac{\omega\sin(\pi\alpha)}{\pi(s^2+\omega^2)(s^{2\alpha}+2s^\alpha\omega_n^2\cos(\pi\alpha)+\omega_n^{2\alpha})}. \tag{41}$$

An oscillator that follows the oscillation equation below is called a fractional oscillator in Class II.

$$m\frac{d^2y_2(t)}{dt^2} + c\frac{d^\beta y_2(t)}{dt^\beta} + ky_2(t) = 0, 0 < \beta \leq 1. \tag{42}$$

The equation below is called an oscillation equation of a fractional oscillator in Class III.

$$m\frac{d^\alpha y_3(t)}{dt^\alpha} + c\frac{d^\beta y_3(t)}{dt^\beta} + ky_3(t) = 0, 1 < \alpha \leq 2, 0 < \beta \leq 1. \tag{43}$$

2.3. Equivalence of Functions in the Sense of Fourier Transform

Denote by $F_1(\omega)$ and $F_2(\omega)$ the Fourier transforms of $f_1(t)$ and $f_2(t)$, respectively. Then, if

$$F_1(\omega) = F_2(\omega), \tag{44}$$

one says that

$$f_1(t) = f_2(t), \tag{45}$$

in the sense of Fourier transform (Gelfand and Vilenkin [67], Papoulis [76]), implying

$$\int_{-\infty}^{\infty} [f_1(t) - f_2(t)]e^{-i\omega t}dt = 0. \tag{46}$$

The above implies that a null function as a difference between $f_1(t)$ and $f_2(t)$ is allowed for (45). An example relating to oscillation theory is the unit step function.

Denote by $u_1(t)$ in the form

$$u_1(t) = \begin{cases} 1, t \geq 0 \\ 0, t < 0 \end{cases}. \tag{47}$$

Let $u_2(t)$ be

$$u_2(t) = \begin{cases} 1, t > 0 \\ 0, t \leq 0 \end{cases}. \tag{48}$$

Clearly, either $u_1(t)$ or $u_2(t)$ is a unit step function. The difference between two is a null function given by

$$u_1(t) - u_2(t) = \begin{cases} 1, & t = 1 \\ 0, & \text{elsewhere} \end{cases}. \tag{49}$$

Thus, $u_1(t) = u_2(t)$. In fact, the Fourier transform of either $u_1(t)$ or $u_2(t)$ equals to the right side on (23). Similarly, if $f_1(t) = f_2(t)$, we say that (44) holds in the sense of

$$\int_{-\infty}^{\infty} [F_1(\omega) - F_2(\omega)]e^{i\omega t}d\omega = 0. \tag{50}$$

3. Problem Statement and Research Thoughts

We have mentioned three classes of fractional oscillators in Section 2. This section contains two parts. One is the problem statement and the other research thoughts.

3.1. Problem Statement

We first take fractional oscillators in Class I as a case to state the problems this research concerns with.

The analytical expressions with respect to the responses of free, impulse, step, to the oscillators of Class I are mathematically obtained (Mainardi [25], Achar et al. [33], Uchaikin ([38], Chapter 7)), also see Section 2.2 in this article. All noticed that a fractional oscillator of Class I is damping free in form but it is damped in nature due to fractional if $1 < \alpha < 2$. However, there are problems unsolved in this regard.

Problem 1. *How to analytically represent the damping of Class I oscillators?*

In this article, we call the damping of fractional oscillators in Class I equivalent damping denoted by c_{eq1}.

It is known that damping relates to mass. Therefore, if we find c_{eq1} in a fractional oscillator in Class I, its intrinsic mass must be different from the primary one m unless $\alpha = 2$. We call it equivalent mass and denote it by m_{eq1}.

Problem 2. *How to analytically represent* m_{eq1}?

Because a fractional oscillator in Class I is damped in nature for $\alpha \neq 2$, there must exist a damped natural frequency. We call it equivalent damped natural frequency, denoted by $\omega_{eqd,1}$. Then, comes the problem below.

Problem 3. *What is the representation of* $\omega_{eqd,1}$?

As there exists m_{eq1} that differs from m if $\alpha \neq 2$, the equivalent damping free natural frequency, we denote it by $\omega_{eqn,1}$, is different from the primary damping free natural frequency $\omega_n = \sqrt{\frac{k}{m}}$. Consequently, the following problem appears.

Problem 4. *What is the expression of* $\omega_{eqn,1}$?

If we find the solutions to the above four, a consequent problem is as follows.

Problem 5. *How to represent response (free, or impulse, or step, or sinusoidal) with* m_{eq1}, c_{eq1}, $\omega_{eqn,1}$, *and* $\omega_{eqd,1}$ *to a fractional oscillator in Class I?*

If we solve the above problems, the solution to the following problem is ready.

Problem 6. *What is the physical mechanism of a fractional oscillator in Class I?*

Note that the intrinsic damping for a Class II fractional oscillator must differ from its primary damping c owing to the fractional friction $c\frac{d^\beta y_2(t)}{dt^\beta}$ for $\beta \neq 1$. We call it the equivalent damping denoted by c_{eq2}. Because $c_{eq2} \neq c$ if $\beta \neq 1$, the equivalent mass of a fractional oscillator in Class II, denoted by m_{eq2}, is not equal to the primary m for $\beta \neq 1$. Thus, the six stated above are also unsolved problems for fractional oscillators in Class II. They are, consequently, the problems unsolved for Class III fractional oscillators.

Note that there are other problems regarding with three classes of fractional oscillators. For example, the explicit expression of the sinusoidal response (37) in closed form needs investigation because of the difficulty in finding the solution to $\int_0^\infty e^{-st} K_\alpha(s) ds$. We shall deal with them in separate sections. The solutions to the problems described above constitute main highlights of this research.

We note that the damping nature of a fractional oscillator in Class I was also observed by other researchers, not explicitly stated though, as can be seen from, e.g., Zurigat ([26], Figure 1), Blaszczyk et al. ([28], Figure 2), Al-rabtah et al. ([29], Figure 2), Ryabov and Puzenko ([36], Equation (5)), Uchaikin ([38], Chapter 7), Duan et al. ([39], Equation (4.3), Figure 2), Gomez-Aguilar et al. ([53], Equation (15), Figures 2 and 3), Chung and Jung ([77], Figure 1). One thing remarkable is by Tofighi, who explored the intrinsic damping of an oscillator in Class I, see ([35], pp. 32–33). That was an advance regarding with the damping implied in (31) but it may be unsatisfactory if one desires its closed form of analytic expression.

3.2. Research Thoughts

Let us qualitatively consider possible performances of equivalent mass and damping. In engineering, people may purposely connect an auxiliary mass m_a to the primary mass m so that the equivalent mass of the total system is related to the oscillation frequency ω (Harris ([4], p. 6.4)). In the field of ship hull vibrations, added mass has to be taken into account in the equivalent mass (i.e., total mass) of a ship hull (Korotkin [78]) so that the equivalent mass is ω-varying. In fact, the three dimensional fluid coefficient with respect to the added mass to a ship hull relates to the oscillation frequency, see, e.g., Jin and Xia ([79], pp. 135–136), Nakagawa et al. [80].

In addition, damping may be also ω-varying. A well-known case of ω-varying damping is the Coulomb damping (Timoshenko ([2], Chapter 1), Harris ([4], Equation (30.4))). Frequency varying damping is a technique used in damping treatments, see, e.g., Harris ([4], Equation (37.8)). Besides, commonly used damping assumptions in ship hull vibrations, such as the Copoknh's, the Voigt's, the Rayleigh's, are all ω-varying (Jin and Xia ([79], pp. 157–158)). Therefore, with the concept of ω-varying mass and damping, I purposely generalize the simple oscillation model expressed by (1) in the form

$$\begin{cases} m_{eq}(\omega)\frac{d^2q(t)}{dt^2} + c_{eq}(\omega)\frac{dq(t)}{dt} + kq(t) = e(t) \\ q(0) = q_0, q'(0) = v_0. \end{cases} \tag{51}$$

The above second-order equation may not be equivalent to a fractional oscillator unless m_{eq} and or c_{eq} are appropriately expressed and properly related to the fractional order α for Class I oscillators, or β for Class II oscillators, or (α, β) for oscillators in Class III. For those reasons, we further generalize (51) by

$$\begin{cases} m_{eq1}(\omega, \alpha)\frac{d^2x_1(t)}{dt^2} + c_{eq1}(\omega, \alpha)\frac{dx_1(t)}{dt} + kx_1(t) = e(t) \\ x_1(0) = x_{10}, \dot{x}_1(0) = v_{10}, \end{cases} \tag{52}$$

for Class I oscillators. As for Class II oscillators, (51) should be generalized by

$$\begin{cases} m_{eq2}(\omega, \beta)\frac{d^2x_2(t)}{dt^2} + c_{eq2}(\omega, \beta)\frac{dx_2(t)}{dt} + kx_2(t) = e(t) \\ x_2(0) = x_{10}, \dot{x}_2(0) = v_{20}. \end{cases} \tag{53}$$

Similarly, for Class III oscillators, we generalize (51) to be the form

$$\begin{cases} m_{eq3}(\omega, \alpha, \beta)\frac{d^2x_3(t)}{dt^2} + c_{eq3}(\omega, \alpha, \beta)\frac{dx_2(t)}{dt} + kx_3(t) = e(t) \\ x_3(0) = x_{30}, \dot{x}_3(0) = v_{30}. \end{cases} \tag{54}$$

Three generalized oscillation Equations (52)–(54), can be unified in the form

$$\begin{cases} m_{eqj}\frac{d^2x_j(t)}{dt^2} + c_{eqj}\frac{dx_j(t)}{dt} + kx_j(t) = e(t) \\ x_j(0) = x_{j0}, \dot{x}_j(0) = v_{30} \end{cases}, j = 1, 2, 3. \tag{55}$$

By introducing the symbols $\omega_{eqn,j} = \sqrt{\frac{k}{m_{eqj}}}$ and $\varsigma_{eqj} = \frac{c_{eqj}}{2\sqrt{m_{eqj}k}}$ for $j = 1, 2, 3$, we rewrite the above by

$$\begin{cases} \frac{d^2x_j(t)}{dt^2} + 2\varsigma_{eqj}\omega_{eqn,j}\frac{dx_j(t)}{dt} + \omega_{eqn,j}^2 x_j(t) = \frac{e(t)}{m_{eqj}} \\ x_j(0) = x_{j0}, \dot{x}_j(0) = v_{30} \end{cases}, j = 1, 2, 3. \tag{56}$$

Let $Y_j(\omega)$ be the Fourier transform of $y_j(t)$, where $y_j(t)(j = 1, 2, 3)$ respectively corresponds to the one in (31), (42), and (43). Denote by $X_j(\omega)$ the Fourier transform of $x_j(t)$. Then, if we find proper m_{eqj} and c_{eqj} such that

$$Y_j(\omega) = X_j(\omega), \quad j = 1, 2, 3, \tag{57}$$

the second-order equation (52), or (53), or (54) is equal to the fractional oscillation Equation (31), or (42), or (43), respectively.

Obviously, once we discover the equivalent equations of the fractional oscillation Equations (52), or (53), and (54), all problems stated previously can be readily solved.

4. Equivalent Systems of Three Classes of Fractional Oscillators

In this section, we first present an equivalent system and then its equivalent mass and damping in Sections 4.1–4.3, respectively for each class of fractional oscillators.

4.1. Equivalent System for Fractional Oscillators in Class I

4.1.1. Equivalent Oscillation Equation of Fractional Oscillators in Class I

Theorem 1 gives the equivalent oscillator with the integer order for the fractional oscillators in Class I.

Theorem 1 (Equivalent oscillator I). *Denote a fractional oscillator in Class I by*

$$m\frac{d^\alpha y_1(t)}{dt^\alpha} + ky_1(t) = 0, 1 < \alpha \le 2. \tag{58}$$

Then, its equivalent oscillator with the equation of order 2 is in the form

$$-m\omega^{\alpha-2}\cos\frac{\alpha\pi}{2}\frac{d^2 x_1(t)}{dt^2} + m\omega^{\alpha-1}\sin\frac{\alpha\pi}{2}\frac{dx_1(t)}{dt} + kx_1(t) = 0, 1 < \alpha \le 2. \tag{59}$$

Proof. Consider the frequency response of (58) with the excitation of the Dirac-delta function $\delta(t)$. In doing so, we study

$$m\frac{d^\alpha h_{y1}(t)}{dt^\alpha} + kh_{y1}(t) = \delta(t), 1 < \alpha \le 2. \tag{60}$$

Doing the Fourier transform on the both sides of (60) produces

$$\left[m(i\omega)^\alpha + k\right] H_{y1}(\omega) = 1, 1 < \alpha \le 2, \tag{61}$$

where $H_{y1}(\omega)$ is the Fourier transform of $h_{y1}(t)$. Using the principal value of i, we have

$$i^\alpha = \cos\frac{\alpha\pi}{2} + i\sin\frac{\alpha\pi}{2}. \tag{62}$$

Thus, (61) implies

$$\begin{aligned}
\left[m(i\omega)^\alpha + k\right] H_{y1}(\omega) &= \left\{m\left(\cos\frac{\alpha\pi}{2} + i\sin\frac{\alpha\pi}{2}\right)\omega^\alpha + k\right\} H_{y1}(\omega) \\
&= \left(m\omega^\alpha \cos\frac{\alpha\pi}{2} + im\omega^\alpha \sin\frac{\alpha\pi}{2} + k\right) H_{y1}(\omega) = 1.
\end{aligned} \tag{63}$$

Therefore, we have the frequency response of (60) in the form

$$H_{y1}(\omega) = \frac{1}{m\omega^\alpha \cos\frac{\alpha\pi}{2} + im\omega^\alpha \sin\frac{\alpha\pi}{2} + k}. \tag{64}$$

On the other hand, for $1 < \alpha \le 2$, we consider (59) by

$$-m\omega^{\alpha-2}\cos\frac{\alpha\pi}{2}\frac{d^2 h_{x1}(t)}{dt^2} + m\omega^{\alpha-1}\sin\frac{\alpha\pi}{2}\frac{dh_{x1}(t)}{dt} + kh_{x1}(t) = \delta(t). \tag{65}$$

Performing the Fourier transform on the both sides of (65) yields

$$\begin{aligned}
&\left[-m\omega^{\alpha-2}\cos\frac{\alpha\pi}{2}(-\omega^2) + m\omega^{\alpha-1}\sin\frac{\alpha\pi}{2}(i\omega) + k\right] H_{x1}(\omega) \\
&= \left(m\omega^\alpha \cos\frac{\alpha\pi}{2} + im\omega^\alpha \sin\frac{\alpha\pi}{2} + k\right) H_{x1}(\omega) = 1,
\end{aligned} \tag{66}$$

where $H_{x1}(\omega)$ is the Fourier transform of $h_{x1}(t)$. Therefore, we have

$$H_{x1}(\omega) = \frac{1}{m\omega^{\alpha}\cos\frac{\alpha\pi}{2} + im\omega^{\alpha}\sin\frac{\alpha\pi}{2} + k}.$$

(67)

By comparing (64) with (67), we see that

$$H_{y1}(\omega) = H_{x1}(\omega).$$

(68)

Thus, (59) is the equivalent equation of (58). The proof completes. $\square$

4.1.2. Equivalent Mass of Fractional Oscillators in Class I

From the first item on the left side of (59), we obtain the equivalent mass for the fractional oscillators of Class I type.

Theorem 2 (Equivalent mass I). *The equivalent mass of the fractional generators in Class I, denoted by m_{eq1}, is expressed by*

$$m_{eq1} = m_{eq1}(\omega, \alpha) = -\left(\omega^{\alpha-2}\cos\frac{\alpha\pi}{2}\right)m, 1 < \alpha \leq 2.$$

(69)

Proof. According to the Newton's second law, the inertia force in the system of the fractional oscillator (58) corresponds to the first item on the left side of its equivalent system (59). That is, $-m\omega^{\alpha-2}\cos\frac{\alpha\pi}{2}\frac{d^2x_1(t)}{dt^2}$. Thus, the coefficient of $\frac{d^2x_1(t)}{dt^2}$ is an equivalent mass expressed by (69). Hence, the proof finishes. $\square$

From Theorem 2, we reveal a power law phenomenon with respect to m_{eq1} in terms of ω.

Remark 1. *The equivalent mass I, m_{eq1}, follows the power law in terms of oscillation frequency ω in the form*

$$m_{eq1}(\omega, \alpha) \sim \omega^{\alpha-2}m, 1 < \alpha \leq 2.$$

(70)

The equivalent mass m_{eq1} relates to the oscillation frequency ω, the fractional order α, and the primary mass m. Denote by

$$R_{m1}(\omega, \alpha) = -\omega^{\alpha-2}\cos\frac{\alpha\pi}{2}, 1 < \alpha \leq 2.$$

(71)

Then, we have

$$m_{eq1} = m_{eq1}(\omega, \alpha) = R_{m1}(\omega, \alpha)m, 1 < \alpha \leq 2.$$

(72)

Note 4.1: Since

$$R_{m1}(\omega, 2) = 1,$$

(73)

$m_{eq1}(\omega, \alpha)$ reduces to the primary mass m when $\alpha = 2$. That is,

$$m_{eq1}(\omega, 2) = m.$$

(74)

In the case of $\alpha = 2$, therefore, both (58) and (59) reach the conventional harmonic oscillation with damping free in the form

$$m\frac{d^2x_1(t)}{dt^2} + kx_1(t) = 0.$$

Note 4.2: If $\alpha \to 1$, we have

$$\lim_{\alpha \to 1} m_{eq1}(\omega, \alpha) = 0 \text{ for } \omega \neq 0.$$

(75)

The above implies that m_{eq1} vanishes if $\alpha \to 1$. Consequently, any oscillation disappears in that case.

Note 4.3: When $1 < \alpha \leq 2$, we attain

$$0 < R_{m1}(\omega, \alpha) \leq 1 \text{ for } \omega > 1. \tag{76}$$

Thus, we reveal an interesting phenomenon expressed by

$$m_{eq1}(\omega, \alpha) \leq m \text{ for } 1 < \alpha \leq 2, \omega > 1. \tag{77}$$

The coefficient $R_{m1}(\omega, \alpha)$ is plotted in Figure 1.

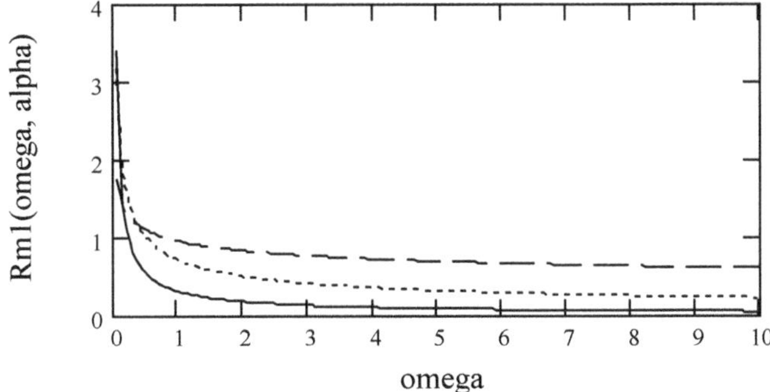

Figure 1. Plots of $R_{m1}(\omega, \alpha)$. Solid line: $\alpha = 1.2$. Dot line: $\alpha = 1.5$. Dash line: $\alpha = 1.8$.

Remark 2. *For $\alpha \in (0, 2)$, we have*

$$\lim_{\omega \to \infty} m_{eq1}(\omega, \alpha) = 0. \tag{78}$$

The interesting and novel behavior, described above, implies that a fractional oscillator in Class I does not oscillate for $\omega \to \infty$ because it is equivalently massless in that case.

Remark 3. *For $\alpha \in (0, 2)$, we have*

$$\lim_{\omega \to 0} m_{eq1}(\omega, \alpha) = \infty. \tag{79}$$

The interesting behavior, revealed above, says that a fractional oscillator of Class I type does not oscillate at $\omega = 0$ because its mass is equivalently infinity in addition to the explanation of static status conventionally described by $\omega = 0$.

4.1.3. Equivalent Damping of Fractional Oscillators of Class I

We now propose the equivalent damping.

Theorem 3 (Equivalent damping I). *The equivalent damping of a fractional oscillator in Class I, denoted by c_{eq1}, is expressed by*

$$c_{eq1} = c_{eq1}(\omega, \alpha) = \left(\omega^{\alpha-1} \sin \frac{\alpha\pi}{2}\right)m, 1 < \alpha \leq 2. \tag{80}$$

Proof. The second term on the left side of (59) is the friction with the linear viscous damping coefficient denoted by (80). The proof completes. $\square$

Denote

$$R_{c1}(\omega, \alpha) = \omega^{\alpha-1} \sin \frac{\alpha\pi}{2}, 1 < \alpha \leq 2. \tag{81}$$

Then, we have

$$c_{eq1}(\omega, \alpha) = R_{c1}(\omega, \alpha)m. \tag{82}$$

The coefficient $R_{c1}(\omega, \alpha)$ is indicated in Figure 2.

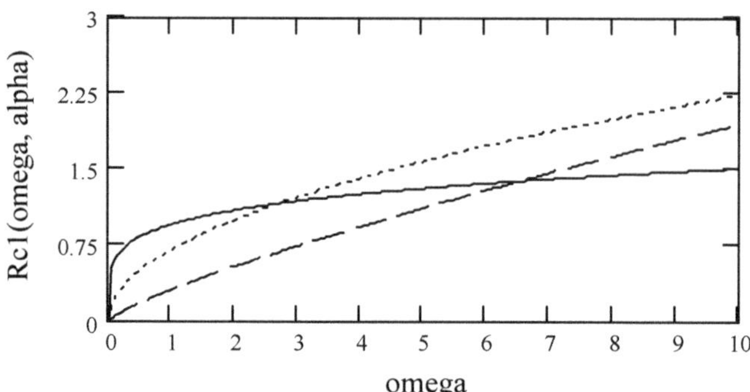

Figure 2. $R_{c1}(\omega, \alpha)$. Solid line: $\alpha = 1.2$. Dot line: $\alpha = 1.5$. Dash line: $\alpha = 1.8$.

Remark 4. *The equivalent damping I relies on ω, m, and α. It obeys the power law in terms of ω in the form*

$$c_{eq1}(\omega, \alpha) \sim \omega^{\alpha-1}m, 1 < \alpha \le 2. \tag{83}$$

Note 4.4: Because

$$c_{eq1}(\omega, \alpha)\big|_{\alpha=2} = 0, \tag{84}$$

we see again that a fractional oscillator of Class I type reduces to the conventional harmonic one when $\alpha = 2$.

Remark 5. *An interesting behavior of c_{eq1}, we found, is expressed by*

$$\lim_{\omega \to \infty} c_{eq1}(\omega, \alpha) = \infty, 1 < \alpha < 2. \tag{85}$$

The above says that the equivalent oscillator (59), as well as the fractional oscillator (58), never oscillates at $\omega \to \infty$ for $1 < \alpha < 2$ because its damping is infinitely large in that case. Due to

$$\lim_{\omega \to 0} c_{eq1}(\omega, \alpha) = 0, 1 < \alpha < 2, \tag{86}$$

we reveal a new damping behavior of a fractional oscillator in Class I in that it is equivalently dampingless for $1 < \alpha < 2$ at $\omega = 0$.

4.2. Equivalent Oscillation System for Fractional Oscillators of Class II Type

4.2.1. Equivalent Oscillation Equation of Fractional Oscillators in Class II

Theorem 4 below describes the equivalent oscillator for the fractional oscillators of Class II type.

Theorem 4 (Equivalent oscillator II). *Denote a fractional oscillator in Class II by*

$$m\frac{d^2 y_2(t)}{dt^2} + c\frac{d^\beta y_2(t)}{dt^\beta} + ky_2(t) = 0, 0 < \beta \le 1. \tag{87}$$

Then, its equivalent 2-order oscillation equation is given by

$$\left(m - c\omega^{\beta-2}\cos\frac{\beta\pi}{2}\right)\frac{d^2x_2(t)}{dt^2} + \left(c\omega^{\beta-1}\sin\frac{\beta\pi}{2}\right)\frac{dx_2(t)}{dt} + kx_2(t) = 0, 0 < \beta \leq 1. \tag{88}$$

Proof. Consider the following equation:

$$m\frac{d^2h_{y2}(t)}{dt^2} + c\frac{d^\beta h_{y2}(t)}{dt^\beta} + kh_{y2}(t) = \delta(t), 0 < \beta \leq 1. \tag{89}$$

Denote by $H_{y2}(\omega)$ the Fourier transform of $h_{y2}(t)$. Then, it is its frequency transfer function. Taking the Fourier transform on the both sides of (89) yields

$$\left[-m\omega^2 + c(i\omega)^\beta + k\right]H_{y2}(\omega) = 1, 0 < \beta \leq 1. \tag{90}$$

With the principal value of i^β, (90) becomes

$$\left[-m\omega^2 + c(i\omega)^\beta + k\right]H_{y2}(\omega) = \left\{-m\omega^2 + c\left(\cos\frac{\beta\pi}{2} + i\sin\frac{\beta\pi}{2}\right)\omega^\beta + k\right\}H_{y2}(\omega)$$
$$= \left(-m\omega^2 + c\omega^\beta\cos\frac{\beta\pi}{2} + k + ic\omega^\beta\sin\frac{\beta\pi}{2}\right)H_{y2}(\omega) = 1. \tag{91}$$

The above means

$$H_{y2}(\omega) = \frac{1}{-m\omega^2 + c\omega^\beta\cos\frac{\beta\pi}{2} + k + ic\omega^\beta\sin\frac{\beta\pi}{2}}. \tag{92}$$

On the other hand, we consider the equivalent oscillation equation II with the Dirac-δ excitation by

$$\left(m - c\omega^{\beta-2}\cos\frac{\beta\pi}{2}\right)\frac{d^2h_{x2}(t)}{dt^2} + \left(c\omega^{\beta-1}\sin\frac{\beta\pi}{2}\right)\frac{dh_{x2}(t)}{dt} + k\frac{d^2h_{x2}(t)}{dt^2} = \delta(t), 0 < \beta \leq 1. \tag{93}$$

Performing the Fourier transform on the both sides of the above produces

$$\left[-m\omega^2 + c\omega^\beta\cos\frac{\beta\pi}{2} + ic\omega^\beta\sin\frac{\beta\pi}{2}(i\omega) + k\right]H_{x2}(\omega)$$
$$= \left(-m\omega^2 + c\omega^\beta\cos\frac{\beta\pi}{2} + k + ic\omega^\beta\sin\frac{\beta\pi}{2}\right)H_{x2}(\omega) = 1, \tag{94}$$

where $H_{x2}(\omega)$ the Fourier transform of $h_{x2}(t)$. Thus, from the above, we have

$$H_{x2}(\omega) = \frac{1}{-m\omega^2 + c\omega^\beta\cos\frac{\beta\pi}{2} + k + ic\omega^\beta\sin\frac{\beta\pi}{2}}. \tag{95}$$

Equations (92) and (95) imply

$$H_{y2}(\omega) = H_{x2}(\omega). \tag{96}$$

Hence, (88) is the equivalent oscillation equation of the fractional oscillators of Class II. This completes the proof. $\square$

4.2.2. Equivalent Mass of Fractional Oscillators of Class II

The equivalent mass of the fractional oscillators of Class II type is presented in Theorem 5.

Theorem 5 (Equivalent mass II). *Let m_{eq2} be the equivalent mass of the fractional oscillators of Class II type. Then,*

$$m_{eq2} = m_{eq2}(\omega, \beta) = m - c\omega^{\beta-2}\cos\frac{\beta\pi}{2}, 0 < \beta \leq 1. \tag{97}$$

Proof. Consider the Newton's second law. Then, we see that the inertia force in the equivalent oscillator II is $\left(m - c\omega^{\beta-2}\cos\frac{\beta\pi}{2}\right)\frac{d^2 x_2}{dt^2}$. Therefore, (97) holds. The proof completes. $\square$

From Theorem 5, we reveal a power law phenomenon with respect to the equivalent mass II.

Remark 6. *The equivalent mass m_{eq2} obeys the power law in terms of ω in the form*

$$m_{eq2} \sim -c\omega^{\beta-2}, 0 < \beta \leq 1. \tag{98}$$

Note 4.5: Equation (97) exhibits that m_{eq2} is related to the oscillation frequency ω, the fractional order β, the primary mass m, and the primary damping c.

Remark 7. *For $0 < \beta \leq 1$, we have*

$$\lim_{\omega\to\infty} m_{eq2}(\omega, \beta) = m. \tag{99}$$

Figure 3 shows its plots for $m = c = 1$ with the part of $m_{eq2}(\omega, \beta) > 0$.

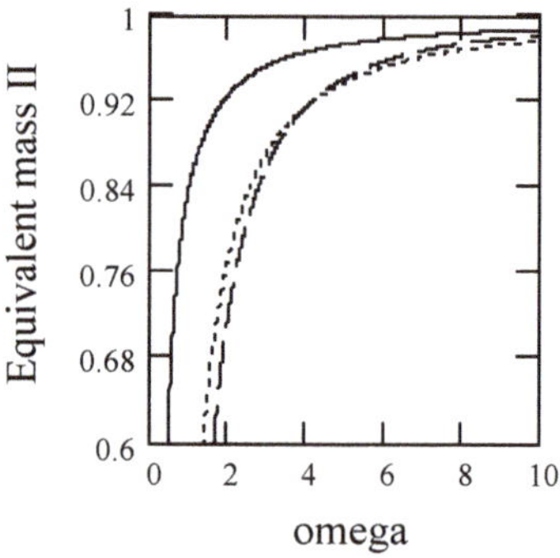

Figure 3. Plots of $m_{eq2}(\omega, \beta) > 0$ for $m = c = 1$.

Remark 8. *For $0 < \beta < 1$, we have*

$$\lim_{\omega\to 0} m_{eq2}(\omega, \beta) = -\infty. \tag{100}$$

Note 4.6: The equivalent mass II is negative if ω is small enough.

Figure 4 exhibits the negative part of $m_{eq2}(\omega, \beta)$.

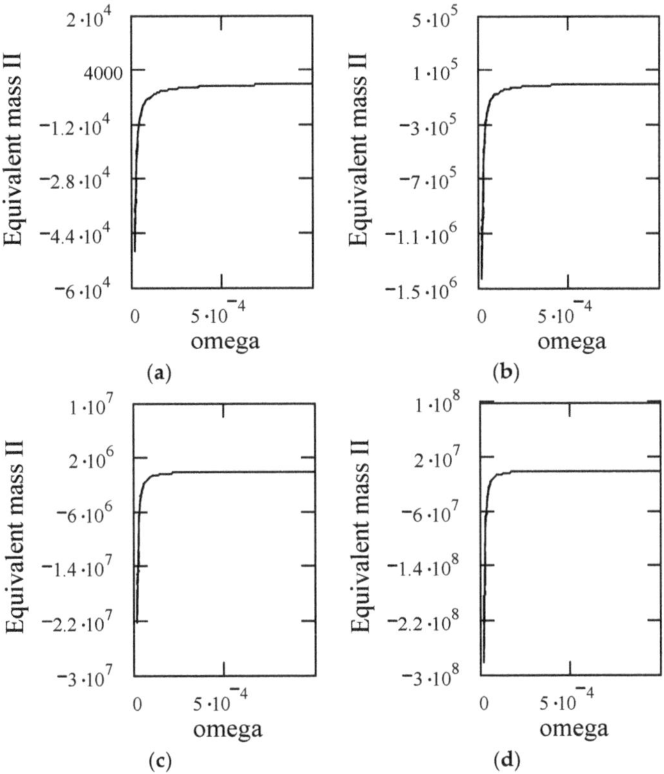

Figure 4. Illustrating negative part of $m_{eq2}(\omega, \beta)$ for $m = c = 1$. (a) $\beta = 0.9$. (b) $\beta = 0.7$. (c) $\beta = 0.5$. (d) $\beta = 0.3$.

Remark 9. *We restrict our research for $m_{eq2}(\omega, \beta) > 0$.*

Note 4.7: The equivalent mass II reduces to the primary mass m for $\beta = 1$ as indicated below.

$$m_{eq2}(\omega, \beta)\big|_{\beta=1} = m. \tag{101}$$

In fact, a fractional oscillator in Class II reduces to the conventional oscillator below if $\beta = 1$

$$m\frac{d^2x_2}{dt^2} + c\frac{dx_2}{ct} + kx_2 = 0.$$

4.2.3. Equivalent Damping of Fractional Oscillators in Class II

Let c_{eq2} be the equivalent damping of a fractional oscillator in Class II. Then, we put forward the expression of c_{eq2} with Theorem 6.

Theorem 6 (Equivalent damping II). *The equivalent damping of the fractional oscillators in Class II is in the form*

$$c_{eq2} = c_{eq2}(\omega, \beta) = c\omega^{\beta-1} \sin\frac{\beta\pi}{2}, 0 < \beta \leq 1. \tag{102}$$

Proof. The second term on the left side of (88) is the friction force with the linear viscous damping coefficient denoted by (102). The proof completes. $\square$

Denote by

$$R_{c2}(\omega,\beta) = \omega^{\beta-1}\sin\frac{\beta\pi}{2}, 0 < \beta \le 1. \tag{103}$$

Then, we have

$$c_{eq2}(\omega,\beta) = R_{c2}(\omega,\beta)c. \tag{104}$$

Figure 5 indicates $R_{c2}(\omega,\beta)$.

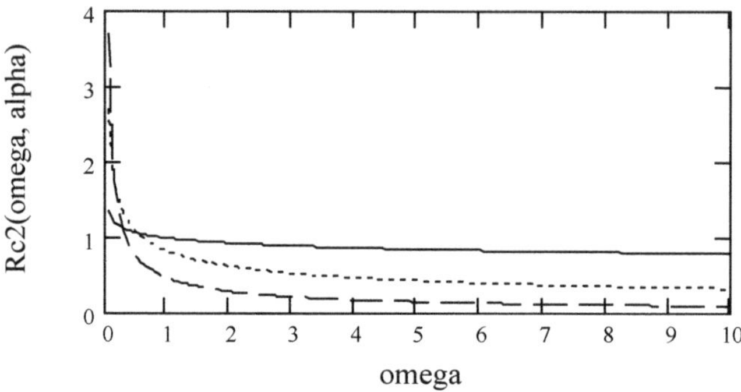

Figure 5. Indication of $R_{c2}(\omega,\beta)$ Solid line: $\beta = 0.9$. Dot line: $\beta = 0.6$. Dash line: $\beta = 0.3$.

Remark 10. *The equivalent damping c_{eq2} is associated with the oscillation frequency ω, the primary damping c, and the fractional order β. It follows the power law in terms of ω in the form*

$$c_{eq2}(\omega,\beta) \sim \omega^{\beta-1}c, 0 < \beta \le 1. \tag{105}$$

Note 4.8: The following says that c_{eq2} reduces to the primary damping c if $\beta = 1$.

$$c_{eq2}(\omega,\beta)\big|_{\beta=1} = c. \tag{106}$$

Remark 11. *The equivalent damping c_{eq2} has, for $\beta \in (0, 1)$, the property given by*

$$\lim_{\omega\to\infty} c_{eq2}(\omega,\beta) = 0. \tag{107}$$

Note 4.9: The equivalent oscillation equation of Class II fractional oscillators reduces to $m\frac{d^2x_2(t)}{dt^2} + kx_2(t) = 0$ in the two cases. One is $\omega \to \infty$, see Remark 7 and Remark 12. The other is $c = 0$.

Note 4.10: Remark 5 for $\lim_{\omega\to\infty} c_{eq1}(\omega,\beta) = \infty$ and Remark 11 just above suggest a substantial difference between two types of fractional oscillators from the point of view of the damping at $\omega \to \infty$.

Remark 12. *The equivalent damping c_{eq2} has, for $\beta \in (0, 1)$, the asymptotic property for $\omega \to 0$ in the form*

$$\lim_{\omega\to 0} c_{eq2}(\omega,\beta) = \infty. \tag{108}$$

The above property implies that a fractional oscillator in Class II does not oscillate at $\omega \to 0$ because not only it is in static status but also its equivalent damping is infinitely large.

4.3. Equivalent Oscillation System for Fractional Oscillators of Class III

4.3.1. Equivalent Oscillation Equation of Fractional Oscillators in Class III

We present Theorem 7 below to explain the equivalent oscillation equation for the fractional oscillators of Class III.

Theorem 7 (Equivalent oscillator III). *Denote a fractional oscillation equation in Class III by*

$$m\frac{d^\alpha y_3(t)}{dt^\alpha} + c\frac{d^\beta y_3(t)}{dt^\beta} + ky_3(t) = 0, 1 < \alpha \le 2, 0 < \beta \le 1. \tag{109}$$

Then, its equivalent oscillator of order 2 for $1 < \alpha \le 2$ and $0 < \beta \le 1$ is in the form

$$-\left(m\omega^{\alpha-2}\cos\tfrac{\alpha\pi}{2} + c\omega^{\beta-2}\cos\tfrac{\beta\pi}{2}\right)\frac{d^2 x_3(t)}{dt^2}$$
$$+\left(m\omega^{\alpha-1}\sin\tfrac{\alpha\pi}{2} + c\omega^{\beta-1}\sin\tfrac{\beta\pi}{2}\right)\frac{dx_3(t)}{dt} + kx_3(t) = 0. \tag{110}$$

Proof. Let us consider the equation

$$m\frac{d^\alpha h_{y3}(t)}{dt^\alpha} + c\frac{d^\beta h_{y3}(t)}{dt^\beta} + kh_{y3}(t) = \delta(t), 1 < \alpha \le 2, 0 < \beta \le 1. \tag{111}$$

Let $H_{y3}(\omega)$ be the Fourier transform of $h_{y3}(t)$. Doing the Fourier transform on the both sides of the above results in

$$\left[m(i\omega)^\alpha + c(i\omega)^\beta + k\right]H_{y3}(\omega) = 1, 1 < \alpha \le 2, 0 < \beta \le 1. \tag{112}$$

Taking into account the principal values of i^α and i^β, (112) becomes

$$\left[m(i\omega)^\alpha + c(i\omega)^\beta + k\right]H_{y3}(\omega)$$
$$= \left[m\left(\cos\tfrac{\alpha\pi}{2} + i\sin\tfrac{\alpha\pi}{2}\right)\omega^\alpha + c\left(\cos\tfrac{\beta\pi}{2} + i\sin\tfrac{\beta\pi}{2}\right)\omega^\beta + k\right]H_{y3}(\omega) \tag{113}$$
$$= \left[m\omega^\alpha\cos\tfrac{\alpha\pi}{2} + c\omega^\beta\cos\tfrac{\beta\pi}{2} + k + i\left(m\omega^\alpha\sin\tfrac{\alpha\pi}{2} + c\omega^\beta\sin\tfrac{\beta\pi}{2}\right)\right]H_{y3}(\omega) = 1.$$

Consequently, we have

$$H_{y3}(\omega) = \frac{1}{m\omega^\alpha\cos\tfrac{\alpha\pi}{2} + c\omega^\beta\cos\tfrac{\beta\pi}{2} + k + i\left(m\omega^\alpha\sin\tfrac{\alpha\pi}{2} + c\omega^\beta\sin\tfrac{\beta\pi}{2}\right)}. \tag{114}$$

On the other hand, considering the equivalent oscillator III driven by the Dirac-δ function, we have

$$-\left(m\omega^{\alpha-2}\cos\tfrac{\alpha\pi}{2} + c\omega^{\beta-2}\cos\tfrac{\beta\pi}{2}\right)\frac{d^2 h_{x3}(t)}{dt^2} + \left(m\omega^{\alpha-1}\sin\tfrac{\alpha\pi}{2} + c\omega^{\beta-1}\sin\tfrac{\beta\pi}{2}\right)\frac{dh_{x3}(t)}{dt} + kh_{x3}(t) = \delta(t). \tag{115}$$

When doing the Fourier transform on the both sides of the above, we obtain

$$\left(m\omega^\alpha\cos\tfrac{\alpha\pi}{2} + c\omega^\beta\cos\tfrac{\beta\pi}{2}\right)H_{x3}(\omega) + i\left(m\omega^\alpha\sin\tfrac{\alpha\pi}{2} + c\omega^\beta\sin\tfrac{\beta\pi}{2}\right)H_{x3}(\omega) + kH_{x3}(\omega) = 1, \tag{116}$$

where $H_{x3}(\omega)$ is the Fourier transform of $h_{x3}(t)$. Therefore, from the above, we get

$$H_{x3}(\omega) = \frac{1}{m\omega^\alpha\cos\tfrac{\alpha\pi}{2} + c\omega^\beta\cos\tfrac{\beta\pi}{2} + k + i\left(m\omega^\alpha\sin\tfrac{\alpha\pi}{2} + c\omega^\beta\sin\tfrac{\beta\pi}{2}\right)}. \tag{117}$$

Two expressions, (114) and (117), imply that

$$H_{x3}(\omega) = H_{y3}(\omega).\tag{118}$$

Thus, Theorem 7 holds. $\square$

4.3.2. Equivalent Mass of Fractional Oscillators in Class III

From Section 4.3.1, we propose the equivalent mass of the fractional oscillators in Class III type by Theorem 8.

Theorem 8 (Equivalent mass III). *Let m_{eq3} be the equivalent mass of the fractional oscillators in Class III. Then, for $1 < \alpha \leq 2$ and $0 < \beta \leq 1$,*

$$m_{eq3} = m_{eq3}(\omega, \alpha, \beta) = -\left(m\omega^{\alpha-2} \cos \frac{\alpha\pi}{2} + c\omega^{\beta-2} \cos \frac{\beta\pi}{2} \right).\tag{119}$$

Proof. When considering the Newton's second law in the equivalent oscillator III (110), we immediately see that Theorem 8 holds. $\square$

Remark 13. *The equivalent mass m_{eq3} obeys the power law in terms of ω.*

Note 4.11: The equivalent mass m_{eq3} is related to ω, m, and c, as well as a pair of fractional orders (α, β).

Note 4.12: If $\alpha = 2$ and $\beta = 1$, m_{eq3} reduces to the primary m, i.e.,

$$m_{eq3}(\omega, \alpha, \beta)\big|_{\alpha=2, \beta=1} = m.\tag{120}$$

As a matter of fact, a fractional oscillator of Class III reduces to the ordinary oscillator when $\alpha = 2$ and $\beta = 1$.

Remark 14. *In the case of $\omega \to \infty$, we obtain*

$$\lim_{\omega \to \infty} m_{eq3}(\omega, \alpha, \beta) = 0, 1 < \alpha < 2, \, 0 < \beta < 1.\tag{121}$$

Therefore, we suggest that a fractional oscillator in Class III does not oscillate for $\omega \to \infty$ because its equivalent mass disappears in that case. Figure 6 shows its positive part for $\alpha = 1.5$, $\beta = 0.9$, $m = c = 1$.

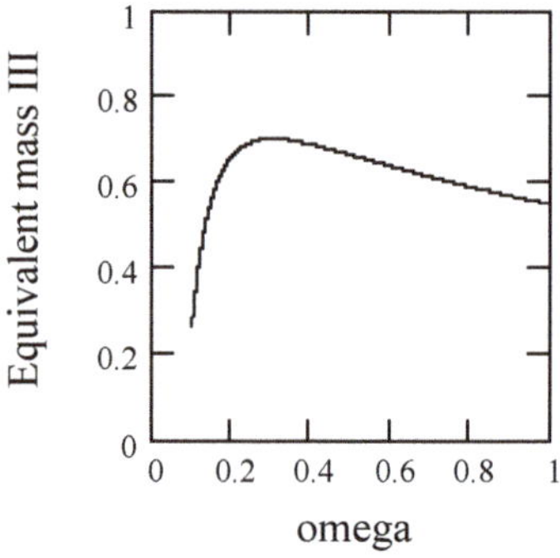

Figure 6. Indicating the positive part of $m_{eq3}(\omega, \alpha, \beta)$ for $\alpha = 1.5$, $\beta = 0.9$, $m = c = 1$.

Remark 15. *In the case of $\omega \to 0$, we obtain*

$$\lim_{\omega \to 0} m_{eq3}(\omega, \alpha, \beta) = -\infty, 1 < \alpha < 2, 0 < \beta < 1. \tag{122}$$

In fact, if ω is small enough, $m_{eq3}(\omega, \alpha, \beta)$ will be negative, see Figure 7.

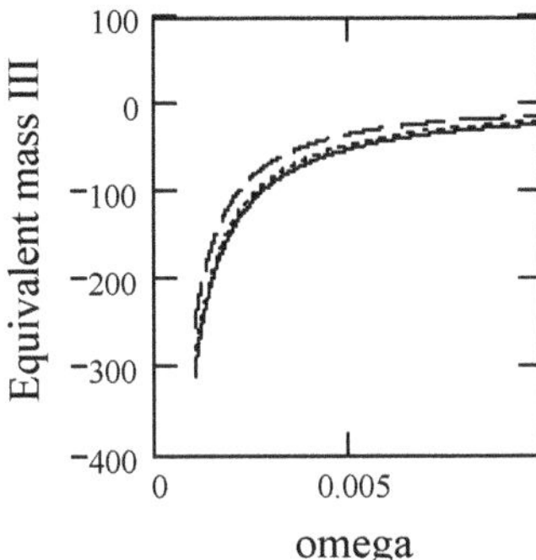

Figure 7. Negative part of $m_{eq3}(\omega, \alpha, \beta)$ for $m = c = 1$ and $\beta = 0.9$. Solid line: $\alpha = 1.9$. Dot line: $\alpha = 1.6$. Dash line: $\alpha = 1.3$.

Remark 16. *This research restricts $m_{eq3}(\omega, \alpha, \beta) \in (0, \infty)$.*

4.3.3. Equivalent Damping of Fractional Oscillators in Class III

Let c_{eq3} be the equivalent damping of a fractional oscillator of Class III type. Then, we propose its expression with Theorem 9.

Theorem 9 (Equivalent damping III). *The equivalent damping of the fractional oscillators in Class III is given by, for $1 < \alpha \leq 2$ and $0 < \beta \leq 1$,*

$$c_{eq3} = c_{eq3}(\omega, \alpha, \beta) = m\omega^{\alpha-1} \sin \frac{\alpha \pi}{2} + c\omega^{\beta-1} \sin \frac{\beta \pi}{2}. \tag{123}$$

Proof. The second term on the left side of the equivalent oscillator III is the friction force with the linear viscous damping coefficient denoted by (123). Thus, the proof completes. $\square$

Remark 17. *The equivalent damping c_{eq3} relates to ω, m, c, and a pair of fractional orders (α, β). It obeys the power law in terms of ω. It contains two terms. The first term is hyperbolically increasing in $\omega^{\alpha-1}$ as $\alpha > 1$ and the second hyperbolically decayed with $\omega^{\beta-1}$ since $\beta < 1$.*

Note 4.13: From (123), we see that c_{eq3} reduces to the primary damping c for $\alpha = 2$ and $\beta = 1$. That is,

$$c_{eq3}(\omega, \alpha, \beta)\big|_{\alpha=2, \beta=1} = c. \tag{124}$$

Remark 18. *One asymptotic property of c_{eq3} for $\omega \to \infty$, due to $\lim_{\omega \to \infty} \omega^{\alpha-1} = \infty$ for $1 < \alpha \leq 2$, is given by*

$$\lim_{\omega \to \infty} c_{eq3}(\omega, \alpha, \beta) = \infty. \tag{125}$$

The above says that a fractional oscillator of Class III does not vibrate for $\omega \to \infty$.

Remark 19. *Another asymptotic property of c_{eq3} in terms of ω for $\omega \to 0$, owing to $\lim\limits_{\omega \to 0} \omega^{\beta-1} = \infty$ for $0 < \beta < 1$, is expressed by*

$$\lim_{\omega \to 0} c_{eq3}(\omega, \alpha, \beta) = \infty. \tag{126}$$

A system does not vibrate obviously in the case of $\omega \to 0$ but Remark 19 suggests a new view about that. Precisely, its equivalent damping is infinitely large at $\omega \to 0$. Figures 8 and 9 illustrate $c_{eq3}(\omega, \alpha, \beta)$ for $m = c = 1$.

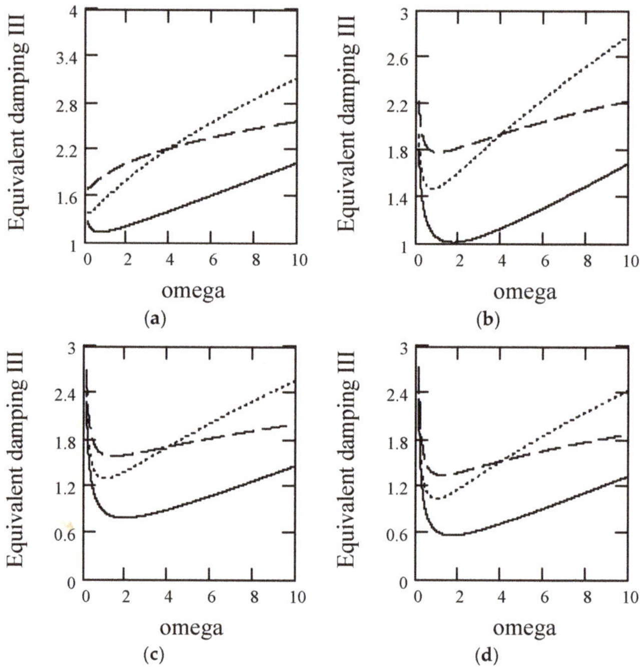

Figure 8. Plots of equivalent damping III for $m = c = 1$. Solid line: $\alpha = 1.9$. Dot line: $\alpha = 1.6$. Dash line: $\alpha = 1.3$. (**a**) For $\beta = 0.9$. (**b**) For $\beta = 0.7$. (**c**) For $\beta = 0.5$. (**d**) For $\beta = 0.3$.

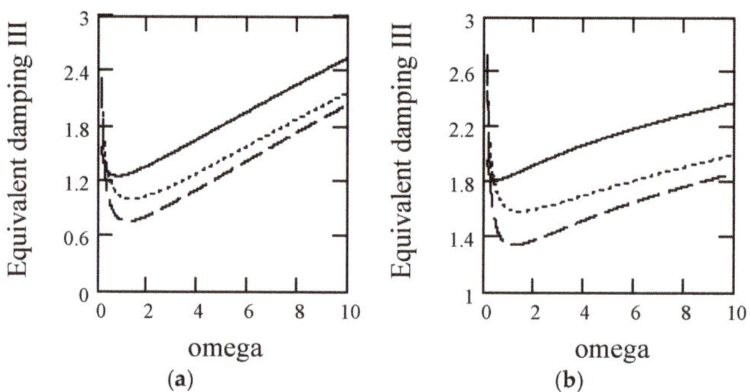

Figure 9. Plots of $c_{eq3}(\omega, \alpha, \beta)$ for $m = c = 1$. Solid line: $\beta = 0.8$. Dot line: $\beta = 0.5$. Dash line: $\beta = 0.3$. (**a**) For $\alpha = 1.8$. (**b**) For $\alpha = 1.3$.

Note 4.14: The equivalent damping $c_{eq3} = 0$ if both $\alpha = 2$ and $c = 0$:

$$c_{eq3}(\omega, \alpha, \beta)\big|_{\alpha=2, c=0} = 0. \tag{127}$$

4.4. Summary

We have proposed three equivalent oscillation equations with order 2 to equivalently characterize three classes of fractional oscillators, opening a novel way of studying fractional oscillators. The analytic expressions of equivalent mass m_{eqj} and damping c_{eqj} ($j = 1, 2, 3$) for each equivalent oscillator have been presented. One general thing regarding m_{eqj} and damping c_{eqj} is that they follow power laws. Another thing in common is that they are dependent on oscillation frequency ω and fractional order.

5. Equivalent Natural Frequencies and Damping Ratio of Three Classes of Fractional Oscillators

We have presented three equivalent oscillation equations corresponding to three classes of fractional oscillators in the last section. Functionally, they are abstracted in a unified form

$$m_{eqj}\frac{d^2x_j(t)}{dt^2} + c_{eqj}\frac{dx_j(t)}{dt} + kx_j(t) = f(t), j = 1, 2, 3. \tag{128}$$

In each equivalent oscillator, either m_{eqj} or c_{eqj} is not a constant in general. Instead, either is a function of the oscillation frequency ω and the fractional order α for m_{eq1} and c_{eq1}, β for m_{eq2} and c_{eq2}, (α, β) for m_{eq3} and c_{eq3}. Consequently, natural frequencies and damping ratios of fractional oscillators should rely on ω and fractional order. We shall propose their analytic expressions in this section.

5.1. Equivalent Natural Frequency I

Definition 1. *Denote by $\omega_{eqn,j}$ a natural frequency of a fractional oscillator in the jth class ($j = 1, 2, 3$). It takes the form*

$$\omega_{eqn,j} = \sqrt{\frac{k}{m_{eqj}}}, j = 1, 2, 3, \tag{129}$$

where m_{eqj} is the equivalent mass of the fractional oscillator in the jth class.

With the above definition, we write (128) by

$$\frac{d^2x_j(t)}{dt^2} + \frac{c_{eqj}}{m_{eqj}}\frac{dx_j(t)}{dt} + \frac{k}{m_{eqj}}x_j(t) = \frac{d^2x_j(t)}{dt^2} + \frac{c_{eqj}}{m_{eqj}}\frac{dx_j(t)}{dt} + \omega_{eqn,j}^2 x_j(t) = \frac{f(t)}{m_{eqj}}, j = 1, 2, 3. \tag{130}$$

Note 5.1: $\omega_{eqn,j}$ may take the conventional natural frequency, denoted by

$$\omega_n = \sqrt{\frac{k}{m}}, \tag{131}$$

as a special case.

Corollary 1 (Equivalent natural frequency I1). *The equivalent natural frequency I1, which we denote it by $\omega_{eqn,1}$, of a fractional oscillator in Class I is given by*

$$\omega_{eqn,1} = \frac{\omega_n}{\sqrt{-\omega^{\alpha-2}\cos\frac{\alpha\pi}{2}}}, 1 < \alpha \leq 2. \tag{132}$$

Proof. According to (129), we have, for $1 < \alpha \leq 2$,

$$\omega_{eqn,1} = \sqrt{\frac{k}{m_{eq1}}} = \sqrt{\frac{k}{-m\omega^{\alpha-2}\cos\frac{\alpha\pi}{2}}} = \sqrt{\frac{1}{-\omega^{\alpha-2}\cos\frac{\alpha\pi}{2}}}\sqrt{\frac{k}{m}} = \frac{\omega_n}{\sqrt{-\omega^{\alpha-2}\cos\frac{\alpha\pi}{2}}}. \tag{133}$$

The proof finishes. $\square$

Figure 10 shows the plots of $\omega_{eqn,1}$.

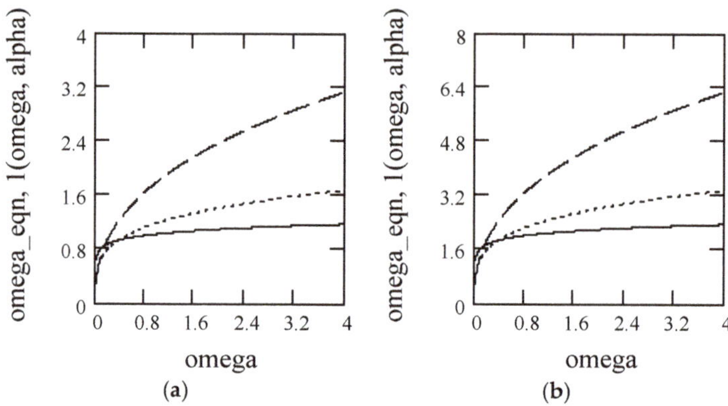

Figure 10. Natural frequency $\omega_{eqn,1}$. Solid line: $\alpha = 1.8$. Dot line: $\alpha = 1.5$. Dash line: $\alpha = 1.2$. (a) $\omega_n = 1$. (b) $\omega_n = 2$.

Note 5.2: From Figure 10, we see that $\omega_{eqn,1}$ is an increasing function with ω. Besides, the greater the value of α the smaller the $\omega_{eqn,1}$.

Note 5.3: $\omega_{eqn,1}$ becomes ω_n if $\alpha = 2$. In fact,

$$\omega_{eqn,1}\big|_{\alpha=2} = \sqrt{\frac{k}{m_{eq1}}}\bigg|_{\alpha=2} = \frac{\omega_n}{\sqrt{-\omega^{\alpha-2}\cos\frac{\alpha\pi}{2}}}\bigg|_{\alpha=2} = \omega_n. \tag{134}$$

Corollary 2 (Equivalent natural frequency I2). *The natural frequency I2, $\omega_{eqn,2}$, of a fractional oscillator in Class II is given by*

$$\omega_{eqn,2} = \frac{\omega_n}{\sqrt{1 - \frac{c}{m}\omega^{\beta-2}\cos\frac{\beta\pi}{2}}}. \tag{135}$$

Proof. Following (129), we have

$$\omega_{eqn,2} = \sqrt{\frac{k}{m_{eq2}}} = \sqrt{\frac{k}{m - c\omega^{\beta-2}\cos\frac{\beta\pi}{2}}} = \sqrt{\frac{k}{m\left(1 - \frac{c}{m}\omega^{\beta-2}\cos\frac{\beta\pi}{2}\right)}} = \frac{\omega_n}{\sqrt{1 - \frac{c}{m}\omega^{\beta-2}\cos\frac{\beta\pi}{2}}}.$$

Hence, the proof completes. $\square$

Figure 11 indicates the curves of $\omega_{eqn,2}$.

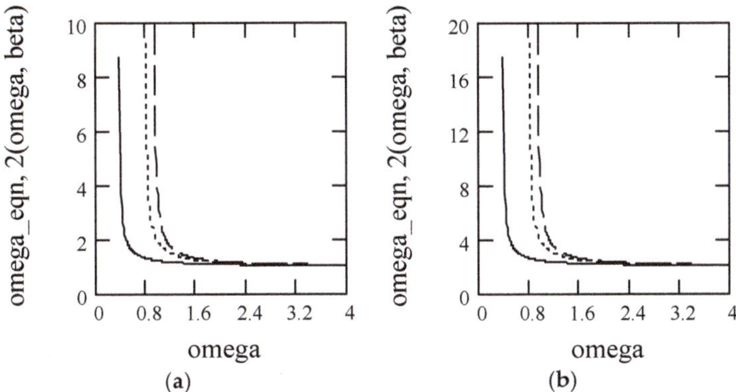

Figure 11. Curves of $\omega_{eqn,2}$ for $m = c = 1$. Solid line: $\beta = 0.8$. Dot line: $\beta = 0.5$. Dash line: $\beta = 0.3$.
(a) $\omega_n = 1$. (b) $\omega_n = 2$.

Note 5.4: Figure 11 shows that $\omega_{eqn,2}$ is a decreasing function with ω. The greater the value of β the smaller the $\omega_{eqn,2}$.

Note 5.5: $\omega_{eqn,2}$ takes ω_n as a special case for $\beta = 1$. As a matter of fact,

$$\omega_{eqn,2}\big|_{\beta=1} = \left. \frac{\omega_n}{\sqrt{1 - \frac{c}{m}\omega^{\beta-2} \cos\frac{\beta\pi}{2}}} \right|_{\beta=1} = \omega_n. \tag{136}$$

Corollary 3 (Equivalent natural frequency I3). *The natural frequency I3, denoted by $\omega_{eqn,3}$, of a fractional oscillator in Class III is given by*

$$\omega_{eqn,3} = \frac{\omega_n}{\sqrt{-\left(\omega^{\alpha-2}\cos\frac{\alpha\pi}{2} + \frac{c}{m}\omega^{\beta-2}\cos\frac{\beta\pi}{2}\right)}}. \tag{137}$$

Proof. With (129), we write

$$\omega_{eqn,3} = \sqrt{\frac{k}{m_{eq3}}} = \sqrt{\frac{k}{-\left(m\omega^{\alpha-2}\cos\frac{\alpha\pi}{2} + c\omega^{\beta-2}\cos\frac{\beta\pi}{2}\right)}} = \frac{\omega_n}{\sqrt{-\left(\omega^{\alpha-2}\cos\frac{\alpha\pi}{2} + \frac{c}{m}\omega^{\beta-2}\cos\frac{\beta\pi}{2}\right)}}. \tag{138}$$

The above completes the proof. $\square$

Figure 12 gives the illustrations of $\omega_{eqn,3}$.

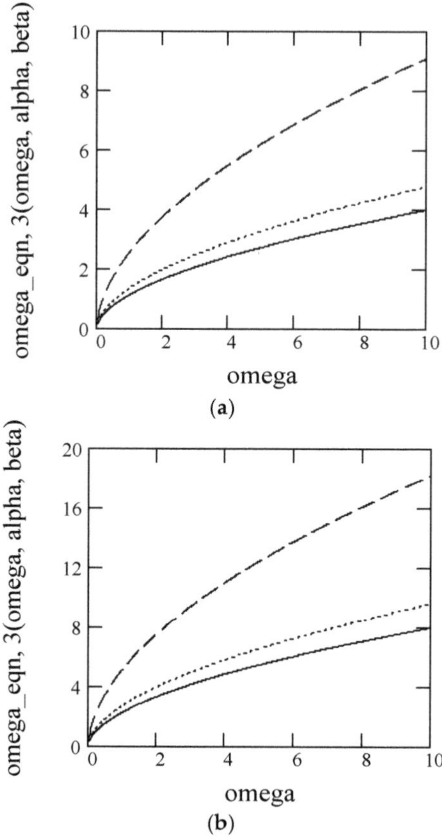

Figure 12. Illustrations of $\omega_{eqn,3}$ for $m = c = 1$. Solid line: $\alpha = 1.8$, $\beta = 0.9$. Dot line: $\alpha = 1.5$, $\beta = 0.9$. Dash line: $\alpha = 1.2$, $\beta = 0.9$. (**a**) $\omega_n = 1$. (**b**) $\omega_n = 2$.

Note 5.6: Figure 12 exhibits that $\omega_{eqn,3}$ is an increasing function in terms of ω.

Note 5.7: $\omega_{eqn,3}$ takes ω_n as a special case for $\alpha = 2$ and $\beta = 1$. Indeed,

$$\omega_{eqn,3}\Big|_{\alpha=2,\beta=0} = \sqrt{\left.\frac{k}{-\left(m\omega^{\alpha-2}\cos\frac{\alpha\pi}{2} + c\omega^{\beta-2}\cos\frac{\beta\pi}{2}\right)}\right|_{\alpha=2,\beta=1}} = \sqrt{\frac{k}{m}} = \omega_n. \tag{139}$$

5.2. Equivalent Damping Ratio

Definition 2. *Let ς_{eqj} be the equivalent damping ratio of the equivalent system of a fractional oscillator in Class j. It is defined by*

$$\varsigma_{eqj} = \frac{c_{eqj}}{2\sqrt{m_{eqj}k}}, j = 1, 2, 3. \tag{140}$$

Corollary 4 (Equivalent damping ratio I). *The equivalent damping ratio of a fractional oscillator in Class I is expressed by*

$$\varsigma_{eq1} = \varsigma_{eq1}(\omega,\alpha) = \frac{\omega^{\frac{\alpha}{2}}\sin\frac{\alpha\pi}{2}}{2\omega_n\sqrt{-\cos\frac{\alpha\pi}{2}}}, 1 < \alpha \le 2. \tag{141}$$

Proof. Replacing m_{eq1} and c_{eq1} in the expression below with the equivalent mass I and the equivalent damping I described in Section 4

$$\varsigma_{eq1} = \frac{c_{eq1}}{2\sqrt{m_{eq1}k}} \tag{142}$$

yields

$$\varsigma_{eq1} = \frac{m\omega^{\alpha-1}\sin\frac{\alpha\pi}{2}}{2\sqrt{\left(-m\omega^{\alpha-2}\cos\frac{\alpha\pi}{2}\right)k}} = \frac{\omega^{\frac{\alpha}{2}}\sin\frac{\alpha\pi}{2}}{2\sqrt{-\cos\frac{\alpha\pi}{2}}}\sqrt{\frac{m}{k}} = \frac{\omega^{\frac{\alpha}{2}}\sin\frac{\alpha\pi}{2}}{2\omega_n\sqrt{-\cos\frac{\alpha\pi}{2}}}, 1 < \alpha \le 2. \tag{143}$$

The proof finishes. $\square$

Remark 20. *The damping ratio ς_{eq1} follows the power law in terms of ω.*

Remark 21. *The damping ratio of fractional oscillators in Class I relates to the oscillation frequency ω and the fractional order α. It is increasing with respect to ω.*

$$\varsigma_{eq1}(0,\alpha) = 0 \text{ and } \varsigma_{eq1}(\infty,\alpha) = \infty. \tag{144}$$

Figure 13 shows the curves of $\varsigma_{eq1}(\omega,\alpha)$.

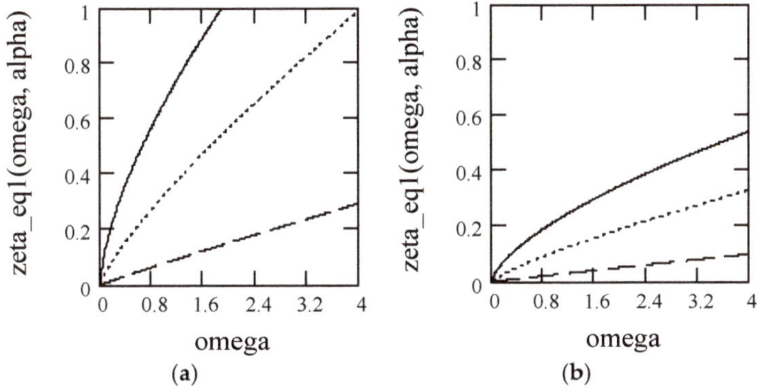

Figure 13. Illustrations of $\varsigma_{eq1}(\omega,\alpha)$. Solid line: $\alpha = 1.3$. Dot line: $\alpha = 1.6$. Dash line: $\alpha = 1.9$. (a) For $\omega_n = 1$. (b) For $\omega_n = 3$.

Note 5.8: Figure 13 indicates that the smaller the α the greater the ς_{eq1}.

Corollary 5 (Equivalent damping ratio II). *The damping ratio of a fractional oscillator in Class II is given by*

$$\varsigma_{eq2} = \varsigma_{eq2}(\omega,\beta) = \frac{\varsigma\omega^{\beta-1}\sin\frac{\beta\pi}{2}}{\sqrt{1-\frac{c}{m}\omega^{\beta-2}\cos\frac{\beta\pi}{2}}}, 0 < \beta \le 1, \tag{145}$$

where $\varsigma = \frac{c}{2\sqrt{mk}}$.

Proof. When replacing the m_{eq2} and c_{eq2} in the following expression by the equivalent mass II and the equivalent damping II proposed in Section 4, we attain

$$\varsigma_{eq2} = \frac{c_{eq2}}{2\sqrt{m_{eq2}k}} = \frac{c\omega^{\beta-1}\sin\frac{\beta\pi}{2}}{2\sqrt{\left(m-c\omega^{\beta-2}\cos\frac{\beta\pi}{2}\right)k}} = \frac{c\omega^{\beta-1}\sin\frac{\beta\pi}{2}}{2\sqrt{\left(1-\frac{c}{m}\omega^{\beta-2}\cos\frac{\beta\pi}{2}\right)mk}}$$

$$= \frac{c\omega^{\beta-1}\sin\frac{\beta\pi}{2}}{2\sqrt{mk}\sqrt{1-\frac{c}{m}\omega^{\beta-2}\cos\frac{\beta\pi}{2}}} = \frac{\varsigma\omega^{\beta-1}\sin\frac{\beta\pi}{2}}{\sqrt{1-\frac{c}{m}\omega^{\beta-2}\cos\frac{\beta\pi}{2}}}, 0 < \beta \leq 1.$$

This finishes the proof. $\square$

Remark 22. *The damping ratio ς_{eq2} obeys the power law in terms of ω.*

Remark 23. *The damping ratio ς_{eq2} is associated with ω and the fractional order β. It is decreasing in terms of ω.*

Note 5.9: ς_{eq2} takes ς as a special case for $\beta = 1$. In fact,

$$\varsigma_{eq2}(\omega,\beta)\big|_{\beta=1} = \left.\frac{\varsigma\omega^{\beta-1}\sin\frac{\beta\pi}{2}}{\sqrt{1-\frac{c}{m}\omega^{\beta-2}\cos\frac{\beta\pi}{2}}}\right|_{\beta=1} = \varsigma. \tag{146}$$

Figure 14 indicates the plots of $\varsigma_{eq2}(\omega,\beta)$ in the case of $m=1$, $c=1$, and $k=1$.

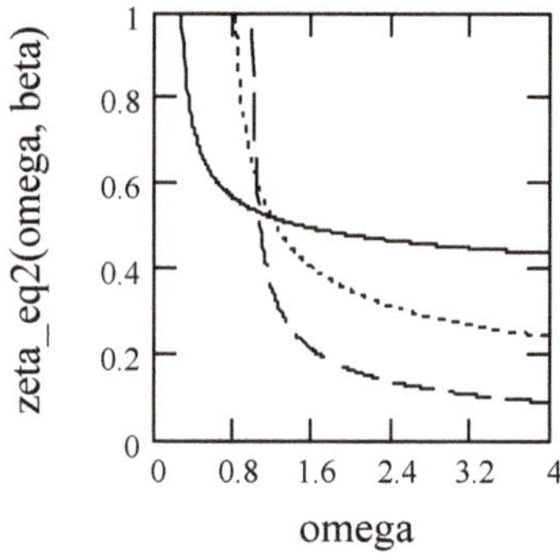

Figure 14. Plots of $\varsigma_{eq2}(\omega,\beta)$ for $m = c = k = 1$. Solid line: $\beta = 0.9$. Dot line: $\beta = 0.6$. Dash line: $\beta = 0.3$.

Corollary 6 (Equivalent damping ratio III). *Let ς_{eq3} be the damping ratio of a fractional oscillator in Class III. Then, for $1 < \alpha \leq 2, 0 < \beta \leq 1$,*

$$\varsigma_{eq3} = \varsigma_{eq3}(\omega,\alpha,\beta) = \frac{\omega^{\alpha-1}\sin\frac{\alpha\pi}{2} + 2\varsigma\omega_n\omega^{\beta-1}\sin\frac{\beta\pi}{2}}{2\omega_n\sqrt{-\left(\omega^{\alpha-2}\cos\frac{\alpha\pi}{2} + 2\varsigma\omega_n\omega^{\beta-2}\cos\frac{\beta\pi}{2}\right)}}. \tag{147}$$

Proof. If replacing the m_{eq3} and c_{eq3} below with the equivalent mass III and the equivalent damping III presented in Section 4, we obtain

$$\varsigma_{eq3} = \frac{c_{eq3}}{2\sqrt{m_{eq3}k}} = \frac{m\omega^{\alpha-1}\sin\frac{\alpha\pi}{2}+c\omega^{\beta-1}\sin\frac{\beta\pi}{2}}{2\sqrt{-\left(m\omega^{\alpha-2}\cos\frac{\alpha\pi}{2}+c\omega^{\beta-2}\cos\frac{\beta\pi}{2}\right)k}}$$

$$= \frac{m\left(\omega^{\alpha-1}\sin\frac{\alpha\pi}{2}+\frac{c}{m}\omega^{\beta-1}\sin\frac{\beta\pi}{2}\right)}{2\sqrt{-\left(\omega^{\alpha-2}\cos\frac{\alpha\pi}{2}+\frac{c}{m}\omega^{\beta-2}\cos\frac{\beta\pi}{2}\right)mk}} = \frac{m\left(\omega^{\alpha-1}\sin\frac{\alpha\pi}{2}+2\varsigma\omega_n\omega^{\beta-1}\sin\frac{\beta\pi}{2}\right)}{2\sqrt{mk}\sqrt{-\left(\omega^{\alpha-2}\cos\frac{\alpha\pi}{2}+2\varsigma\omega_n\omega^{\beta-2}\cos\frac{\beta\pi}{2}\right)}}$$

$$= \frac{\omega^{\alpha-1}\sin\frac{\alpha\pi}{2}+2\varsigma\omega_n\omega^{\beta-1}\sin\frac{\beta\pi}{2}}{2\omega_n\sqrt{-\left(\omega^{\alpha-2}\cos\frac{\alpha\pi}{2}+2\varsigma\omega_n\omega^{\beta-2}\cos\frac{\beta\pi}{2}\right)}}.$$

Thus, we finish the proof. $\square$

Remark 24. *The damping ratio ς_{eq3} follows the power law in terms of ω.*

Remark 25. *ς_{eq3} relates to ω and a pair of fractional orders (α, β).*

Note 5.10: ς_{eq3} regards ζ as a special case for $\alpha = 2$ and $\beta = 1$. As a matter of fact,

$$\varsigma_{eq3}(\omega, 2, 1) = \left.\frac{\omega^{\alpha-1}\sin\frac{\alpha\pi}{2}+2\varsigma\omega_n\omega^{\beta-1}\sin\frac{\beta\pi}{2}}{2\omega_n\sqrt{-\left(\omega^{\alpha-2}\cos\frac{\alpha\pi}{2}+2\varsigma\omega_n\omega^{\beta-2}\cos\frac{\beta\pi}{2}\right)}}\right|_{\alpha=2,\beta=1} = \varsigma. \tag{148}$$

Figure 15 demonstrates the figures of $\varsigma_{eq3}(\omega, \alpha, \beta)$ in the case of $m = 1$, $c = 1$, and $k = 1$.

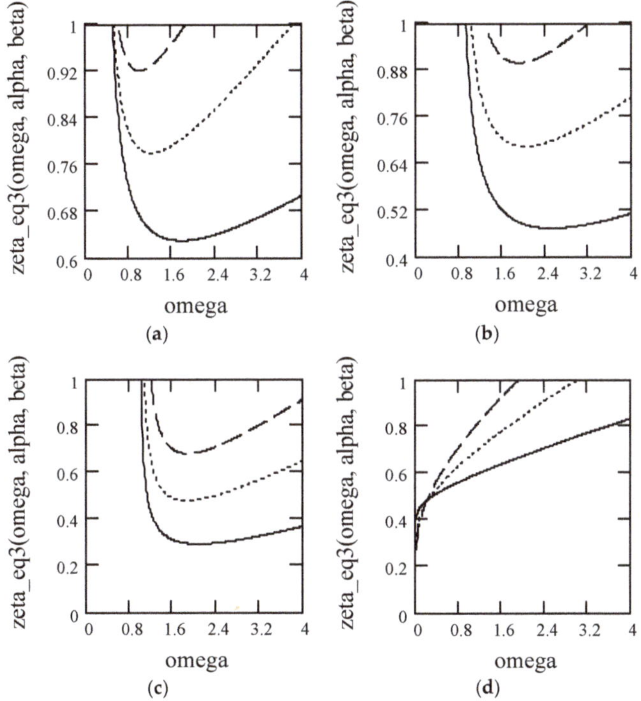

Figure 15. Demonstrations of $\varsigma_{eq3}(\omega, \alpha, \beta)$ in $m = c = k = 1$. Solid line: $\alpha = 1.9$. Dot line: $\alpha = 1.8$. Dash line: $\alpha = 1.7$. (**a**) For $\beta = 0.8$. (**b**) For $\beta = 0.5$. (**c**) For $\beta = 0.2$. (**d**) For $\beta = 1$.

5.3. Equivalent Natural Frequency II

Now, with two parameters $\omega_{eqn,j}$ and ς_{eqj} presented above, we rewrite the equivalent oscillator (130) by

$$\frac{d^2 x_j(t)}{dt^2} + 2\varsigma_{eqj}\omega_{eqn,j}\frac{dx_j(t)}{dt} + \omega_{eqn,j}^2 x_j(t) = \frac{f(t)}{m_{eqj}}, j = 1,2,3. \tag{149}$$

The characteristic equation of (149) is given by

$$s_j^2 + 2\varsigma_{eqj}\omega_{eqn,j}s_j + \omega_{eqn,j}^2 = 0, j = 1,2,3. \tag{150}$$

The characteristic roots are in the form

$$s_{j,1,2} = -\varsigma_{eqj}\omega_{eqn,j} \pm \sqrt{\varsigma_{eqj}^2\omega_{eqn,j}^2 - \omega_{eqn,j}^2} = -\varsigma_{eqj}\omega_{eqn,j} \pm i\omega_{eqn,j}\sqrt{1 - \varsigma_{eqj}^2}, j = 1,2,3. \tag{151}$$

Functionally, we utilize the symbol $\omega_{eqd,j}$ for

$$\omega_{eqd,j} = \omega_{eqn,j}\sqrt{1 - \varsigma_{eqj}^2}, j = 1,2,3. \tag{152}$$

Thus, the characteristic roots are

$$s_{j,1,2} = -\varsigma_{eqj}\omega_{eqn,j} \pm i\omega_{eqd,j}, j = 1,2,3. \tag{153}$$

Note that, in practice, $0 \le \varsigma_{eqj} < 1$ because $1 \le \varsigma_{eqj}$ means no oscillation at all.

We write those above for the sake of applying the theory of linear oscillations to fractional ones. Now, we discuss $\omega_{eqd,j}$.

Corollary 7 (Equivalent natural frequency II1). *Let $\omega_{eqd,1}$ be the functional damped natural frequency of a fractional oscillator in Class I. It may be termed the equivalent natural frequency II1. Then,*

$$\omega_{eqd,1} = \omega_{eqd,1}(\omega,\alpha) = \frac{\omega_n}{\sqrt{-\omega^{\alpha-2}\cos\frac{\alpha\pi}{2}}}\sqrt{1 - \frac{\omega^\alpha \sin^2\frac{\alpha\pi}{2}}{4\omega_n^2\left|\cos\frac{\alpha\pi}{2}\right|}}, 1 < \alpha \le 2. \tag{154}$$

Proof. Note that

$$\omega_{eqd,1} = \omega_{eqn,1}\sqrt{1 - \varsigma_{ed1}^2}. \tag{155}$$

Using the above ς_{ed1}, we have

$$\omega_{eqd,1} = \omega_{eqn,1}\sqrt{1 - \varsigma_{ed1}^2} = \frac{\omega_n}{\sqrt{-\omega^{\alpha-2}\cos\frac{\alpha\pi}{2}}}\sqrt{1 - \left(\frac{\omega^{\frac{\alpha}{2}}\sin\frac{\alpha\pi}{2}}{2\omega_n\sqrt{\left|\cos\frac{\alpha\pi}{2}\right|}}\right)^2}.$$

This finishes the proof. $\square$

The parameter $\omega_{eqd,1}$ functionally takes the form of damped natural frequency as in the conventional linear oscillation theory. In this research, we do not distinguish the natural frequencies with damped or damping free. At most, we just say that it is a functional damped one. It relates to the oscillation frequency ω and the fractional order α.

Remark 26. $\omega_{eqd,1}$ *is not a monotonic function of ω.*

Note 5.11: ω_n is a special case of $\omega_{eqd,1}$ when $\alpha = 2$:

$$\omega_{eqd,1}(\omega,2) = \frac{\omega_n}{\sqrt{-\omega^{\alpha-2}\cos\frac{\alpha\pi}{2}}}\sqrt{1 - \frac{\omega^\alpha \sin^2\frac{\alpha\pi}{2}}{4\omega_n^2\left|\cos\frac{\alpha\pi}{2}\right|}}\,\Bigg|_{\alpha=2} = \omega_n. \tag{156}$$

As a matter of fact, fractional oscillators of Class I are damping free for $\alpha = 2$. Figure 16 illustrates the plots of $\omega_{eqd,1}(\omega,\alpha)$.

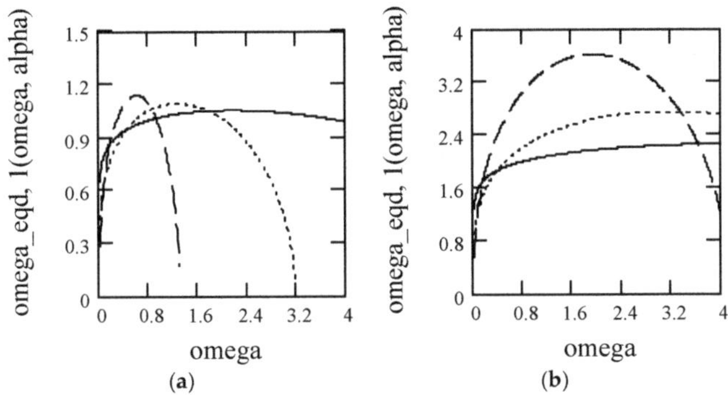

Figure 16. Plots of $\omega_{eqd,1}(\omega,\alpha)$. Solid line: $\alpha = 1.8$. Dot line: $\alpha = 1.5$. Dash line: $\alpha = 1.2$. **(a)** For $\omega_n = 1$. **(b)** For $\omega_n = 2$.

Corollary 8 (Equivalent natural frequency II2). *Let $\omega_{eqd,2}$ be the functional damped natural frequency of a fractional oscillator in Class II. Term it with the equivalent natural frequency II2. Then, for $0 < \beta \leq 1$,*

$$\omega_{eqd,2} = \omega_{eqd,2}(\omega,\beta) = \frac{\omega_n}{\sqrt{1 - \frac{c}{m}\omega^{\beta-2}\cos\frac{\beta\pi}{2}}}\sqrt{1 - \frac{\varsigma^2\omega^{2(\beta-1)}\sin^2\frac{\beta\pi}{2}}{1 - \frac{c}{m}\omega^{\beta-2}\cos\frac{\beta\pi}{2}}}. \tag{157}$$

Proof. Consider

$$\omega_{eqd,2} = \omega_{eqn,2}\sqrt{1 - \varsigma_{ed2}^2}. \tag{158}$$

Replacing $\omega_{eqn,2}$ and ς_{ed2} in the above yields

$$\omega_{eqd,2} = \omega_{eqn,2}\sqrt{1 - \varsigma_{ed2}^2} = \frac{\omega_n}{\sqrt{1 - \frac{c}{m}\omega^{\beta-2}\cos\frac{\beta\pi}{2}}}\sqrt{1 - \left(\frac{\varsigma\omega^{\beta-1}\sin\frac{\beta\pi}{2}}{2\sqrt{1 - \frac{c}{m}\omega^{\beta-2}\cos\frac{\beta\pi}{2}}}\right)^2}$$

$$= \frac{\omega_n}{\sqrt{1 - \frac{c}{m}\omega^{\beta-2}\cos\frac{\beta\pi}{2}}}\sqrt{1 - \frac{\varsigma^2\omega^{2(\beta-1)}\sin^2\frac{\beta\pi}{2}}{1 - \frac{c}{m}\omega^{\beta-2}\cos\frac{\beta\pi}{2}}}.$$

Thus, Corollary 8 holds. $\square$

Remark 27. *$\omega_{eqd,2}$ is related to ω and the fractional order β.*

Note 5.12: The conventional damped natural frequency, say,

$$\omega_d = \omega_n\sqrt{1 - \varsigma^2} \tag{159}$$

is a special case of $\omega_{eqd,2}(\omega,\beta)$ for $\beta = 1$.

Figure 17 gives the plots of $\omega_{eqd,2}(\omega,\beta)$.

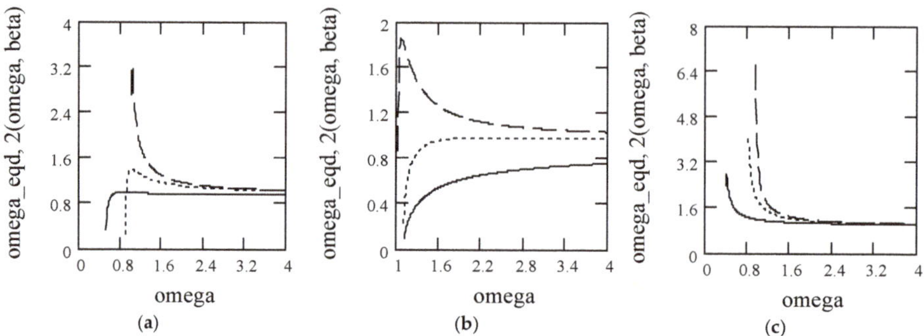

Figure 17. Illustrating $\omega_{eqd,2}(\omega,\beta)$ for $m = c = 1$. Solid line: $\beta = 0.8$. Dot line: $\beta = 0.5$. Dash line: $\beta = 0.2$. (a) For $\omega_n = 1$. (b) For $\omega_n = 0.3$. (c) For $\omega_n = 10$.

Corollary 9 (Equivalent natural frequency II3). *Let* $\omega_{eqd,3} = \omega_{eqd,3}(\omega,\alpha,\beta)$ *be the functional damped natural frequency of a fractional oscillator in Class III. Call it the equivalent natural frequency II3. Then, for* $1 < \alpha \leq 2$ *and* $0 < \beta \leq 1$, *we have*

$$\omega_{eqd,3} = \frac{\omega_n\sqrt{1 - \left[\dfrac{\left(\omega^{\alpha-1}\sin\frac{\alpha\pi}{2} + 2\varsigma\omega_n\omega^{\beta-1}\sin\frac{\beta\pi}{2}\right)^2}{4\omega_n^2\left[-\left(\omega^{\alpha-2}\cos\frac{\alpha\pi}{2} + 2\varsigma\omega_n\omega^{\beta-2}\cos\frac{\beta\pi}{2}\right)\right]}\right]^2}}{\sqrt{-\left(\omega^{\alpha-2}\cos\frac{\alpha\pi}{2} + \frac{c}{m}\omega^{\beta-2}\cos\frac{\beta\pi}{2}\right)}}. \tag{160}$$

Proof. In the expression below

$$\omega_{eqd,3} = \omega_{eqn,3}\sqrt{1 - \varsigma_{ed3}^2}, \tag{161}$$

we replace $\omega_{eqd,3}$ and ς_{eq3} by those expressed above. Then, we have

$$\omega_{eqd,3} = \omega_{eqn,3}\sqrt{1 - \varsigma_{ed3}^2} = \sqrt{\frac{k}{m_{eq3}}}\sqrt{1 - \varsigma_{ed3}^2}$$

$$= \frac{\omega_n\sqrt{1 - \left[\dfrac{\omega^{\alpha-1}\sin\frac{\alpha\pi}{2} + 2\varsigma\omega_n\omega^{\beta-1}\sin\frac{\beta\pi}{2}}{2\omega_n\sqrt{-\left(\omega^{\alpha-2}\cos\frac{\alpha\pi}{2} + 2\varsigma\omega_n\omega^{\beta-2}\cos\frac{\beta\pi}{2}\right)}}\right]^2}}{\sqrt{-\left(\omega^{\alpha-2}\cos\frac{\alpha\pi}{2} + \frac{c}{m}\omega^{\beta-2}\cos\frac{\beta\pi}{2}\right)}}$$

$$= \frac{\omega_n\sqrt{1 - \left[\dfrac{\left(\omega^{\alpha-1}\sin\frac{\alpha\pi}{2} + 2\varsigma\omega_n\omega^{\beta-1}\sin\frac{\beta\pi}{2}\right)^2}{4\omega_n^2\left[-\left(\omega^{\alpha-2}\cos\frac{\alpha\pi}{2} + 2\varsigma\omega_n\omega^{\beta-2}\cos\frac{\beta\pi}{2}\right)\right]}\right]^2}}{\sqrt{-\left(\omega^{\alpha-2}\cos\frac{\alpha\pi}{2} + \frac{c}{m}\omega^{\beta-2}\cos\frac{\beta\pi}{2}\right)}}.$$

Therefore, the corollary holds. $\square$

Note 5.13: The conventional damped natural frequency ω_d is a special case of $\omega_{eqd,3}$ for $(\alpha,\beta) = (2,1)$. Indeed,

$$\omega_{eqd,3}(\omega,2,1) = \left.\frac{\omega_n\sqrt{1 - \left[\dfrac{\left(\omega^{\alpha-1}\sin\frac{\alpha\pi}{2}+2\varsigma\omega_n\omega^{\beta-1}\sin\frac{\beta\pi}{2}\right)^2}{4\omega_n^2\left[-\left(\omega^{\alpha-2}\cos\frac{\alpha\pi}{2}+2\varsigma\omega_n\omega^{\beta-2}\cos\frac{\beta\pi}{2}\right)\right]}\right]^2}}{\sqrt{-\left(\omega^{\alpha-2}\cos\frac{\alpha\pi}{2}+\frac{c}{m}\omega^{\beta-2}\cos\frac{\beta\pi}{2}\right)}}\right|_{\alpha=1,\beta=1} = \omega_n\sqrt{1-\varsigma^2}. \tag{162}$$

Remark 28. *The natural frequency $\omega_{eqd,3}$ is associated with ω and a pair of fractional orders $(\alpha,\,\beta)$.*

Figures 18 and 19 indicate its plots.

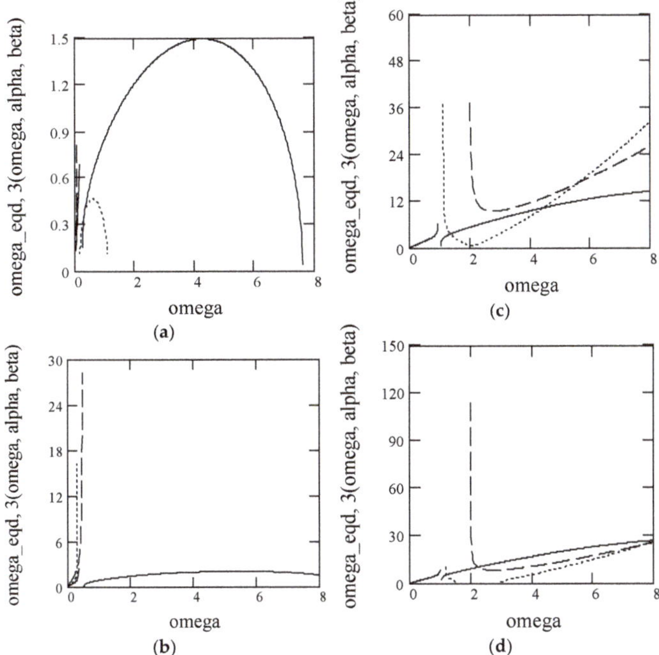

Figure 18. Demonstrations of $\omega_{eqd,3}(\omega,\alpha,\beta)$ for $m = c = k = 1$. Solid line: $\alpha = 1.9$. Dot line: $\alpha = 1.6$. Dash line: $\alpha = 1.3$. (**a**) For $\beta = 0.9$. (**b**) For $\beta = 0.8$. (**c**) For $\beta = 0.3$. (**d**) For $\beta = 0.2$.

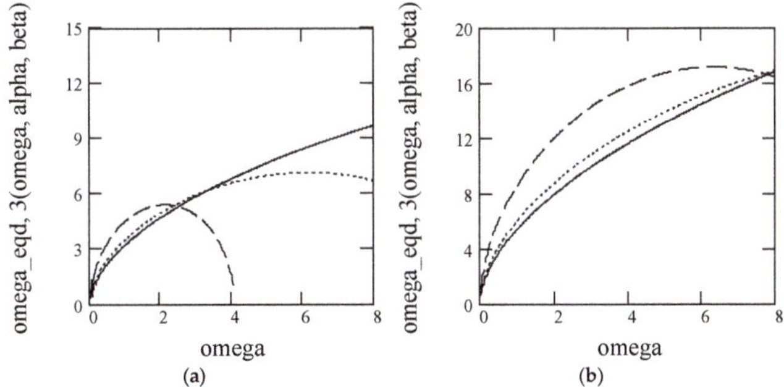

Figure 19. $\omega_{eqd,3}(\omega,\alpha,\beta)$ for $m = c = 1$, $\beta = 0.9$. Solid line: $\alpha = 1.9$. Dot line: $\alpha = 1.6$. Dash line: $\alpha = 1.3$. (**a**) For $\omega_n = 3$. (**b**) For $\omega_n = 5$.

5.4. There Exists Infinity of Natural Frequencies of a Fractional Oscillator

The previous discussions imply that there exists infinity of natural frequencies, for either $\omega_{eqn,j}$ or $\omega_{eqd,j}$, because each is dependent on $\omega \in (0, \infty)$. We functionally derived the two characteristic roots of the frequency equation (151), namely, $s_{j,1,2}$, actually stand for infinity of roots owing to $\omega \in (0, \infty)$.

Taking a fractional oscillator in Class I into account, its frequency equation is given by

$$s^{\alpha} + \omega_n^2 = 0, 1 < \alpha \le 2. \tag{163}$$

Then, it is easy to see that there exists infinitely many characteristic roots in the above, also see Li et al. [18].

A contribution in this work in representing characteristic roots of three classes of fractional oscillators is that they are expressed analytically. Moreover, functionally, they take the form as that in the theory of conventional linear oscillations, making it possible to represent solutions to three classes of fractional oscillators by using elementary functions, which are easier for use in both engineering applications and theoretic analysis of fractional oscillators.

6. Free Responses to Three Classes of Fractional Oscillators

We put forward the free responses in this section to three classes of fractional oscillators based on their equivalent oscillators presented in Section 4. Since the equivalent oscillators are expressed by using second-order differential equations in form, in methodology, therefore, it is easy for us to find the responses we concern with. Note that the equivalence explained in Section 4 says that

$$Y_j(\omega) = X_j(\omega), \; j = 1, \, 2, \, 3, \tag{164}$$

where the subscript j stands for the Class I to III. Consequently,

$$y_j(t) = x_j(t), \; j = 1, \, 2, \, 3. \tag{165}$$

Therefore, our research implies three advances.

- First, proposing the free responses to three classes of fractional oscillators using the way of solving conventional oscillators.
- Then, since the responses to conventional oscillators are represented by elementary functions while those to fractional ones are expressed by special functions, such as the Mittag-Leffler function and its generalizations, we shall present novel representation to a certain special functions by elementary ones.
- Finally, analytic expressions of the logarithmic decrements, which are useful in practice, of three classes of fractional oscillators are proposed.

6.1. General Form of Free Responses

Consider the free response to the functional equivalent oscillator in Class j in the form

$$\begin{cases} m_{eqj}\dfrac{d^2 x_j(t)}{dt^2} + c_{eqj}\dfrac{dx_j(t)}{dt} + kx_j(t) = 0 \\ x_j(0) = x_{j0}, \; \dfrac{dx_j(t)}{dt}\Big|_{t=0} = v_{j0} \end{cases} , j = 1,2,3. \tag{166}$$

Following the representation style in engineering, we rewrite it by

$$\begin{cases} \dfrac{d^2 x_j(t)}{dt^2} + 2\varsigma_{eqj}\omega_{eqn,j}\dfrac{dx_j(t)}{dt} + \omega_{eqn,j}^2 x_j(t) = 0 \\ x_j(0) = x_{j0}, \; \dfrac{dx_j(t)}{dt}\Big|_{t=0} = v_{j0} \end{cases} , j = 1,2,3. \tag{167}$$

Therefore (Timoshenko ([1], p. 34), Jin and Xia ([79], p. 11)), we have, for $t \geq 0$,

$$x_j(t) = e^{-\varsigma_{eqj}\omega_{eqn,j}t}\left(x_{j0}\cos\omega_{eqd,j}t + \frac{v_{j0} + \varsigma_{eqj}\omega_{eqn,j}x_{j0}}{\omega_{eqd,j}}\sin\omega_{eqd,j}t\right). \tag{168}$$

The above may be rewritten in the form

$$x_j(t) = A_{eqj}e^{-\varsigma_{eqj}\omega_{eqn,j}t}\cos\left(\omega_{eqd,j}t - \theta_{eqj}\right), t \geq 0, \tag{169}$$

where the equivalent amplitude A_{eqj} is given by

$$A_{eqj} = \sqrt{x_{j0}^2 + \left[\frac{v_{j0} + \varsigma_{eqj}\omega_{eqn,j}x_{j0}}{\omega_{eqd,j}}\right]^2}, \tag{170}$$

and the equivalent phase θ_{eqj} is

$$\theta_{eqj} = \tan^{-1}\frac{v_{j0} + \varsigma_{eqj}\omega_{eqn,j}x_{j0}}{\omega_{eqd,j}x_{j0}}. \tag{171}$$

Note that, for $\omega_{eqn,j}$, ς_{eqj}, A_{eqj}, and θ_{eqj}, each is not constant for fractional oscillators. Instead, each is generally a function of oscillation frequency ω and fractional order.

6.2. Free Response to Fractional Oscillators in Class I

We state the free response to a fractional oscillator in Class I by Theorem 10.

Theorem 10 (Free response I). *Let $x_1(t)$ be the free response to a fractional oscillator in Class I. Then, for $1 < \alpha \leq 2$, $x_1(t)$ is given by*

$$x_1(t) = e^{-\frac{\omega\sin\frac{\alpha\pi}{2}}{2|\cos\frac{\alpha\pi}{2}|}t}\left[\begin{array}{l} x_{10}\cos\left(\dfrac{\omega_n}{\sqrt{\omega^{\alpha-2}}|\cos\frac{\alpha\pi}{2}|}\sqrt{1 - \dfrac{\omega^\alpha\sin^2\frac{\alpha\pi}{2}}{4\omega_n^2|\cos\frac{\alpha\pi}{2}|}}\,t\right) \\[2em] + \dfrac{v_{10} + \dfrac{\omega^{\frac{\alpha}{2}}\sin\frac{\alpha\pi}{2}}{2\omega_n\sqrt{|\cos\frac{\alpha\pi}{2}|}}x_{10}}{\sqrt{1 - \dfrac{\omega^\alpha\sin^2\frac{\alpha\pi}{2}}{4\omega_n^2|\cos\frac{\alpha\pi}{2}|}}}\sin\left(\dfrac{\omega_n}{\sqrt{\omega^{\alpha-2}}|\cos\frac{\alpha\pi}{2}|}\sqrt{1 - \dfrac{\omega^\alpha\sin^2\frac{\alpha\pi}{2}}{4\omega_n^2|\cos\frac{\alpha\pi}{2}|}}\,t\right)\end{array}\right]. \tag{172}$$

Proof. For $t \geq 0$, consider

$$x_1(t) = e^{-\varsigma_{eq1}\omega_{eqn,1}t}\left(x_{10}\cos\omega_{eqd,1}t + \frac{v_{10} + \varsigma_{eq1}\omega_{eqn,1}x_{10}}{\omega_{eqd,1}}\sin\omega_{eqd,1}t\right). \tag{173}$$

In the above, replacing $\omega_{eqn,1}$ by the one in (132), ς_{eq1} with that in (141), and $\omega_{eqd,1}$ by the one in (154) yields

$$x_1(t) = e^{-\frac{\omega^{\frac{\alpha}{2}}\sin\frac{\alpha\pi}{2}}{2\omega_n\sqrt{\left|\cos\frac{\alpha\pi}{2}\right|}}\frac{\omega_n}{\sqrt{\omega^{\alpha-2}\left|\cos\frac{\alpha\pi}{2}\right|}}t}\left[\begin{array}{c} x_{10}\cos\left(\frac{\omega_n\sqrt{1-\frac{\omega^{\alpha}\sin^2\frac{\alpha\pi}{2}}{4\omega_n^2\left|\cos\frac{\alpha\pi}{2}\right|}}}{\sqrt{\omega^{\alpha-2}\left|\cos\frac{\alpha\pi}{2}\right|}}t\right) \\[2em] +\frac{v_{10}+\frac{\omega^{\frac{\alpha}{2}}\sin\frac{\alpha\pi}{2}}{2\omega_n\sqrt{\left|\cos\frac{\alpha\pi}{2}\right|}}\frac{\omega_n}{\sqrt{\omega^{\alpha-2}\left|\cos\frac{\alpha\pi}{2}\right|}}x_{10}}{\frac{\omega_n}{\sqrt{\omega^{\alpha-2}\left|\cos\frac{\alpha\pi}{2}\right|}}\sqrt{1-\frac{\omega^{\alpha}\sin^2\frac{\alpha\pi}{2}}{4\omega_n^2\left|\cos\frac{\alpha\pi}{2}\right|}}}\sin\left(\frac{\omega_n\sqrt{1-\frac{\omega^{\alpha}\sin^2\frac{\alpha\pi}{2}}{4\omega_n^2\left|\cos\frac{\alpha\pi}{2}\right|}}}{\sqrt{\omega^{\alpha-2}\left|\cos\frac{\alpha\pi}{2}\right|}}t\right) \end{array}\right]$$

$$= e^{-\frac{\omega\sin\frac{\alpha\pi}{2}}{2\left|\cos\frac{\alpha\pi}{2}\right|}t}\left[x_{10}\cos\left(\frac{\omega_n\sqrt{1-\frac{\omega^{\alpha}\sin^2\frac{\alpha\pi}{2}}{4\omega_n^2\left|\cos\frac{\alpha\pi}{2}\right|}}}{\sqrt{\omega^{\alpha-2}\left|\cos\frac{\alpha\pi}{2}\right|}}t\right) + \frac{v_{10}+\frac{\omega^{\frac{\alpha}{2}}\sin\frac{\alpha\pi}{2}}{2\omega_n\sqrt{\left|\cos\frac{\alpha\pi}{2}\right|}}x_{10}}{\sqrt{1-\frac{\omega^{\alpha}\sin^2\frac{\alpha\pi}{2}}{4\omega_n^2\left|\cos\frac{\alpha\pi}{2}\right|}}}\sin\left(\frac{\omega_n\sqrt{1-\frac{\omega^{\alpha}\sin^2\frac{\alpha\pi}{2}}{4\omega_n^2\left|\cos\frac{\alpha\pi}{2}\right|}}}{\sqrt{\omega^{\alpha-2}\left|\cos\frac{\alpha\pi}{2}\right|}}t\right)\right].$$

This completes the proof. $\square$

Figure 20 indicates $x_1(t)$ with fixed ω.

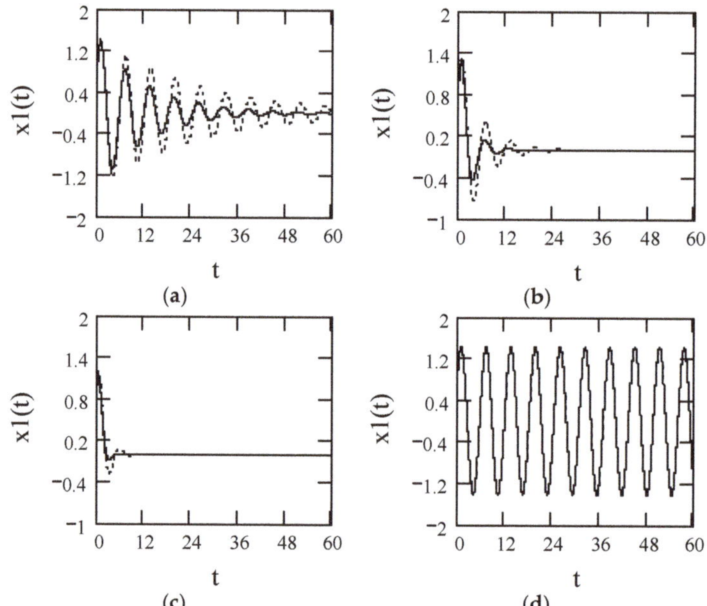

Figure 20. Indicating free response $x_1(t)$ for $x_{10} = v_{10} = \omega_n = 1$. (**a**) $\alpha = 1.9$. Solid line: $\omega = 1$ ($\varsigma_{eq1} = 0.08$). Dot line: $\omega = 0.7$ ($\varsigma_{eq1} = 0.04$). (**b**) $\alpha = 1.6$. Solid line: $\omega = 1$ ($\varsigma_{eq1} = 0.33$). Dot line: $\omega = 0.7$ ($\varsigma_{eq1} = 0.16$). (**c**) $\alpha = 1.3$. Solid line: $\omega = 1$ ($\varsigma_{eq1} = 0.66$). Dot line: $\omega = 0.7$ ($\varsigma_{eq1} = 0.42$). (**d**) $\alpha = 2$. Solid line: $\omega = 1$ ($\varsigma_{eq1} = 0$). Dot line: $\omega = 0.7$ ($\varsigma_{eq1} = 0$).

Note 6.1: As indicated in Figure 20, both oscillation frequency ω and the fractional order α have affects on the damping $\varsigma_{eq1}(\omega, \alpha)$, also see Figure 10. When $\alpha = 2$, $x_1(t)$ reduces to the free response to the ordinary harmonic oscillation with damping free in the form (also see Figure 20d)

$$x_1(t) = \left(x_{10} \cos \omega_n t + \frac{v_{10}}{\omega_n} \sin \omega_n t \right), t \geq 0.$$

The free response to a fractional oscillator in Class I is presented in (172). It uses elementary functions instead of special functions.

Since there exists infinity of natural frequencies for a fractional oscillator, as we explained in Section 5, $x_1(t)$ is actually a function of both t and ω as can be seen from (172). In Figure 20, plots are only specifically with fixed ω. Its plots with varying ω are viewed by Figure 21.

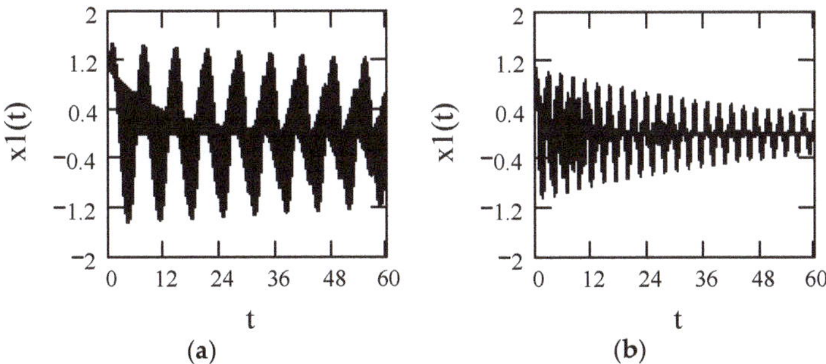

Figure 21. Illustrating free response $x_1(t)$ with variable ω (= 0, 0.2, 0.4, ..., 10) for $x_{10} = v_{10} = 1$. (**a**) For $\omega_n = 1$ and $\alpha = 1.9$ ($0 \leq \varsigma_{eq1} \leq 0.64$). (**b**) For $\omega_n = 3$ and $\alpha = 1.6$ ($0 \leq \varsigma_{eq1} \leq 0.63$).

When emphasizing the point of time-frequency behavior, we view it in t-ω plane as Figure 22 shows.

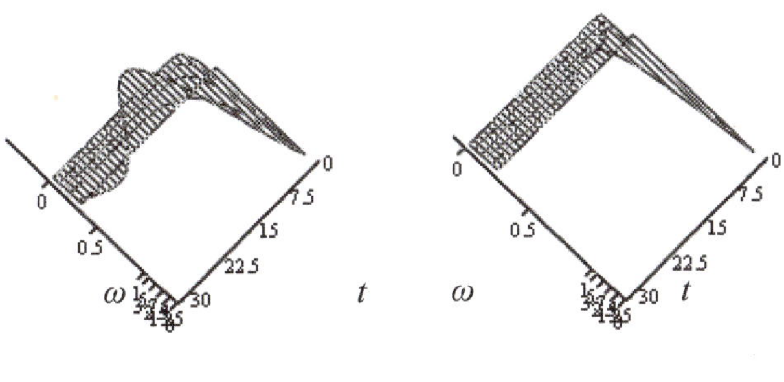

Figure 22. *Cont.*

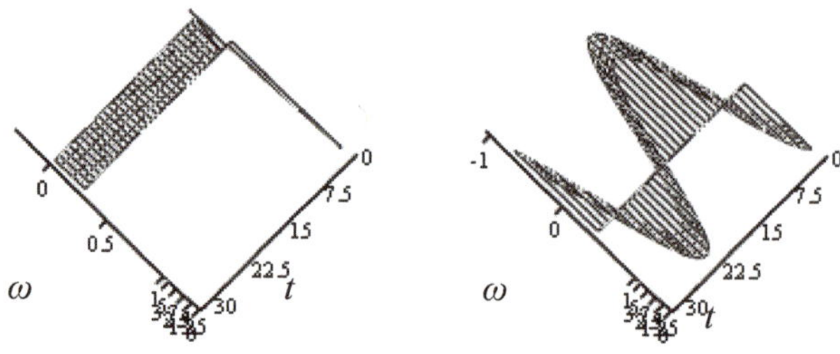

Figure 22. Indicating free response $x_1(t)$ in t-ω plane for t = 0, 1, …, 30 and ω = 1, 2, …, 5, with $x_{10} = v_{10} = 1$, $\omega_n = 6$. (a) α = 1.9 (0.01 $\leq \varsigma_{eq1} \leq$ 0.28). (b) α = 1.6 (0.05 $\leq \varsigma_{eq1} \leq$ 0.72). (c) α = 1.3 (0.11 $\leq \varsigma_{eq1} \leq$ 0.89). (d) α = 2 (ς_{eq1} = 0).

Let t_i and t_{i+1} be two time points where $x_j(t_i)$ reaches its successive peak values of $x_j(t_i)$ and $x_j(t_{i+1})$, respectively. Let Δ_{eqj} be the logarithmic decrement of $x_j(t_i)$. Then, from (178), we immediately obtain

$$\Delta_{eqj} = \ln \frac{x_j(t_i)}{x_j(t_{i+1})} = \frac{2\pi \varsigma_{eqj}}{\sqrt{1 - \varsigma_{eqj}^2}}. \tag{174}$$

Corollary 10 (Decrement I). *Let $x_1(t)$ be the free response of a fractional oscillator in Class I. Then, its logarithmic decrement is given in the form*

$$\Delta_{eq1} = \frac{\pi}{\sqrt{1 - \left(\dfrac{\omega^{\frac{\alpha}{2}} \sin \frac{\alpha\pi}{2}}{2\omega_n \sqrt{-\cos \frac{\alpha\pi}{2}}}\right)^2}} \frac{\omega^{\frac{\alpha}{2}} \sin \frac{\alpha\pi}{2}}{\omega_n \sqrt{-\cos \frac{\alpha\pi}{2}}}, 1 < \alpha \leq 2. \tag{175}$$

Proof. According to (174), we have

$$\Delta_{eq1} = \ln \frac{x_1(t_i)}{x_1(t_{i+1})} = \frac{2\pi \varsigma_{eq1}}{\sqrt{1 - \varsigma_{eq1}^2}} = \frac{\pi}{\sqrt{1 - \left(\dfrac{\omega^{\frac{\alpha}{2}} \sin \frac{\alpha\pi}{2}}{2\omega_n \sqrt{-\cos \frac{\alpha\pi}{2}}}\right)^2}} \frac{\omega^{\frac{\alpha}{2}} \sin \frac{\alpha\pi}{2}}{\omega_n \sqrt{-\cos \frac{\alpha\pi}{2}}}, 1 < \alpha \leq 2.$$

The proof finishes. $\square$

Since Δ_{eq1} is a function of ω and α, we may write it with $\Delta_{eq1}(\omega, \alpha)$. Figure 23 indicates Δ_{eq1}.

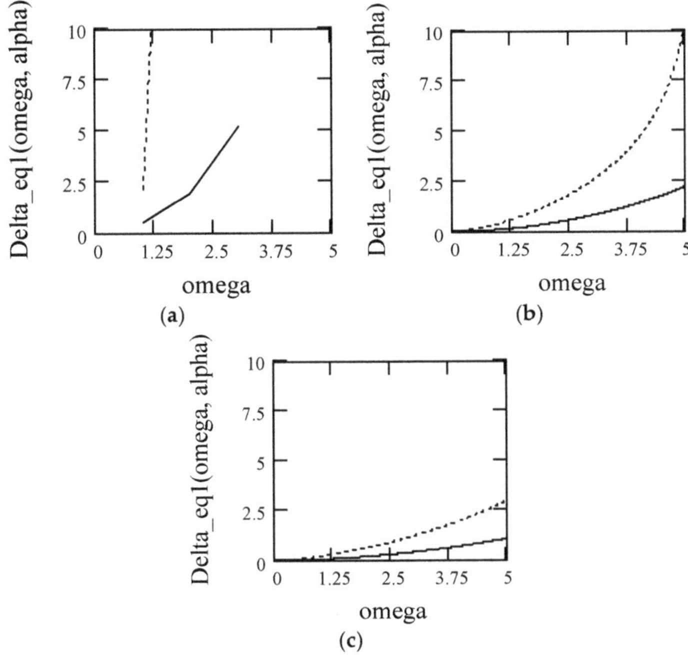

Figure 23. Plots of Δ_{eq1}. Solid line: $\alpha = 1.9$. Dot line: $\alpha = 1.6$. (**a**) For $\omega_n = 1$. (**b**) For $\omega_n = 5$. (**c**) For $\omega_n = 10$.

Note 6.2: $\Delta_{eq1} = 0$ for $\alpha = 2$. As a matter of fact, a fractional oscillator in Class I reduces to a harmonic one if $\alpha = 2$. Accordingly, $\Delta_{eq1} = 0$ in that case.

6.3. Free Response to Fractional Oscillators in Class II

We state the free response to a fractional oscillator in Class II by Theorem 11.

Theorem 11 (Free response II). *Denote by $x_2(t)$ the free response to a fractional oscillator of Class II type. Then, it is, for $t \geq 0$ and $1 < \beta \leq 2$, in the form*

$$
x_2(t) = e^{-\frac{\varsigma \omega_n \omega^{\beta-1} \sin\frac{\beta\pi}{2}}{1-\frac{c}{m}\omega^{\beta-2}\cos\frac{\beta\pi}{2}}t}
\left[
x_{20}\cos\left(\frac{\omega_n\sqrt{1-\frac{c^2\omega^{2(\beta-1)}\sin^2\frac{\beta\pi}{2}}{4\left(m-c\omega^{\beta-2}\cos\frac{\beta\pi}{2}\right)k}}\,t}{\sqrt{\left(1-\frac{c}{m}\omega^{\beta-2}\cos\frac{\beta\pi}{2}\right)}}\right)
+ \frac{v_{20}+\frac{c\omega^{\beta-1}\sin\frac{\beta\pi}{2}}{2\left(m-c\omega^{\beta-2}\cos\frac{\beta\pi}{2}\right)}x_{20}}{\omega_n\sqrt{1-\frac{c^2\omega^{2(\beta-1)}\sin^2\frac{\beta\pi}{2}}{4\left(m-c\omega^{\beta-2}\cos\frac{\beta\pi}{2}\right)k}}}
\sin\left(\frac{\omega_n\sqrt{1-\frac{c^2\omega^{2(\beta-1)}\sin^2\frac{\beta\pi}{2}}{4\left(m-c\omega^{\beta-2}\cos\frac{\beta\pi}{2}\right)k}}\,t}{\sqrt{\left(1-\frac{c}{m}\omega^{\beta-2}\cos\frac{\beta\pi}{2}\right)}}\right)
\right].
\tag{176}
$$

Proof. Note that, for $t \geq 0$,

$$x_2(t) = e^{-\varsigma_{eq2}\omega_{eqn,2}t}\left(x_{20}\cos\omega_{eqd,2}t + \frac{v_{20} + \varsigma_{eq2}\omega_{eqn,2}x_{20}}{\omega_{eqd,2}}\sin\omega_{eqd,2}t\right). \tag{177}$$

In the above expression, we replace ς_{eq2}, $\omega_{eqd,2}$, and $\omega_{eqn,2}$ by those expressed in Section 5. Then, we have (176). Thus, Theorem 11 holds. $\square$

Note 6.3: If $\beta = 1$, $x_2(t)$ degenerates to the ordinary free response to an oscillator with the viscous damper c. In fact,

$$x_2(t)|_{\beta=1} = e^{-\frac{\varsigma\omega_n\omega^{\beta-1}\sin\frac{\beta\pi}{2}}{1-\frac{\varsigma}{m}\omega^{\beta-2}\cos\frac{\beta\pi}{2}}t}\left[\begin{array}{c} x_{20}\cos\left(\dfrac{\omega_n\sqrt{1-\frac{c^2\omega^{2(\beta-1)}\sin^2\frac{\beta\pi}{2}}{4\left(m-c\omega^{\beta-2}\cos\frac{\beta\pi}{2}\right)k}}}{\sqrt{\left(1-\frac{c}{m}\omega^{\beta-2}\cos\frac{\beta\pi}{2}\right)}}t\right) \\[2em] + \dfrac{v_{20}+\dfrac{c\omega^{\beta-1}\sin\frac{\beta\pi}{2}}{2\left(m-c\omega^{\beta-2}\cos\frac{\beta\pi}{2}\right)}x_{20}}{\omega_n\sqrt{1-\frac{c^2\omega^{2(\beta-1)}\sin^2\frac{\beta\pi}{2}}{4\left(m-c\omega^{\beta-2}\cos\frac{\beta\pi}{2}\right)k}}} \\[2em] \sin\left(\dfrac{\omega_n\sqrt{1-\frac{c^2\omega^{2(\beta-1)}\sin^2\frac{\beta\pi}{2}}{4\left(m-c\omega^{\beta-2}\cos\frac{\beta\pi}{2}\right)k}}}{\sqrt{\left(1-\frac{c}{m}\omega^{\beta-2}\cos\frac{\beta\pi}{2}\right)}}t\right) \end{array}\right]_{\beta=1}$$

$$= e^{-\varsigma\omega_n t}\left[x_{20}\cos\left(\omega_n\sqrt{1-\frac{c^2}{4mk}}t\right) + \frac{v_{20}+\frac{c}{2m}x_{20}}{\omega_n\sqrt{1-\frac{c^2}{4mk}}}\sin\left(\omega_n\sqrt{1-\frac{c^2}{4mk}}t\right)\right]$$

$$= e^{-\varsigma\omega_n t}\left[x_{20}\cos\left(\omega_n\sqrt{1-\varsigma^2}t\right) + \frac{v_{20}+\varsigma\omega_n x_{20}}{\omega_n\sqrt{1-\varsigma^2}}\sin\left(\omega_n\sqrt{1-\varsigma^2}t\right)\right].$$

Note 6.4: As far as a fractional oscillator in Class II was concerned, its free response in the closed form is rarely reported. Theorem 11 gives it by using elementary functions.

Let $m = c = k = x_{10} = v_{10} = 1$, and $\omega = 30$. We use Figure 24 to illustrate $x_2(t)$.

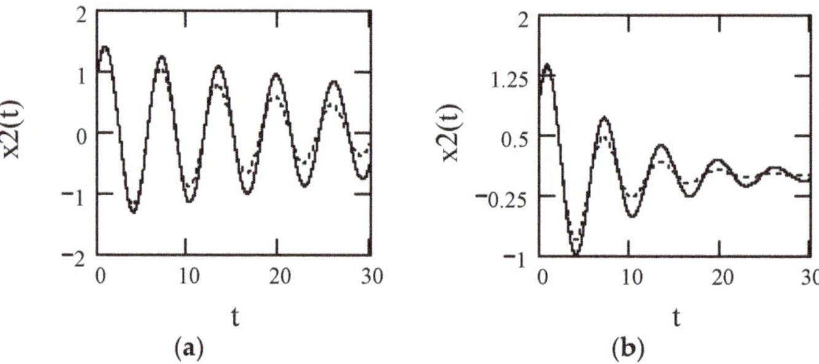

Figure 24. *Cont.*

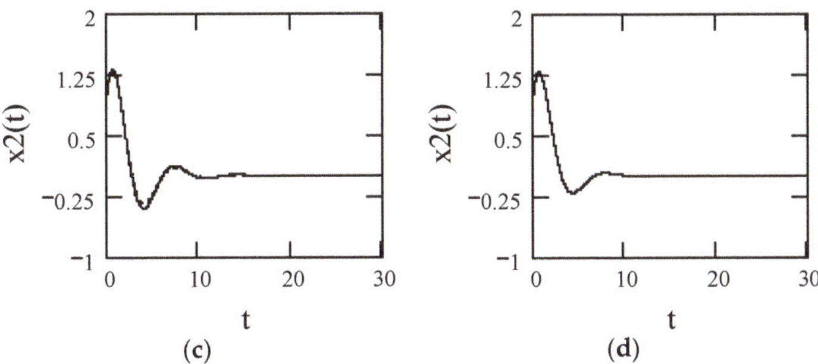

Figure 24. Illustrating free response $x_2(t)$ with fixed ω when $m = c = k = x_{10} = v_{10} = 1$. (**a**) $\beta = 0.3$. Solid line: $\omega = 30$ ($\varsigma_{eq2} = 0.02$). Dot line: $\omega = 10$ ($\varsigma_{eq2} = 0.05$). (**b**) $\beta = 0.6$. Solid line: $\omega = 30$ ($\varsigma_{eq2} = 0.10$). Dot line: $\omega = 10$ ($\varsigma_{eq2} = 0.16$). (**c**) $\beta = 0.9$ Solid line: $\omega = 30$ ($\varsigma_{eq2} = 0.35$). Dot line: $\omega = 10$ ($\varsigma_{eq2} = 0.40$). (**d**) $\beta = 1$. Solid line: $\omega = 30$ ($\varsigma_{eq2} = 0.50$). Dot line: $\omega = 10$ ($\varsigma_{eq2} = 0.50$).

Similar to $x_1(t)$, $x_2(t)$ is also with the argument ω. Its plots with variable ω are demonstrated in Figure 25. Figure 26 shows its plots in t-ω plane.

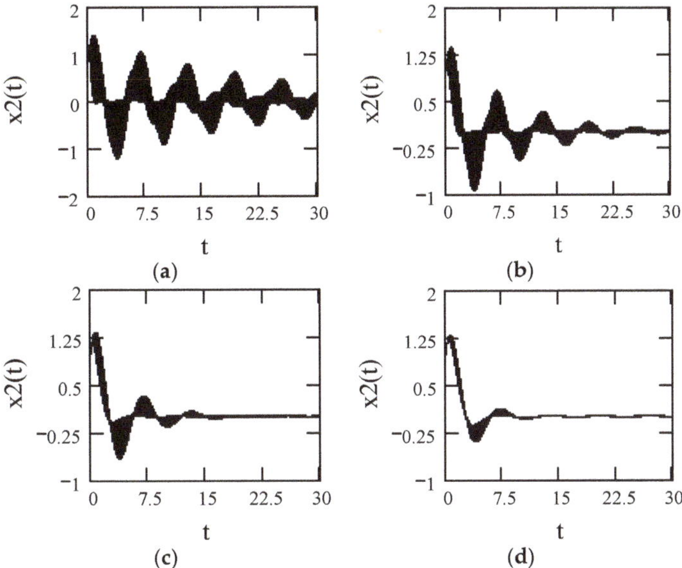

Figure 25. Plots of free response $x_2(t)$ with ω (= 1, 2, ..., 5), $m = c = k = 1 = x_{10} = v_{10} = 1$. (**a**) For $\beta = 0.2$ ($0.04 \leq \varsigma_{eq2} \leq 0.70$). (**b**) For $\beta = 0.4$ ($0.12 \leq \varsigma_{eq2} \leq 0.67$). (**c**) For $\beta = 0.6$ ($0.22 \leq \varsigma_{eq2} \leq 0.63$). (**d**) For $\beta = 0.8$ ($0.35 \leq \varsigma_{eq2} \leq 0.57$).

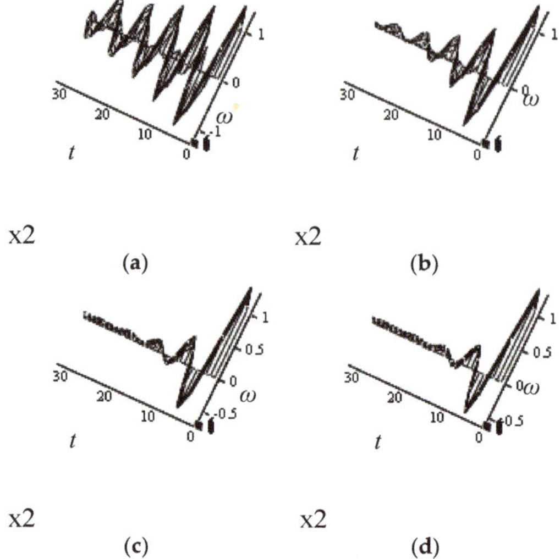

Figure 26. Plots of free response $x_2(t)$ in t-ω plane with $m = k = x_{10} = v_{10} = 1$, $c = 0.5$, for $t = 0, 1, ..., 30$; $\omega = 1, 2, ..., 5$. (**a**) $\beta = 0.3$ ($0.04 \leq \varsigma_{eq2} \leq 0.15$). (**b**) $\beta = 0.6$ ($0.11 \leq \varsigma_{eq2} \leq 0.24$). (**c**) $\beta = 0.9$ ($0.21 \leq \varsigma_{eq2} \leq 0.26$). (**d**) $\beta = 1$ ($\varsigma_{eq2} = 0.25$).

Corollary 11 (Decrement II). *Denote by $x_2(t)$ the free response to a fractional oscillator in Class II. Then, for $0 < \beta \leq 1$, its logarithmic decrement Δ_{eq2}, is in the form*

$$\Delta_{eq2} = \frac{2\pi}{\sqrt{1 - \left(\dfrac{\varsigma\omega^{\beta-1}\sin\frac{\beta\pi}{2}}{\sqrt{1 - \frac{c}{m}\omega^{\beta-2}\cos\frac{\beta\pi}{2}}} \right)^2}} \frac{\varsigma\omega^{\beta-1}\sin\frac{\beta\pi}{2}}{\sqrt{1 - \frac{c}{m}\omega^{\beta-2}\cos\frac{\beta\pi}{2}}}. \tag{178}$$

Proof. According to (174), we have

$$\Delta_{eq2} = \ln\frac{x_2(t_i)}{x_2(t_{i+1})} = \frac{2\pi\varsigma_{eq2}}{\sqrt{1 - \varsigma_{eq2}^2}}. \tag{179}$$

Replacing the above ς_{eq2} with that in (145) produces

$$\Delta_{eq2} = \frac{2\pi\varsigma_{eq2}}{\sqrt{1 - \varsigma_{eq2}^2}} = \frac{2\pi}{\sqrt{1 - \left(\dfrac{\varsigma\omega^{\beta-1}\sin\frac{\beta\pi}{2}}{\sqrt{1 - \frac{c}{m}\omega^{\beta-2}\cos\frac{\beta\pi}{2}}} \right)^2}} \frac{\varsigma\omega^{\beta-1}\sin\frac{\beta\pi}{2}}{\sqrt{1 - \frac{c}{m}\omega^{\beta-2}\cos\frac{\beta\pi}{2}}}.$$

This finishes the proof. $\square$

Similar to Δ_{eq1}, we may write Δ_{eq2} with $\Delta_{eq2}(\omega, \beta)$. Figure 27 indicates its plots.

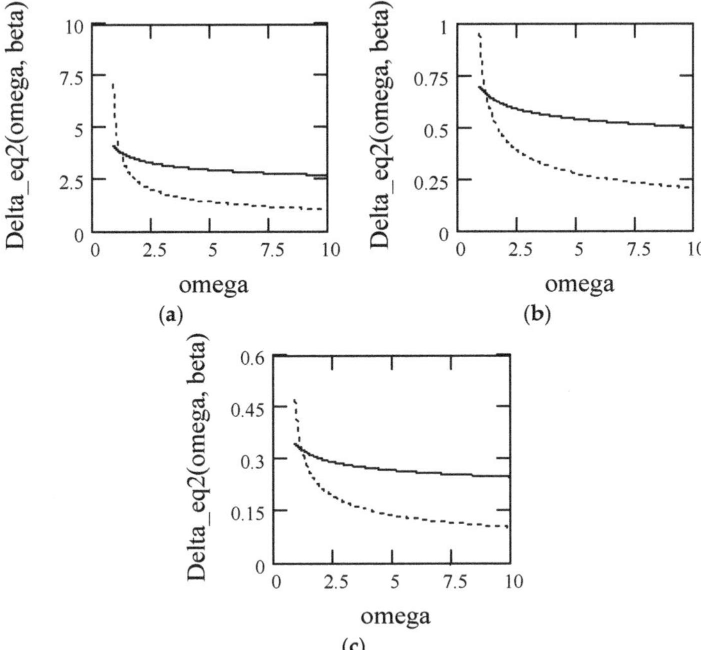

Figure 27. Logarithmic decrement Δ_{eq2} of fractional oscillator in Class II for $m = c = 1$. Solid line: $\beta = 0.9$. Dot line: $\beta = 0.6$. (a) For $\omega_n = 1$. (b) For $\omega_n = 5$. (c) For $\omega_n = 10$.

Note 6.5: Δ_{eq2} reduces to the conventional logarithmic decrement if $\beta = 1$, because

$$\Delta_{eq2}\big|_{\beta=1} = \left[\frac{\dfrac{2\pi\varsigma\omega^{\beta-1}\sin\frac{\beta\pi}{2}}{\sqrt{1-\frac{c}{m}\omega^{\beta-2}\cos\frac{\beta\pi}{2}}}}{\sqrt{1-\left(\dfrac{\varsigma\omega^{\beta-1}\sin\frac{\beta\pi}{2}}{\sqrt{1-\frac{c}{m}\omega^{\beta-2}\cos\frac{\beta\pi}{2}}}\right)^2}} \right]_{\beta=1} = \frac{2\pi\varsigma}{\sqrt{1-\varsigma^2}}. \tag{180}$$

6.4. Free Response to Fractional Oscillators in Class III

We now present the free response to a fractional oscillator in Class III by Theorem 12.

Theorem 12 (Free response III). *Let $x_3(t)$ be the free response to a fractional oscillator in Class III. Then, for $t \geq 0$, $1 < \alpha \leq 2$, $0 < \beta \leq 1$, it is given by*

$$x_3(t) = e^{-\frac{m\omega^{\alpha-1}\sin\frac{\alpha\pi}{2}+c\omega^{\beta-1}\sin\frac{\beta\pi}{2}}{2(m\omega^{\alpha-2}|\cos\frac{\alpha\pi}{2}|-c\omega^{\beta-2}\cos\frac{\beta\pi}{2})}t}\left[\begin{aligned}&x_{30}\cos\left(\frac{\omega_n\sqrt{1-\frac{\left(m\omega^{\alpha-1}\sin\frac{\alpha\pi}{2}+c\omega^{\beta-1}\sin\frac{\beta\pi}{2}\right)^2}{4\left[-\left(m\omega^{\alpha-2}\cos\frac{\alpha\pi}{2}+c\omega^{\beta-2}\cos\frac{\beta\pi}{2}\right)k\right]}}}{\sqrt{-\left(\omega^{\alpha-2}\cos\frac{\alpha\pi}{2}+\frac{c}{m}\omega^{\beta-2}\cos\frac{\beta\pi}{2}\right)}}t\right)\\[6pt]&+\left\{\frac{v_{30}+\frac{m\omega^{\alpha-1}\sin\frac{\alpha\pi}{2}+c\omega^{\beta-1}\sin\frac{\beta\pi}{2}}{2\left(m\omega^{\alpha-2}\left|\cos\frac{\alpha\pi}{2}\right|-c\omega^{\beta-2}\cos\frac{\beta\pi}{2}\right)}x_{30}}{\dfrac{\omega_n}{\sqrt{\left(\omega^{\alpha-2}\left|\cos\frac{\alpha\pi}{2}\right|-\frac{c}{m}\omega^{\beta-2}\cos\frac{\beta\pi}{2}\right)}\sqrt{1-\frac{\left(m\omega^{\alpha-1}\sin\frac{\alpha\pi}{2}+c\omega^{\beta-1}\sin\frac{\beta\pi}{2}\right)^2}{4\left[-\left(m\omega^{\alpha-2}\cos\frac{\alpha\pi}{2}+c\omega^{\beta-2}\cos\frac{\beta\pi}{2}\right)k\right]}}}}\right\}\\[6pt]&\times\sin\left(\frac{\omega_n\sqrt{1-\frac{\left(m\omega^{\alpha-1}\sin\frac{\alpha\pi}{2}+c\omega^{\beta-1}\sin\frac{\beta\pi}{2}\right)^2}{4\left[-\left(m\omega^{\alpha-2}\cos\frac{\alpha\pi}{2}+c\omega^{\beta-2}\cos\frac{\beta\pi}{2}\right)k\right]}}}{\sqrt{-\left(\omega^{\alpha-2}\cos\frac{\alpha\pi}{2}+\frac{c}{m}\omega^{\beta-2}\cos\frac{\beta\pi}{2}\right)}}t\right)\end{aligned}\right]. \tag{181}$$

Proof. Note that, for $t \geq 0$,

$$x_3(t) = e^{-\varsigma_{eq3}\omega_{eqn,3}t}\left(x_{30}\cos\omega_{eqd,3}t + \frac{v_{30}+\varsigma_{eq3}\omega_{eqn,3}x_{30}}{\omega_{eqd,3}}\sin\omega_{eqd,3}t\right). \tag{182}$$

In (182), when substituting ς_{eq3}, $\omega_{eqd,3}$, and $\omega_{eqn,3}$ with those explained in Section 5, we have (181). The proof finishes. $\square$

Note 6.6: If $(\alpha, \beta) = (2, 1)$, $x_3(t)$ returns to be the free response to an ordinary oscillator with the viscous damping. As a matter of fact,

$$x_3(t)\big|_{\alpha=2,\beta=1} = e^{-\varsigma\omega_n t}\left(x_{30}\cos\omega_d t + \frac{v_{30}+\varsigma\omega_n x_{30}}{\omega_d}\sin\omega_d t\right), \tag{183}$$

where $\omega_d = \omega_n\sqrt{1-\varsigma^2}$.

Figure 28 indicates $x_3(t)$ for $m = c = k = x_{10} = v_{10} = 1$.

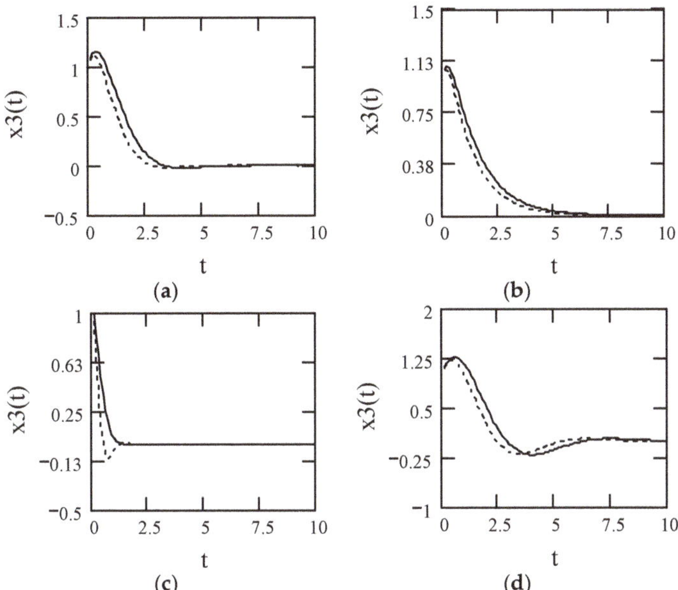

Figure 28. Plots of free response $x_3(t)$ for $m = c = k = x_{10} = v_{10} = 1$. Solid line: $\omega = 1.1$. Dot line: $\omega = 1.5$. (**a**) $\alpha = 1.8$, $\beta = 0.8$ ($\varsigma_{eq3} = 0.78$). (**b**) $\alpha = 1.5$, $\beta = 0.8$ ($\varsigma_{eq3} = 1.33$). (**c**) $\alpha = 1.8$, $\beta = 0.3$ ($\varsigma_{eq3} = 0.91$). (**d**) $\alpha = 2$, $\beta = 1$ ($\varsigma_{eq3} = 0.50$).

Note that the plots regarding with $x_3(t)$ in Figure 28 are with fixed ω. However, actual $x_3(t)$ is frequency varying. Figure 29 shows its frequency varying pictures in time domain and Figure 30 in t-ω plane.

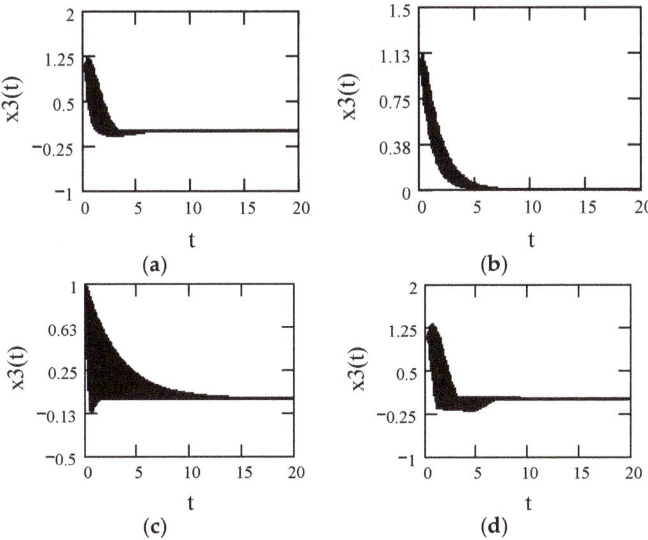

Figure 29. Free response $x_3(t)$ with variable ω (=1, 2, ..., 10) for $m = c = k = 1$. (**a**) $\alpha = 1.9$, $\beta = 0.8$ ($\varsigma_{eq3} = 0.66$). (**b**) $\alpha = 1.5$, $\beta = 0.8$ ($\varsigma_{eq3} = 1.33$). (**c**) $\alpha = 1.8$, $\beta = 0.3$ ($\varsigma_{eq3} = 0.91$). (**d**) $\alpha = 2$, $\beta = 1$ ($\varsigma_{eq3} = 0.50$).

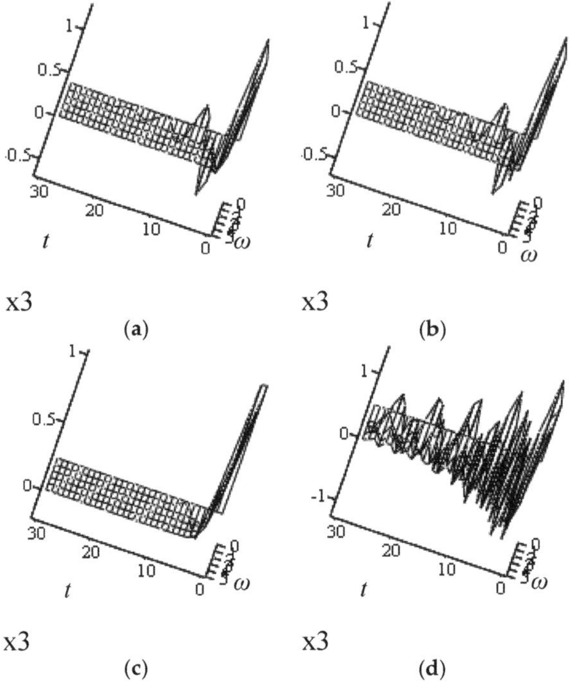

Figure 30. Free response $x_3(t)$ in t-ω plane with $m = k = x_{10} = v_{10} = 1$ and $c = 0.1$ for $t = 0, 1, ..., 30$; $\omega = 1, 2, ..., 5$. (**a**) $\alpha = 1.8$, $\beta = 0.8$ ($0.21 \leq \varsigma_{eq3} \leq 0.72$). (**b**) $\alpha = 1.8$, $\beta = 0.5$ ($0.20 \leq \varsigma_{eq3} \leq 0.70$). (**c**) $\alpha = 1.5$, $\beta = 0.8$ ($0.49 \leq \varsigma_{eq3} \leq 1.49$). (**d**) $\alpha = 2$, $\beta = 1$ ($\varsigma_{eq3} = 0.05$).

Corollary 12 (Decrement III). *Denote by $x_3(t)$ the free response to a fractional oscillator in Class III. Then, for $1 < \alpha \leq 2$ and $0 < \beta \leq 1$, its logarithmic decrement, denoted by Δ_{eq3}, is given by*

$$\Delta_{eq3} = \frac{\dfrac{\pi\left(\omega^{\alpha-1}\sin\frac{\alpha\pi}{2}+2\varsigma\omega_n\omega^{\beta-1}\sin\frac{\beta\pi}{2}\right)}{\omega_n\sqrt{-\left(\omega^{\alpha-2}\cos\frac{\alpha\pi}{2}+2\varsigma\omega_n\omega^{\beta-2}\cos\frac{\beta\pi}{2}\right)}}}{\sqrt{1-\left(\dfrac{\omega^{\alpha-1}\sin\frac{\alpha\pi}{2}+2\varsigma\omega_n\omega^{\beta-1}\sin\frac{\beta\pi}{2}}{2\omega_n\sqrt{-\left(\omega^{\alpha-2}\cos\frac{\alpha\pi}{2}+2\varsigma\omega_n\omega^{\beta-2}\cos\frac{\beta\pi}{2}\right)}}\right)^2}}. \tag{184}$$

Proof. Note that

$$\Delta_{eq3} = \ln\frac{x_3(t_i)}{x_3(t_{i+1})} = \frac{2\pi\varsigma_{eq3}}{\sqrt{1-\varsigma_{eq3}^2}}. \tag{185}$$

Replacing the above ς_{eq3} with that in (147) yields (184). This completes the proof. $\square$

When $\alpha = 2$ and $\beta = 1$, $\Delta_{eq3} = \Delta_{eq3}(\omega, \alpha, \beta)$ becomes the conventional logarithmic decrement. In fact,

$$\Delta_{eq3}\big|_{\alpha=1,\beta=1} = \left. \frac{\dfrac{\pi\left(\omega^{\alpha-1}\sin\frac{\alpha\pi}{2}+2\varsigma\omega_n\omega^{\beta-1}\sin\frac{\beta\pi}{2}\right)}{\omega_n\sqrt{-\left(\omega^{\alpha-2}\cos\frac{\alpha\pi}{2}+2\varsigma\omega_n\omega^{\beta-2}\cos\frac{\beta\pi}{2}\right)}}}{\sqrt{1-\left(\dfrac{\omega^{\alpha-1}\sin\frac{\alpha\pi}{2}+2\varsigma\omega_n\omega^{\beta-1}\sin\frac{\beta\pi}{2}}{2\omega_n\sqrt{-\left(\omega^{\alpha-2}\cos\frac{\alpha\pi}{2}+2\varsigma\omega_n\omega^{\beta-2}\cos\frac{\beta\pi}{2}\right)}}\right)^2}} \right|_{\alpha=1,\beta=1} = \frac{2\pi\varsigma}{\sqrt{1-\varsigma^2}}. \qquad (186)$$

As Δ_{eq3} is a function of ω and (α, β), we write it by $\Delta_{eq3}(\omega, \alpha, \beta)$. Figures 31 and 32 show its plots.

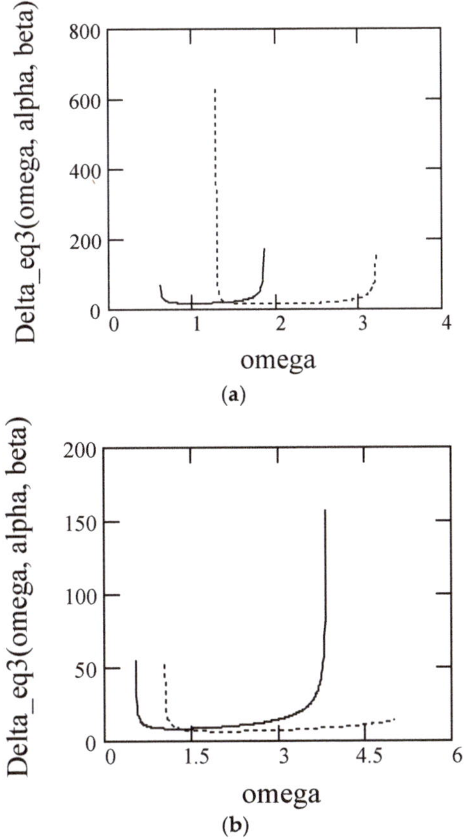

Figure 31. $\Delta_{eq3}(\omega, \alpha, \beta)$: Logarithmic decrement of fractional oscillator in Class III for $m = c = k = 1$. (a) Solid line: $\alpha = 1.7$, $\beta = 0.8$. Dot line: $\alpha = 1.7$, $\beta = 0.5$. (b) Solid line: $\alpha = 1.8$, $\beta = 0.8$. Dot line: $\alpha = 1.8$, $\beta = 0.5$.

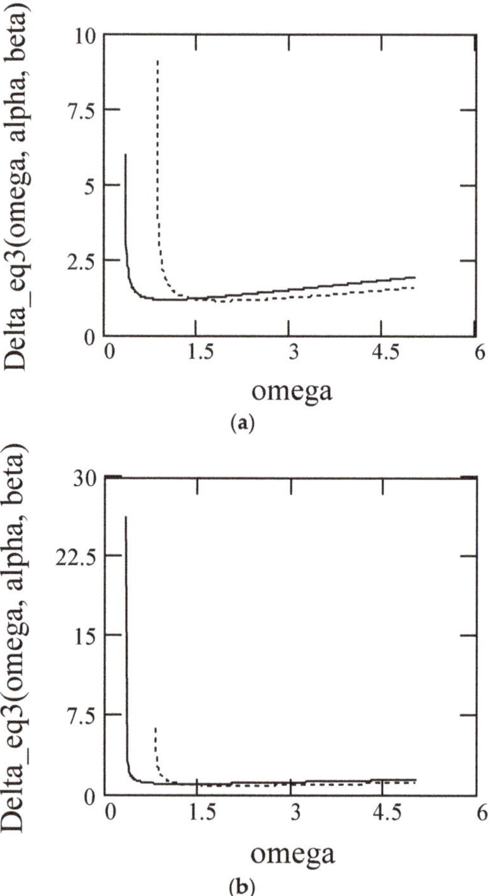

Figure 32. $\Delta_{eq3}(\omega, \alpha, \beta)$ for $m = c = 1$, $k = 25$. (**a**) Solid line: $\alpha = 1.7$, $\beta = 0.8$. Dot line: $\alpha = 1.7$, $\beta = 0.5$. (**b**) Solid line: $\alpha = 1.8$, $\beta = 0.8$. Dot line: $\alpha = 1.8$, $\beta = 0.5$.

6.5. Application to Representing Generalized Mittag-Leffler Function (1)

The previous research (Mainardi [25], Achar et al. [33], Uchaikin ([38], Chapter 7)) presented the free response to fractional oscillators of Class I type by using a kind of special function, called the generalized Mittag-Leffler function, see (32). The novelty of our result presented in Theorem 10 is in that Equation (172) or (173) is consistent with the representation style in engineering by using elementary functions. Thus, we obtain novel representations of the generalized Mittag-Leffler functions as follows.

Corollary 13. *The generalized Mittag-Leffler function in the form*

$$x_1(t) = x_{10}E_{\alpha,1}\left[-(\omega_n t)^\alpha\right] + v_{10}t E_{\alpha,2}\left[-(\omega_n t)^\alpha\right], 1 < \alpha \le 2, t \ge 0, \tag{187}$$

is the solution to fractional oscillators in Class I (Mainardi [25], Achar et al. [33], Uchaikin ([38], Chapter 7)). It can be expressed by the one in (172). That is, for $1 < \alpha \le 2, t \ge 0$,

$$x_1(t) = x_{10}E_{\alpha,1}\left[-(\omega_n t)^\alpha\right] + v_{10}t E_{\alpha,2}\left[-(\omega_n t)^\alpha\right]$$

$$= e^{-\frac{\omega\sin\frac{\alpha\pi}{2}}{2\left|\cos\frac{\alpha\pi}{2}\right|}t}\left[\begin{array}{l} x_{10}\cos\left(\dfrac{\omega_n\sqrt{1-\frac{\omega^\alpha\sin^2\frac{\alpha\pi}{2}}{4\omega_n^2\left|\cos\frac{\alpha\pi}{2}\right|}}}{\sqrt{\omega^{\alpha-2}\left|\cos\frac{\alpha\pi}{2}\right|}}t\right) \\[2em] +\left[\dfrac{v_{10}+\frac{\omega^{\frac{\alpha}{2}}\sin\frac{\alpha\pi}{2}}{2\omega_n\sqrt{\left|\cos\frac{\alpha\pi}{2}\right|}}x_{10}}{\sqrt{1-\frac{\omega^\alpha\sin^2\frac{\alpha\pi}{2}}{4\omega_n^2\left|\cos\frac{\alpha\pi}{2}\right|}}}\right]\sin\left(\dfrac{\omega_n\sqrt{1-\frac{\omega^\alpha\sin^2\frac{\alpha\pi}{2}}{4\omega_n^2\left|\cos\frac{\alpha\pi}{2}\right|}}}{\sqrt{\omega^{\alpha-2}\left|\cos\frac{\alpha\pi}{2}\right|}}t\right) \end{array}\right]. \tag{188}$$

The proof of Corollary 13 is straightforward from (172).

When $v_{10}= 0$ in (187), we obtain a corollary below.

Corollary 14. *The generalized Mittag-Leffler function given by*

$$x_1(t) = x_{10}E_{\alpha,1}\left[-(\omega_n t)^\alpha\right], 1 < \alpha \le 2, t \ge 0, \tag{189}$$

can be expressed by the elementary functions, for $1 < \alpha \le 2$, $t \ge 0$, in the form

$$x_1(t) = x_{10}E_{\alpha,1}\left[-(\omega_n t)^\alpha\right] = e^{-\frac{\omega\sin\frac{\alpha\pi}{2}}{2\left|\cos\frac{\alpha\pi}{2}\right|}t}\left[\begin{array}{l} x_{10}\cos\left(\dfrac{\omega_n}{\sqrt{\omega^{\alpha-2}\left|\cos\frac{\alpha\pi}{2}\right|}}\sqrt{1-\frac{\omega^\alpha\sin^2\frac{\alpha\pi}{2}}{4\omega_n^2\left|\cos\frac{\alpha\pi}{2}\right|}}t\right) \\[2em] +\left(\dfrac{\frac{\omega^{\frac{\alpha}{2}}\sin\frac{\alpha\pi}{2}}{2\omega_n\sqrt{\left|\cos\frac{\alpha\pi}{2}\right|}}x_{10}}{\sqrt{1-\frac{\omega^\alpha\sin^2\frac{\alpha\pi}{2}}{4\omega_n^2\left|\cos\frac{\alpha\pi}{2}\right|}}}\right)\sin\left(\dfrac{\omega_n\sqrt{1-\frac{\omega^\alpha\sin^2\frac{\alpha\pi}{2}}{4\omega_n^2\left|\cos\frac{\alpha\pi}{2}\right|}}}{\sqrt{\omega^{\alpha-2}\left|\cos\frac{\alpha\pi}{2}\right|}}t\right) \end{array}\right]. \tag{190}$$

Proof. If $v_{10} = 0$ in (187), (188) becomes (189). The proof completes. $\square$

If $x_{10}= 0$ in (187), we obtain another corollary as follows.

Corollary 15. *The generalized Mittag-Leffler function expressed by*

$$x_1(t) = v_{10}t E_{\alpha,2}\left[-(\omega_n t)^\alpha\right], 1 < \alpha \le 2, t \ge 0, \tag{191}$$

can be represented, for $1 < \alpha \le 2$, $t \ge 0$, by the elementary functions in the form

$$x_1(t) = v_{10}t E_{\alpha,2}\left[-(\omega_n t)^\alpha\right] = e^{-\frac{\omega\sin\frac{\alpha\pi}{2}}{2\left|\cos\frac{\alpha\pi}{2}\right|}t}\left[\left(\dfrac{v_{10}}{\sqrt{1-\frac{\omega^\alpha\sin^2\frac{\alpha\pi}{2}}{4\omega_n^2\left|\cos\frac{\alpha\pi}{2}\right|}}}\right)\sin\left(\dfrac{\omega_n\sqrt{1-\frac{\omega^\alpha\sin^2\frac{\alpha\pi}{2}}{4\omega_n^2\left|\cos\frac{\alpha\pi}{2}\right|}}}{\sqrt{\omega^{\alpha-2}\left|\cos\frac{\alpha\pi}{2}\right|}}t\right)\right]. \tag{192}$$

Proof. When $x_{10} = 0$ in (187), (188) becomes (192). The proof finishes. $\square$

7. Impulse Responses to Three Classes of Fractional Oscillators

In this section, we shall present the impulse responses to three classes of fractional oscillators using elementary functions.

In Section 4, we have proved that

$$H_{yj}(\omega) = H_{xj}(\omega),\ j = 1,\ 2,\ 3,$$

where $H_{yj}(\omega)$ is the frequency response function solved directly from a jth fractional oscillator while $H_{xj}(\omega)$ is the one derived from its equivalent oscillator. Doing the inverse Fourier transform on the both sides above, therefore, we have

$$h_{yj}(t) = h_{xj}(t),\ j = 1,\ 2,\ 3,$$

where $h_{yj}(t)$ is the impulse response obtained directly from the jth fractional oscillator but $h_{xj}(t)$ is the one solved from its equivalent one. In that way, therefore, we may establish the theoretic foundation for representing the impulse responses to three classes of fractional oscillators by using elementary functions.

The main highlight presented in this section is to propose the impulse responses to three classes of fractional oscillators in the closed analytic form expressed by elementary functions. As a by product, we shall represent a certain generalized Mittag-Leffler functions using elementary functions.

7.1. General Form of Impulse Responses

Given a following functional form of equivalent oscillators for finding their impulse responses, we denote by $h_j(t)$ the impulse response to the equivalent oscillator in Class j in the form

$$m_{eqj}\frac{d^2h_j(t)}{dt^2} + c_{eqj}\frac{dh_j(t)}{dt} + kh_j(t) = \delta(t), j = 1,2,3. \tag{193}$$

Rewrite the above in the form

$$\frac{d^2h_j(t)}{dt^2} + \frac{c_{eqj}}{m_{eqj}}\frac{dh_j(t)}{dt} + \frac{k}{m_{eqj}}h_j(t) = \frac{\delta(t)}{m_{eqj}}, j = 1,2,3. \tag{194}$$

According to the results in the previous sections, we have

$$\frac{d^2h_j(t)}{dt^2} + 2\varsigma_{eqj}\omega_{eqn,j}\frac{dh_j(t)}{dt} + \omega_{eqn,j}^2 h_j(t) = \frac{\delta(t)}{m_{eqj}}, j = 1,2,3. \tag{195}$$

Therefore, functionally, we have

$$h_j(t) = \frac{e^{-\varsigma_{eqj}\omega_{eqn,j}t}}{m_{eqj}\omega_{eqd,j}}\sin\omega_{eqd,j}t, t \geq 0. \tag{196}$$

Equation (196) is a general form of the impulse response to fractional oscillators for Class j ($j = 1, 2, 3$). Its specific form for each Class is discussed as follows.

7.2. Impulse Response to Fractional Oscillators in Class I

Theorem 13 (Impulse response I). *Let $h_1(t)$ be the impulse response to a fractional oscillator in Class I. Then, for $t \geq 0$ and $1 < \alpha \leq 2$, we have*

$$h_1(t) = \frac{e^{-\frac{\omega\sin\frac{\alpha\pi}{2}}{2\left|\cos\frac{\alpha\pi}{2}\right|}t}\sin\left(\frac{\omega_n}{\sqrt{\omega^{\alpha-2}\left|\cos\frac{\alpha\pi}{2}\right|}}\sqrt{1 - \frac{\omega^{2\alpha}\sin^2\frac{\alpha\pi}{2}}{4\omega_n^2\left|\cos\frac{\alpha\pi}{2}\right|}}t\right)}{m\omega_n\sqrt{\omega^{\alpha-2}\left|\cos\frac{\alpha\pi}{2}\right|}\sqrt{1 - \frac{\omega^{2\alpha}\sin^2\frac{\alpha\pi}{2}}{4\omega_n^2\left|\cos\frac{\alpha\pi}{2}\right|}}}. \tag{197}$$

Proof. From (196), we have

$$h_1(t) = \frac{e^{-\varsigma_{eq1}\omega_{eqn,1}t}}{m_{eq1}\omega_{eqd,1}}\sin\omega_{eqd,1}t, t \geq 0. \tag{198}$$

When replacing m_{eq1} by that in Section 4, ς_{eq1} and $\omega_{eqd,1}$ as well as $\omega_{eqn,1}$ with those in Section 5, respectively, we obtain

$$h_1(t) = \frac{e^{-\varsigma_{eq1}\omega_{eqn,1}t}}{m_{eq1}\omega_{eqd,1}}\sin\omega_{eqd,1}t = \frac{-\dfrac{\omega^{\frac{\alpha}{2}}\sin\frac{\alpha\pi}{2}}{2\omega_n\sqrt{|\cos\frac{\alpha\pi}{2}|}\sqrt{\omega^{\alpha-2}|\cos\frac{\alpha\pi}{2}|}}t\ \sin\dfrac{\omega_n\sqrt{1-\dfrac{\omega^{2\alpha}\sin^2\frac{\alpha\pi}{2}}{4\omega_n^2|\cos\frac{\alpha\pi}{2}|}}}{\sqrt{\omega^{\alpha-2}|\cos\frac{\alpha\pi}{2}|}}t}{\omega^{\alpha-2}|\cos\frac{\alpha\pi}{2}|m\dfrac{\omega_n}{\sqrt{\omega^{\alpha-2}|\cos\frac{\alpha\pi}{2}|}}\sqrt{1-\dfrac{\omega^{2\alpha}\sin^2\frac{\alpha\pi}{2}}{4\omega_n^2|\cos\frac{\alpha\pi}{2}|}}}$$

$$= \frac{e^{-\dfrac{\omega\sin\frac{\alpha\pi}{2}}{2|\cos\frac{\alpha\pi}{2}|}t}\sin\dfrac{\omega_n\sqrt{1-\dfrac{\omega^{2\alpha}\sin^2\frac{\alpha\pi}{2}}{4\omega_n^2|\cos\frac{\alpha\pi}{2}|}}}{\sqrt{\omega^{\alpha-2}|\cos\frac{\alpha\pi}{2}|}}t}{m\omega_n\sqrt{\omega^{\alpha-2}|\cos\frac{\alpha\pi}{2}|}\sqrt{1-\dfrac{\omega^{2\alpha}\sin^2\frac{\alpha\pi}{2}}{4\omega_n^2|\cos\frac{\alpha\pi}{2}|}}}.$$

This finishes the proof. □

Figure 33 shows the plots of $h_1(t)$, where the oscillation frequency ω is fixed. Note that ω is an argument of $h_1(t)$. Therefore, its pictures in time domain are indicated in Figure 34. Figure 35 indicates its figures in t-ω plane.

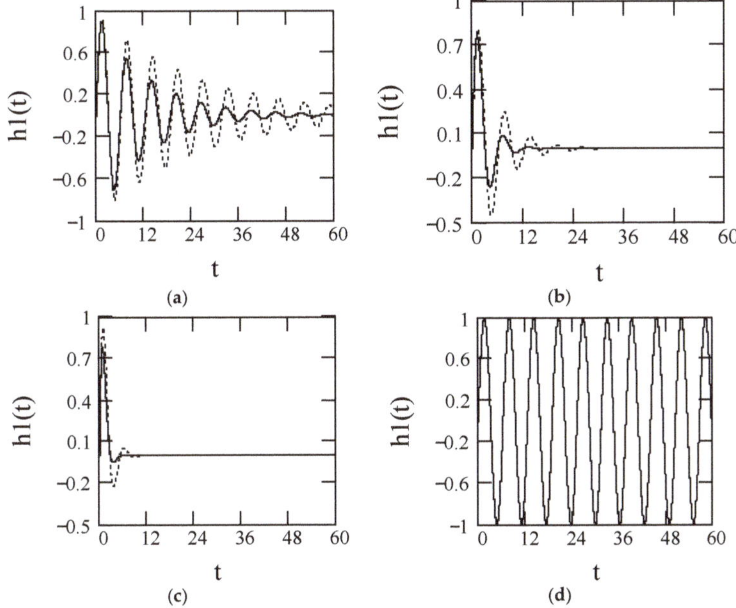

Figure 33. Plots of impulse response $h_1(t)$ with $\omega_n = 1$. (**a**) $\alpha = 1.9$, solid line: $\omega = 1$ ($\varsigma_{eq1} = 0.08$); dot line: $\omega = 0.7$ ($\varsigma_{eq1} = 0.04$). (**b**) $\alpha = 1.6$, solid line: $\omega = 1$ ($\varsigma_{eq1} = 0.33$); dot line: $\omega = 0.7$ ($\varsigma_{eq1} = 0.19$). (**c**) $\alpha = 1.3$, solid line: $\omega = 1$ ($\varsigma_{eq1} = 0.66$); dot line: $\omega = 0.7$ ($\varsigma_{eq1} = 0.42$). (**d**) $\alpha = 2$, solid line: $\omega = 1$ ($\varsigma_{eq1} = 0$); dot line: $\omega = 0.7$ ($\varsigma_{eq1} = 0$).

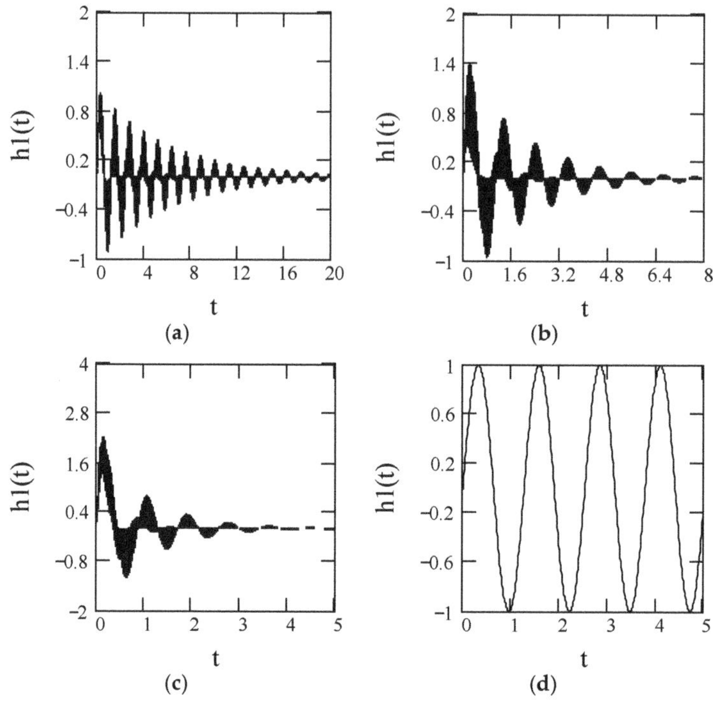

Figure 34. Plots of impulse response $h_1(t)$ for $\omega_n = 5$, $\omega = 0, 1, ..., 5$. (**a**) For $\alpha = 1.8$ ($0 \leq \varsigma_{eq1} \leq 0.57$). (**b**) For $\alpha = 1.5$ ($0 \leq \varsigma_{eq1} \leq 0.94$). (**c**) For $\alpha = 1.3$ ($0 \leq \varsigma_{eq1} \leq 1.07$). (**d**) For $\alpha = 2$ ($\varsigma_{eq1} = 0$).

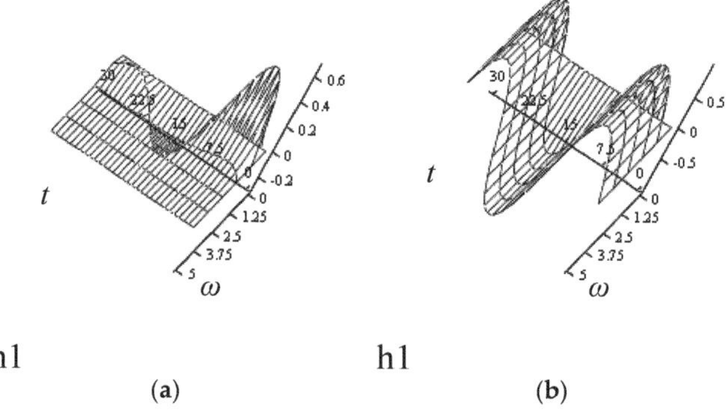

Figure 35. Impulse response $h_1(t)$ in t-ω plane with $m = 1$, $\omega_n = 0.3$ for $t = 0, 1, ..., 30$; $\omega = 1, 2, ..., 5$. (**a**) $\alpha = 1.9$ ($0.26 \leq \varsigma_{eq1} \leq 5.58$). (**b**) $\alpha = 2$ ($\varsigma_{eq1} = 0$).

Note 7.1: The impulse response $h_1(t)$ reduces to the conventional one if $\alpha = 2$. In fact,

$$h_1(t)\big|_{\alpha=2} = \left(\frac{e^{-\frac{\omega \sin \frac{\alpha\pi}{2}}{2\left|\cos \frac{\alpha\pi}{2}\right|}t} \sin \frac{\omega_n}{\sqrt{\omega^{\alpha-2}\left|\cos \frac{\alpha\pi}{2}\right|}}\sqrt{1 - \frac{\omega^{2\alpha}\sin^2 \frac{\alpha\pi}{2}}{4\omega_n^2\left|\cos \frac{\alpha\pi}{2}\right|}}\,t}{m\omega_n \sqrt{\omega^{\alpha-2}\left|\cos \frac{\alpha\pi}{2}\right|}\sqrt{1 - \frac{\omega^{2\alpha}\sin^2 \frac{\alpha\pi}{2}}{4\omega_n^2\left|\cos \frac{\alpha\pi}{2}\right|}}} \right)_{\alpha=2} = \frac{1}{m\omega_n}\sin \omega_n t. \tag{199}$$

7.3. Impulse Response to Fractional Oscillators in Class II

Theorem 14 (Impulse response II)**.** *Denote by $h_2(t)$ the impulse response to a fractional oscillator in Class II. For $t \geq 0$ and $1 < \beta \leq 2$, therefore, it is given by*

$$h_2(t) = \frac{e^{-\frac{\varsigma \omega_n \omega^{\beta-1} \sin \frac{\beta\pi}{2}}{1-\frac{\varsigma}{m}\omega^{\beta-2}\cos\frac{\beta\pi}{2}}t} \sin \frac{\omega_n\sqrt{1-\frac{\varsigma^2\omega^{2(\beta-1)}\sin^2\frac{\beta\pi}{2}}{1-\frac{\varsigma}{m}\omega^{\beta-2}\cos\frac{\beta\pi}{2}}}}{\sqrt{1-\frac{\varsigma}{m}\omega^{\beta-2}\cos\frac{\beta\pi}{2}}}t}{\omega_n m\sqrt{1-\frac{\varsigma}{m}\omega^{\beta-2}\cos\frac{\beta\pi}{2}}\sqrt{1-\frac{\varsigma^2\omega^{2(\beta-1)}\sin^2\frac{\beta\pi}{2}}{1-\frac{\varsigma}{m}\omega^{\beta-2}\cos\frac{\beta\pi}{2}}}}. \tag{200}$$

Proof. From (196), we have

$$h_2(t) = e^{-\varsigma_{eq2}\omega_{eqn,2}t}\frac{1}{m_{eq2}\omega_{eqd,2}}\sin\omega_{eqd,2}t, \, t \geq 0. \tag{201}$$

By replacing m_{eq2} with that in Section 4, ς_{eq2}, $\omega_{eqn,2}$, and $\omega_{eqd,2}$ by those in Section 5, we obtain

$$h_2(t) = \frac{e^{-\varsigma_{eq2}\omega_{eqn,2}t}\sin\omega_{eqd,2}t}{m_{eq2}\omega_{eqd,2}} = \frac{e^{-\frac{\varsigma\omega^{\beta-1}\sin\frac{\beta\pi}{2}}{\sqrt{1-\frac{\varsigma}{m}\omega^{\beta-2}\cos\frac{\beta\pi}{2}}}\frac{\omega_n}{\sqrt{1-\frac{\varsigma}{m}\omega^{\beta-2}\cos\frac{\beta\pi}{2}}}t}\sin\omega_{eqd,2}t}{\left(m-c\omega^{\beta-2}\cos\frac{\beta\pi}{2}\right)\omega_{eqd,2}}$$

$$= e^{-\frac{\varsigma\omega_n\omega^{\beta-1}\sin\frac{\beta\pi}{2}}{1-\frac{\varsigma}{m}\omega^{\beta-2}\cos\frac{\beta\pi}{2}}t}\frac{\sin\frac{\omega_n\sqrt{1-\frac{\varsigma^2\omega^{2(\beta-1)}\sin^2\frac{\beta\pi}{2}}{1-\frac{\varsigma}{m}\omega^{\beta-2}\cos\frac{\beta\pi}{2}}}}{\sqrt{1-\frac{\varsigma}{m}\omega^{\beta-2}\cos\frac{\beta\pi}{2}}}t}{\frac{\left(m-c\omega^{\beta-2}\cos\frac{\beta\pi}{2}\right)\omega_n}{\sqrt{1-\frac{\varsigma}{m}\omega^{\beta-2}\cos\frac{\beta\pi}{2}}}\sqrt{1-\frac{\varsigma^2\omega^{2(\beta-1)}\sin^2\frac{\beta\pi}{2}}{1-\frac{\varsigma}{m}\omega^{\beta-2}\cos\frac{\beta\pi}{2}}}}$$

$$= e^{-\frac{\varsigma\omega_n\omega^{\beta-1}\sin\frac{\beta\pi}{2}}{1-\frac{\varsigma}{m}\omega^{\beta-2}\cos\frac{\beta\pi}{2}}t}\frac{\sin\frac{\omega_n}{\sqrt{1-\frac{\varsigma}{m}\omega^{\beta-2}\cos\frac{\beta\pi}{2}}}\sqrt{1-\frac{\varsigma^2\omega^{2(\beta-1)}\sin^2\frac{\beta\pi}{2}}{1-\frac{\varsigma}{m}\omega^{\beta-2}\cos\frac{\beta\pi}{2}}}t}{\omega_n m\sqrt{1-\frac{\varsigma}{m}\omega^{\beta-2}\cos\frac{\beta\pi}{2}}\sqrt{1-\frac{\varsigma^2\omega^{2(\beta-1)}\sin^2\frac{\beta\pi}{2}}{1-\frac{\varsigma}{m}\omega^{\beta-2}\cos\frac{\beta\pi}{2}}}}.$$

This is (200). Hence, the proof completes. $\square$

Figure 36 illustrates $h_2(t)$ with fixed ω. Its plots with variable ω are shown in Figure 37. Its pictures in t-ω plane are indicated in Figure 38.

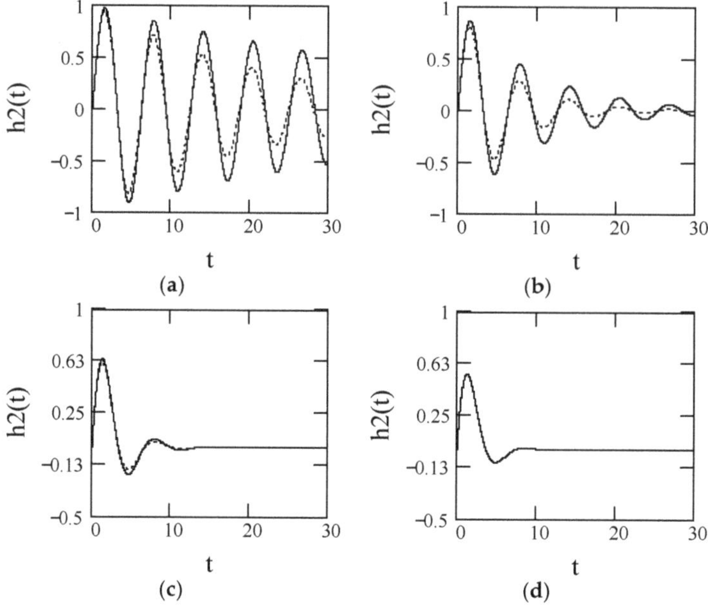

Figure 36. Illustrating impulse response $h_2(t)$ for $m = c = k = 1$. Solid line: $\omega = 30$. Dot line: $\omega = 10$. (**a**) $\beta = 0.3$, solid line: $\omega = 30$ ($\varsigma_{eq2} = 0.02$); dot line: $\omega = 10$ ($\varsigma_{eq2} = 0.05$). (**b**) $\beta = 0.6$, solid line: $\omega = 30$ ($\varsigma_{eq2} = 0.10$); dot line: $\omega = 10$ ($\varsigma_{eq2} = 0.16$). (**c**) $\beta = 0.9$, solid line: $\omega = 30$ ($\varsigma_{eq2} = 0.35$); dot line: $\omega = 10$ ($\varsigma_{eq2} = 0.40$). (**d**) $\beta = 1$, solid line: $\omega = 30$ ($\varsigma_{eq2} = 0.50$); dot line: $\omega = 10$ ($\varsigma_{eq2} = 0.50$).

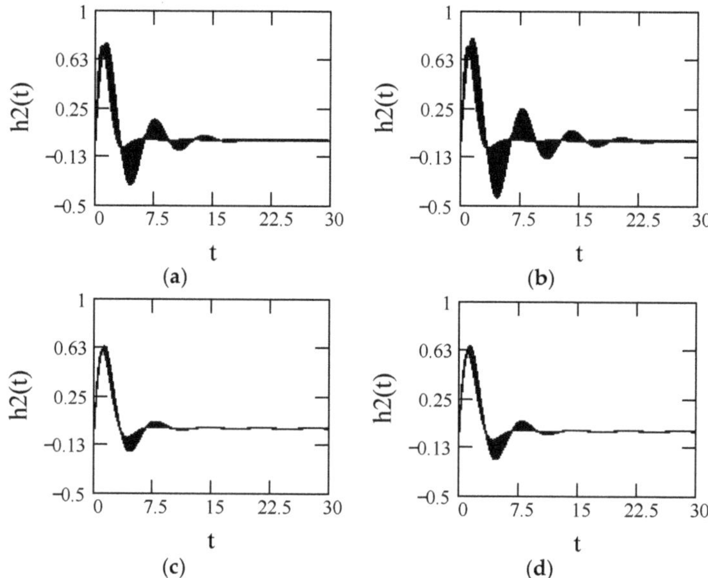

Figure 37. Plots of impulse response $h_2(t)$ with variable ω for $m = c = k = 1$ in time domain. (**a**) For $\beta = 0.63$, $\omega = 1, 2, ..., 5$ ($0.24 \leq \varsigma_{eq2} \leq 0.62$). (**b**) For $\beta = 0.63$, $\omega = 1, 2, ..., 10$ ($0.18 \leq \varsigma_{eq2} \leq 0.62$). (**c**) For $\beta = 0.83$, $\omega = 1, 2, ..., 5$ ($0.37 \leq \varsigma_{eq2} \leq 0.56$). (**d**) For $\beta = 0.83$, $\omega = 1, 2, ..., 10$ ($0.33 \leq \varsigma_{eq2} \leq 0.56$).

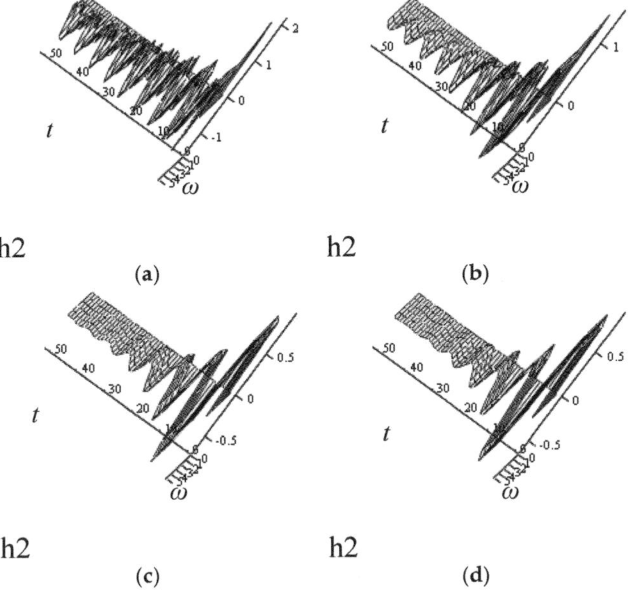

Figure 38. Illustrating impulse response $h_2(t)$ in t-ω plane for $m = c = k = 1$ with $t = 0, 1, \ldots, 50$; $\omega = 1, 2, \ldots, 5$. (a) $\beta = 0.3$ ($0.08 \leq \varsigma_{eq2} \leq 0.69$). (b) $\beta = 0.6$ ($0.22 \leq \varsigma_{eq2} \leq 0.63$). (c) $\beta = 0.9$ ($0.43 \leq \varsigma_{eq2} \leq 0.54$). (d) $\beta = 1$ ($\varsigma_{eq2} = 0.50$).

Note 7.2: The impulse response $h_2(t)$ reduces to the conventional one if $\beta = 1$. As a matter of fact,

$$
h_2(t)|_{\beta=1} = \left[\frac{e^{-\frac{\varsigma\omega_n\omega^{\beta-1}\sin\frac{\beta\pi}{2}}{1-\frac{c}{m}\omega^{\beta-2}\cos\frac{\beta\pi}{2}}t}\sin\omega_n\sqrt{1-\frac{\varsigma^2\omega^{2(\beta-1)}\sin^2\frac{\beta\pi}{2}}{1-\frac{c}{m}\omega^{\beta-2}\cos\frac{\beta\pi}{2}}}t}{\omega_n m\sqrt{1-\frac{c}{m}\omega^{\beta-2}\cos\frac{\beta\pi}{2}}\sqrt{1-\frac{\varsigma^2\omega^{2(\beta-1)}\sin^2\frac{\beta\pi}{2}}{1-\frac{c}{m}\omega^{\beta-2}\cos\frac{\beta\pi}{2}}}} \right]_{\beta=1}
\tag{202}
$$

$$
= \frac{e^{-\varsigma\omega_n t}}{m\omega_n\sqrt{1-\varsigma^2}}\sin\omega_n\sqrt{1-\varsigma^2}\,t.
$$

7.4. Impulse Response to Fractional Oscillators in Class III

We present the impulse response to fractional oscillators in Class III with Theorem 15.

Theorem 15 (Impulse response III). *Let $h_3(t)$ be the impulse response to a fractional oscillator in Class III. For $t \geq 0$, $1 < \alpha \leq 2$, $0 < \beta \leq 1$, it is in the form*

$$
h_3(t) = \frac{e^{-\frac{m\omega^{\alpha-1}\sin\frac{\alpha\pi}{2}+c\omega^{\beta-1}\sin\frac{\beta\pi}{2}}{2\sqrt{-(m\omega^{\alpha-2}\cos\frac{\alpha\pi}{2}+c\omega^{\beta-2}\cos\frac{\beta\pi}{2})k}}\omega_{eqn,3}t}\sin\omega_{eqd,3}t}{-\left(m\omega^{\alpha-2}\cos\frac{\alpha\pi}{2}+c\omega^{\beta-2}\cos\frac{\beta\pi}{2}\right)\omega_{eqd,3}},
\tag{203}
$$

where

$$
\omega_{eqn,3} = \frac{\omega_n}{\sqrt{-\left(\omega^{\alpha-2}\cos\frac{\alpha\pi}{2}+\frac{c}{m}\omega^{\beta-2}\cos\frac{\beta\pi}{2}\right)}},
$$

and

$$\omega_{eqd,3} = \frac{\omega_n \sqrt{1 - \dfrac{\left(m\omega^{\alpha-1} \sin \frac{\alpha\pi}{2} + c\omega^{\beta-1} \sin \frac{\beta\pi}{2}\right)^2}{4\left[-\left(m\omega^{\alpha-2} \cos \frac{\alpha\pi}{2} + c\omega^{\beta-2} \cos \frac{\beta\pi}{2}\right)k\right]}}}{\sqrt{-\left(\omega^{\alpha-2} \cos \frac{\alpha\pi}{2} + \frac{c}{m}\omega^{\beta-2} \cos \frac{\beta\pi}{2}\right)}}.$$

Proof. With (196), we get

$$h_3(t) = e^{-\varsigma_{eq3}\omega_{eqn,3}t} \frac{1}{m_{eq3}\omega_{eqd,3}} \sin \omega_{eqd,3}t, \, t \geq 0. \tag{204}$$

In the above expression, substitute m_{eq3} with the one in Section 4, ς_{eq3}, $\omega_{eqd,3}$, $\omega_{eqn,3}$ by those in Section 5, respectively, we have, for $t \geq 0$,

$$h_3(t) = \frac{e^{-\frac{m\omega^{\alpha-1} \sin \frac{\alpha\pi}{2} + c\omega^{\beta-1} \sin \frac{\beta\pi}{2}}{2\sqrt{-(m\omega^{\alpha-2} \cos \frac{\alpha\pi}{2} + c\omega^{\beta-2} \cos \frac{\beta\pi}{2})k}}\omega_{eqn,3}t}}{-\left(m\omega^{\alpha-2} \cos \frac{\alpha\pi}{2} + c\omega^{\beta-2} \cos \frac{\beta\pi}{2}\right)\omega_{eqd,3}} \sin \omega_{eqd,3}t.$$

The right side on the above is (203). Thus, the proof completes. $\square$

The plots of $h_3(t)$ with fixed ω are shown in Figure 39, with variable ω in Figure 40, and in t-ω plane by Figure 41.

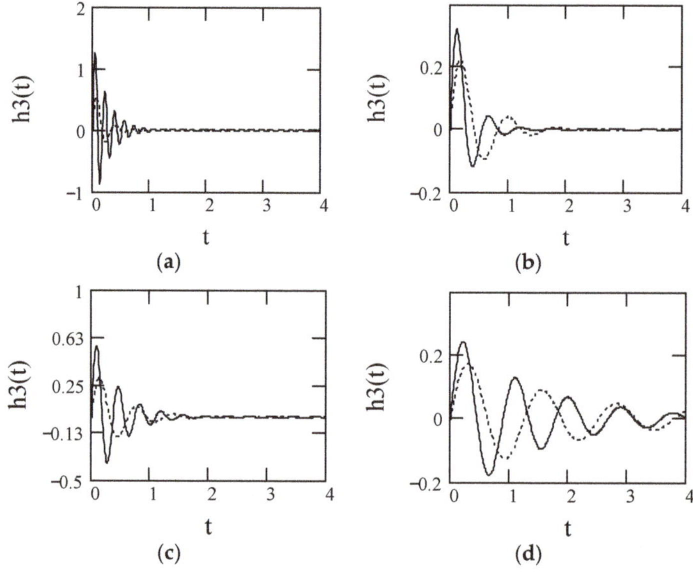

Figure 39. Impulse response $h_3(t)$ for $m = c = 1$, $k = 25$ ($\omega_n = 5$). (**a**) (α, β) = (1.8, 0.3), solid line: $\omega = 2$ ($\varsigma_{eq3} = 0.03$); dot line: $\omega = 1$ ($\varsigma_{eq3} = 0.02$). (**b**) (α, β) = (1.5, 0.8), solid line: $\omega = 2$ ($\varsigma_{eq3} = 0.20$); dot line: $\omega = 1$ ($\varsigma_{eq3} = 0.10$). (**c**) (α, β) = (1.8, 0.5), solid line: $\omega = 2$ ($\varsigma_{eq3} = 0.05$); dot line: $\omega = 1$ ($\varsigma_{eq3} = 0.02$). (**d**) (α, β) = (2, 1), solid line: $\omega = 2$ ($\varsigma_{eq3} = 0$); dot line: $\omega = 1$ ($\varsigma_{eq3} = 0$).

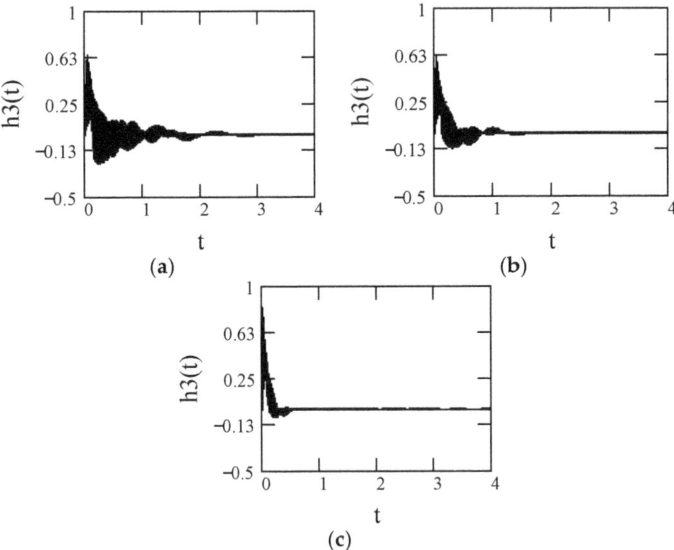

Figure 40. Impulse response $h_3(t)$ to a fractional oscillator in Class III for $m = c = 1$, $k = 25$ ($\omega_n = 5$). (a) $(\alpha, \beta) = (1.8, 0.8)$, $\omega = 1, 2, ..., 5$ ($0.09 \leq \varsigma_{eq3} \leq 0.45$). (b) $(\alpha, \beta) = (1.5, 0.8)$, $\omega = 1, 2, ..., 5$ ($0.09 \leq \varsigma_{eq3} \leq 0.20$). (c) $(\alpha, \beta) = (1.3, 0.8)$, $\omega = 1, 2, ..., 5$ ($0.48 \leq \varsigma_{eq3} \leq 0.67$).

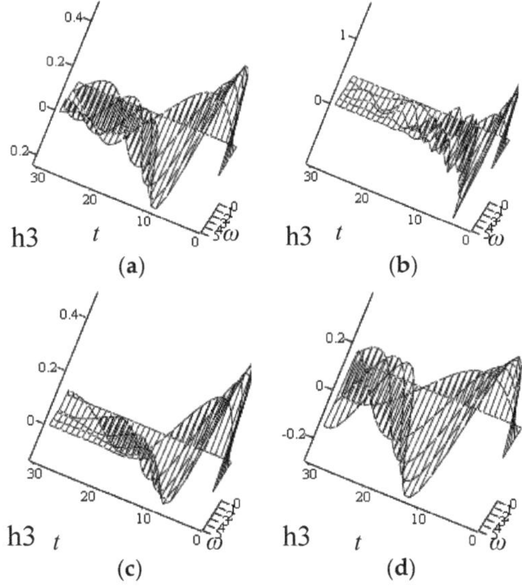

Figure 41. Impulse response to a fractional oscillator in Class III in t-ω plane for $m = c = 1$, $k = 25$ ($\omega_n = 5$) with $t = 0, 1, ..., 30$; $\omega = 1, 2, ..., 5$. (a) $\alpha = 1.8$, $\beta = 0.8$ ($0.09 \leq \varsigma_{eq3} \leq 0.45$). (b) $\alpha = 1.8$, $\beta = 0.4$ ($0.07 \leq \varsigma_{eq3} \leq 0.15$). (c) $\alpha = 1.5$, $\beta = 0.8$ ($0.33 \leq \varsigma_{eq3} \leq 0.91$). (d) $\alpha = 2$, $\beta = 1$ ($\varsigma_{eq3} = 0$).

Note 7.3: The impulse response $h_3(t)$ degenerates to the conventional one when $\alpha = 2$ and $\beta = 1$. Indeed,

$$h_3(t)\big|_{\alpha=2,\beta=1} = \left[\frac{e^{-\frac{m\omega^{\alpha-1}\sin\frac{\alpha\pi}{2}+c\omega^{\beta-1}\sin\frac{\beta\pi}{2}}{2\sqrt{-(m\omega^{\alpha-2}\cos\frac{\alpha\pi}{2}+c\omega^{\beta-2}\cos\frac{\beta\pi}{2})k}}\omega_{eqn,3}t}\sin\omega_{eqd,3}t}{-\left(m\omega^{\alpha-2}\cos\frac{\alpha\pi}{2}+c\omega^{\beta-2}\cos\frac{\beta\pi}{2}\right)\omega_{eqd,3}} \right]_{\alpha=2,\beta=1} = \frac{e^{-\varsigma\omega_n t}}{m\omega_d}\sin\omega_d t. \tag{205}$$

7.5. Application to Represetenting Generalized Mittag-Leffler Function (2)

The impulse response to fractional oscillators in Class I by using the generalized Mittag-Leffler function is in the form (Uchaikin ([38], Chapter 7))

$$h_1(t) = t^{\alpha-1}E_{\alpha,\alpha}\left[-(\omega_n t)^{\alpha}\right], 1 < \alpha \leq 2, t \geq 0. \tag{206}$$

In this section, we propose the representation of (206) by elementary functions.

Corollary 16. *The generalized Mittag-Leffler function in the form (206) can be expressed by the elementary functions in Theorem 13, for $1 < \alpha \leq 2$ and $t \geq 0$, in the form*

$$t^{\alpha-1}E_{\alpha,\alpha}\left[-(\omega_n t)^{\alpha}\right] = \frac{e^{-\frac{\omega\sin\frac{\alpha\pi}{2}}{2|\cos\frac{\alpha\pi}{2}|}t}\sin\frac{\omega_n\sqrt{1-\frac{\omega^{2\alpha}\sin^2\frac{\alpha\pi}{2}}{4\omega_n^2|\cos\frac{\alpha\pi}{2}|}}}{\sqrt{\omega^{\alpha-2}|\cos\frac{\alpha\pi}{2}|}}t}{m\omega_n\sqrt{\omega^{\alpha-2}|\cos\frac{\alpha\pi}{2}|}\sqrt{1-\frac{\omega^{2\alpha}\sin^2\frac{\alpha\pi}{2}}{4\omega_n^2|\cos\frac{\alpha\pi}{2}|}}}. \tag{207}$$

The proof is straightforward from Theorem 13 and (206).

8. Step Responses to Three Classes of Fractional Oscillators

In this section, we shall put forward the unit step responses to three classes of fractional oscillators in the analytic closed forms with elementary functions. Besides, we shall suggest a novel expression of a certain generalized Mittag-Leffler function by using elementary functions.

8.1. General Form of Step Responses

Denote by $g_j(t)$ ($j = 1, 2, 3$) the step response to a fractional oscillator in the jth Class. Then, it is also the step response to the jth equivalent oscillator. Precisely, $g_j(t)$ is the solution to the jth equivalent oscillator expressed by

$$\begin{cases} m_{eqj}\ddot{g}_j(t) + c_{eqj}\dot{g}_j(t) + kg_j(t) = u(t) \\ g_j(0) = 0, \dot{g}_j(0) = 0 \end{cases}, j = 1,2,3. \tag{208}$$

The solution to the above equation is given by

$$g_j(t) = \int_0^t h_j(\tau)d\tau = \frac{1}{k}\left[1 - \frac{e^{-\varsigma_{eqj}\omega_{eqn,j}t}}{\sqrt{1-\varsigma_{eqj}^2}}\cos\left(\omega_{eqd,j}t - \phi_j\right)\right], j = 1,2,3, \tag{209}$$

where

$$\phi_j = \tan^{-1}\frac{\varsigma_{eqj}}{\sqrt{1-\varsigma_{eqj}^2}}, j = 1,2,3. \tag{210}$$

8.2. Step Response to a Fractional Oscillator in Class I

Theorem 16 (Step response I). *Let $g_1(t)$ be the unit step response to a fractional oscillator in Class I. For $t \geq 0$ and $1 < \alpha \leq 2$, it is given by*

$$g_1(t) = \frac{1}{k}\left[1 - \frac{e^{-\frac{\omega \sin \frac{\alpha\pi}{2}}{2|\cos \frac{\alpha\pi}{2}|}t} \cos\left(\frac{\omega_n \sqrt{1 - \frac{\omega^\alpha \sin^2 \frac{\alpha\pi}{2}}{4\omega_n^2 \left|\cos \frac{\alpha\pi}{2}\right|}}}{\sqrt{-\omega^{\alpha-2}\cos \frac{\alpha\pi}{2}}}t - \phi_1 \right)}{\sqrt{1 - \left(\frac{\omega^{\frac{\alpha}{2}} \sin \frac{\alpha\pi}{2}}{2\omega_n \sqrt{-\cos \frac{\alpha\pi}{2}}} \right)^2}} \right], \tag{211}$$

where

$$\phi_1 = \tan^{-1}\frac{\varsigma_{eq1}}{\sqrt{1 - \varsigma_{eq1}^2}} = \tan^{-1}\frac{\frac{\omega^{\frac{\alpha}{2}} \sin \frac{\alpha\pi}{2}}{2\omega_n \sqrt{\left|\cos \frac{\alpha\pi}{2}\right|}}}{\sqrt{1 - \left(\frac{\omega^{\frac{\alpha}{2}} \sin \frac{\alpha\pi}{2}}{2\omega_n \sqrt{\left|\cos \frac{\alpha\pi}{2}\right|}} \right)^2}}. \tag{212}$$

Proof. Note that

$$g_1(t) = \frac{1}{k}\left[1 - \frac{e^{-\varsigma_{eq1}\omega_{eqn,1}t}}{\sqrt{1 - \varsigma_{eq1}^2}}\cos\left(\omega_{eqd,1}t - \phi_1\right) \right]. \tag{213}$$

Substituting ς_{eq1} with the one in (141) into the above produces

$$g_1(t) = \frac{1}{k}\left[1 - \frac{e^{-\varsigma_{eq1}\omega_{eqn,1}t}\cos\left(\omega_{eqd,1}t - \phi_1\right)}{\sqrt{1 - \varsigma_{eq1}^2}} \right] = \frac{1}{k}\left[1 - \frac{e^{-\frac{\omega^{\frac{\alpha}{2}} \sin \frac{\alpha\pi}{2}}{2\omega_n \sqrt{-\cos \frac{\alpha\pi}{2}}}\omega_{eqn,1}t}\cos\left(\omega_{eqd,1}t - \phi_1\right)}{\sqrt{1 - \left(\frac{\omega^{\frac{\alpha}{2}} \sin \frac{\alpha\pi}{2}}{2\omega_n \sqrt{-\cos \frac{\alpha\pi}{2}}} \right)^2}} \right]. \tag{214}$$

Replacing $\omega_{eqn,1}$ and $\omega_{eqd,1}$ with those in Section 5 in the above yields (211) and (212). The proof finishes. $\square$

Figure 42 shows the unit step response $g_1(t)$ with fixed oscillation frequency ω. Note that $g_1(t)$ takes ω as an argument. Thus, we use Figure 43 to indicate $g_1(t)$ with variable ω in time domain. Its plots in t-ω plane are shown in Figure 44.

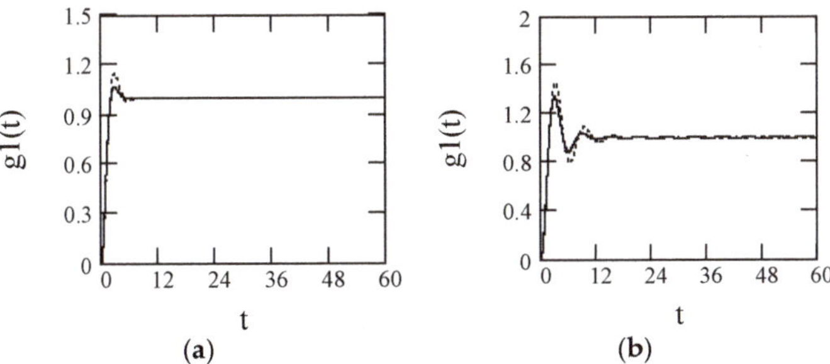

Figure 42. *Cont.*

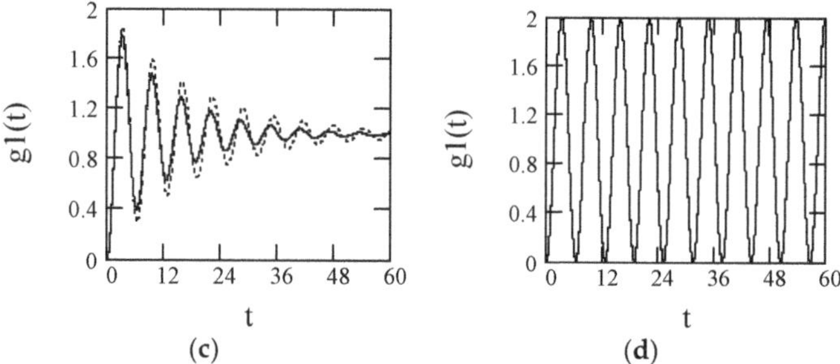

Figure 42. Unit step response $g_1(t)$ to a fractional oscillator in Class I with fixed ω for $m = k = 1$. (a) $\alpha = 1.3$, solid line: $\omega = 1$ ($\varsigma_{eq1} = 0.66$); dot line: $\omega = 0.7$ ($\varsigma_{eq1} = 0.52$). (b) $\alpha = 1.6$, solid line: $\omega = 1$ ($\varsigma_{eq1} = 0.33$); dot line: $\omega = 0.7$ ($\varsigma_{eq1} = 0.25$). (c) $\alpha = 1.9$, solid line: $\omega = 1$ ($\varsigma_{eq1} = 0.08$); dot line: $\omega = 0.7$ ($\varsigma_{eq1} = 0.06$). (d) $\alpha = 2$ ($\varsigma_{eq1} = 0$).

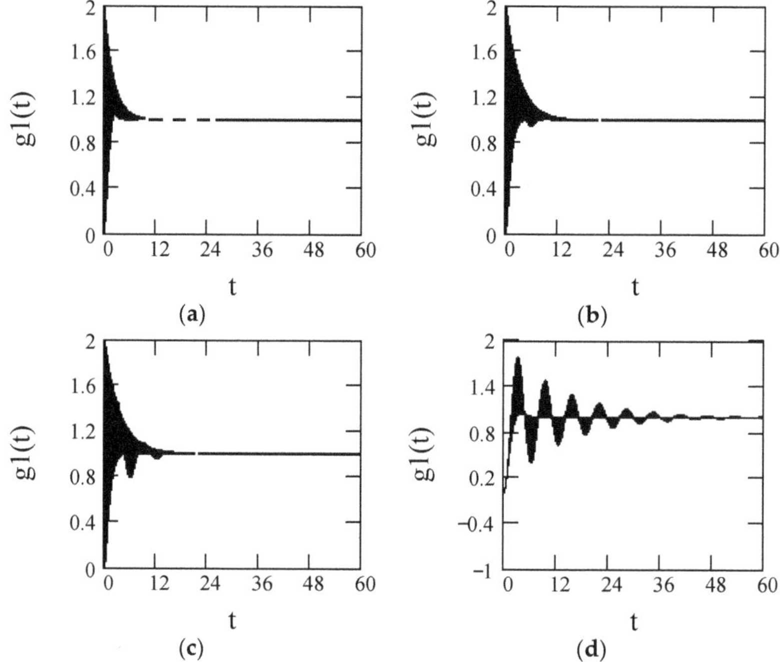

Figure 43. Step response $g_1(t)$ to a fractional oscillator in Class I with variable ω for $m = k = 1$. (a) $\alpha = 1.3$, $\omega = 1, 1.2, 1.4, ..., 5$ ($0.66 \leq \varsigma_{eq1} \leq 1.88$). (b) $\alpha = 1.5$, $\omega = 1, 1.2, 1.4, ..., 10$ ($0.66 \leq \varsigma_{eq1} \leq 2.95$). (c) $\alpha = 1.7$, $\omega = 1, 1.2, 1.4, ..., 10$ ($0.08 \leq \varsigma_{eq1} \leq 0.36$). (d) $\alpha = 1.9$, $\omega = 1, 1.12, 1.14, ..., 10$ ($0.08 \leq \varsigma_{eq1} \leq 0.70$).

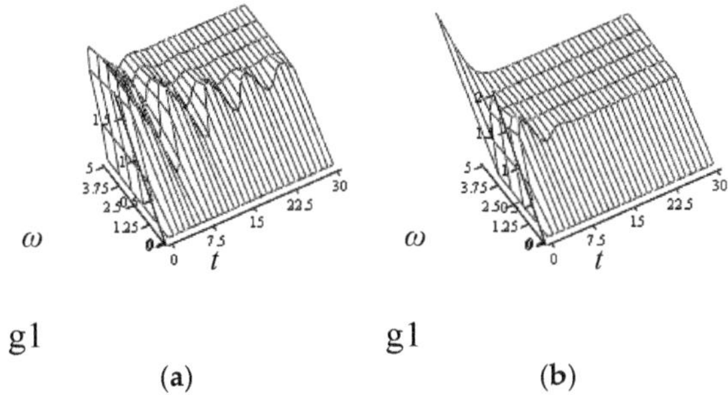

Figure 44. Step response $g_1(t)$ to a fractional oscillator in Class I in t-ω plane for $m = k = 1$, $\omega = 0, 1, ..., 5$. (a) $\alpha = 1.9$ ($0 \leq \varsigma_{eq1} \leq 0.36$). (b) $\alpha = 1.6$ ($0 \leq \varsigma_{eq1} \leq 1.18$).

Note 8.1: If $\alpha = 2$, $g_1(t)$ reduces to the conventional step response with damping free. In fact,

$$g_1(t)|_{\alpha=2} = \frac{1}{k}\left[1 - \frac{e^{-\frac{\omega \sin \frac{\alpha\pi}{2}}{2|\cos \frac{\alpha\pi}{2}|}t}\cos\left(\frac{\omega_n\sqrt{1-\frac{\omega^\alpha \sin^2 \frac{\alpha\pi}{2}}{4\omega_n^2|\cos \frac{\alpha\pi}{2}|}}}{\sqrt{-\omega^{\alpha-2}\cos \frac{\alpha\pi}{2}}}t - \phi_1\right)}{\sqrt{1-\left(\frac{\omega^{\frac{\alpha}{2}}\sin \frac{\alpha\pi}{2}}{2\omega_n\sqrt{-\cos \frac{\alpha\pi}{2}}}\right)^2}}\right]_{\alpha=2} = \frac{1}{k}(1-\cos \omega_n t), \tag{215}$$

and

$$\phi_1|_{\alpha=1} = \tan^{-1}\frac{\frac{\omega^{\frac{\alpha}{2}}\sin \frac{\alpha\pi}{2}}{2\omega_n\sqrt{|\cos \frac{\alpha\pi}{2}|}}}{\sqrt{1-\left(\frac{\omega^{\frac{\alpha}{2}}\sin \frac{\alpha\pi}{2}}{2\omega_n\sqrt{|\cos \frac{\alpha\pi}{2}|}}\right)^2}}\Bigg|_{\alpha=2} = 0. \tag{216}$$

8.3. Step Response to a Fractional Oscillator in Class II

Theorem 17 (Step response II). *Denote by $g_2(t)$ the unit step response to a fractional oscillator in Class II. It is in the form, for $t \geq 0$ and $0 < \beta \leq 1$,*

$$g_2(t) = \frac{1}{k}\left[1 - \frac{e^{-\frac{\varsigma\omega_n\omega^{\beta-1}\sin \frac{\beta\pi}{2}}{1-\frac{\varsigma}{m}\omega^{\beta-2}\cos \frac{\beta\pi}{2}}t}\cos\left(\frac{\omega_n\sqrt{1-\frac{\varsigma^2\omega^{2(\beta-1)}\sin^2 \frac{\beta\pi}{2}}{1-\frac{\varsigma}{m}\omega^{\beta-2}\cos \frac{\beta\pi}{2}}}}{\sqrt{\left(1-\frac{\varsigma}{m}\omega^{\beta-2}\cos \frac{\beta\pi}{2}\right)}}t - \phi_2\right)}{\sqrt{1-\frac{\varsigma^2\omega^{2(\beta-1)}\sin^2 \frac{\beta\pi}{2}}{1-\frac{\varsigma}{m}\omega^{\beta-2}\cos \frac{\beta\pi}{2}}}}\right], \tag{217}$$

where

$$\phi_2 = \tan^{-1}\frac{\varsigma_{eq2}}{\sqrt{1-\varsigma_{eq2}^2}} = \tan^{-1}\frac{\frac{\varsigma\omega^{\beta-1}\sin \frac{\beta\pi}{2}}{\sqrt{1-\frac{\varsigma}{m}\omega^{\beta-2}\cos \frac{\beta\pi}{2}}}}{\sqrt{1-\frac{\varsigma^2\omega^{2(\beta-1)}\sin^2 \frac{\beta\pi}{2}}{1-\frac{\varsigma}{m}\omega^{\beta-2}\cos \frac{\beta\pi}{2}}}}. \tag{218}$$

Proof. ; Substituting ς_{eq2} with that in Section 5 into the following expression

$$g_2(t) = \frac{1}{k}\left[1 - \frac{e^{-\varsigma_{eq2}\omega_{eqn,2}t}}{\sqrt{1-\varsigma_{eq2}^2}}\cos\left(\omega_{eqd,2}t - \phi_2\right)\right] \tag{219}$$

yields

$$g_2(t) = \frac{1}{k}\left[1 - \frac{e^{-\frac{\varsigma\omega^{\beta-1}\sin\frac{\beta\pi}{2}}{\sqrt{1-\frac{\varsigma}{m}\omega^{\beta-2}\cos\frac{\beta\pi}{2}}}\omega_{eqn,2}t}}{\sqrt{1-\frac{\varsigma^2\omega^{2(\beta-1)}\sin^2\frac{\beta\pi}{2}}{1-\frac{\varsigma}{m}\omega^{\beta-2}\cos\frac{\beta\pi}{2}}}}\cos\left(\omega_{eqd,2}t - \phi_2\right)\right]. \tag{220}$$

On the other side, replacing $\omega_{eqn,2}$ by the one in (135) in the above results in

$$g_2(t) = \frac{1}{k}\left[1 - \frac{e^{-\frac{\varsigma\omega^{\beta-1}\sin\frac{\beta\pi}{2}}{\sqrt{1-\frac{\varsigma}{m}\omega^{\beta-2}\cos\frac{\beta\pi}{2}}}\omega_{eqn,2}t}}{\sqrt{1-\frac{\varsigma^2\omega^{2(\beta-1)}\sin^2\frac{\beta\pi}{2}}{1-\frac{\varsigma}{m}\omega^{\beta-2}\cos\frac{\beta\pi}{2}}}}\cos\left(\omega_{eqd,2}t - \phi_2\right)\right] = \frac{1}{k}\left[1 - \frac{e^{-\frac{\varsigma\omega n\omega^{\beta-1}\sin\frac{\beta\pi}{2}}{1-\frac{\varsigma}{m}\omega^{\beta-2}\cos\frac{\beta\pi}{2}}t}}{\sqrt{1-\frac{\varsigma^2\omega^{2(\beta-1)}\sin^2\frac{\beta\pi}{2}}{1-\frac{\varsigma}{m}\omega^{\beta-2}\cos\frac{\beta\pi}{2}}}}\cos\left(\omega_{eqd,2}t - \phi_2\right)\right].$$

Finally, substituting $\omega_{eqd,2}$ by that in (157) in the above produces (217) and (218). Hence, we finish the proof. □

We use Figure 45 to indicate $g_2(t)$ with fixed ω. When considering variable ω, we show $g_2(t)$ in Figure 46 in time domain and Figure 47 in t-ω plane.

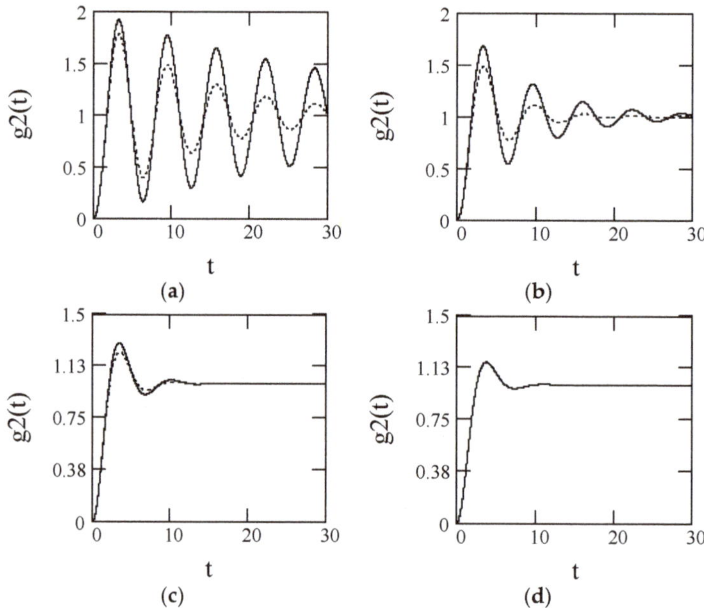

Figure 45. Step response $g_2(t)$ to a fractional oscillator in Class II with fixed ω for $m = c = k = 1$. (a) $\beta = 0.3$, solid line: $\omega = 20$ ($\varsigma_{eq2} = 0.03$); dot line: $\omega = 5$ ($\varsigma_{eq2} = 0.08$). (b) $\beta = 0.6$, solid line: $\omega = 20$ ($\varsigma_{eq2} = 0.12$); dot line: $\omega = 5$ ($\varsigma_{eq2} = 0.22$). (c) $\beta = 0.9$, solid line: $\omega = 20$ ($\varsigma_{eq2} = 0.37$); dot line: $\omega = 5$ ($\varsigma_{eq2} = 0.43$). (d) $\beta = 1$, solid line: $\omega = 20$ ($\varsigma_{eq2} = 0.50$); dot line: $\omega = 5$ ($\varsigma_{eq2} = 0.50$).

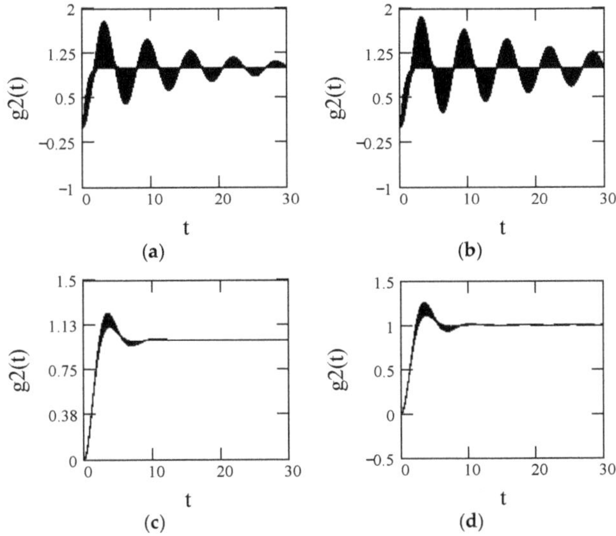

Figure 46. Step response $g_2(t)$ to a fractional oscillator in Class II with variable ω for $m = c = k = 1$. (a) $\beta = 0.3$, $\omega = 1, 2, ..., 5$ ($0.08 \leq \varsigma_{eq2} \leq 0.69$). (b) $\beta = 0.3$, $\omega = 1, 2, ..., 10$ ($0.22 \leq \varsigma_{eq2} \leq 0.63$). (c) $\beta = 0.9$, $\omega = 1, 2, ..., 5$ ($0.43 \leq \varsigma_{eq2} \leq 0.54$). (d) $\beta = 0.9$, $\omega = 1, 2, ..., 10$ ($0.40 \leq \varsigma_{eq2} \leq 0.54$).

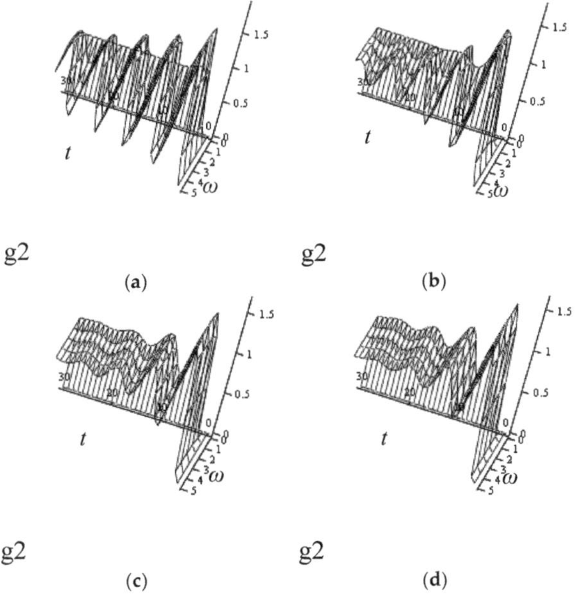

Figure 47. Step response $g_2(t)$ in t-ω plane for $m = c = 1$ and $\omega_n = 0.3$ ($k = 0.09$), with $t = 0, 1, ..., 30$, $\omega = 1, 2, 3, 4$. (a) $\beta = 0.3$ ($0.09 \leq \varsigma_{eq2} \leq 0.69$). (b) $\beta = 0.6$ ($0.24 \leq \varsigma_{eq2} \leq 0.63$). (c) $\beta = 0.9$ ($0.44 \leq \varsigma_{eq2} \leq 0.54$). (d) $\beta = 1$ ($\varsigma_{eq2} = 0.50$).

Note 8.2: When $\beta = 1$, $g_2(t)$ turns to be the ordinary step response. As a matter of fact,

$$g_2(t)|_{\beta=1} = \frac{1}{k}\left[1 - \frac{e^{-\varsigma\omega_{eqn,2}t}}{\sqrt{1-\varsigma^2}}\cos\left(\omega_n\sqrt{1-\varsigma^2}t - \phi_2\right)\right], \tag{221}$$

where

$$\phi_2|_{\beta=1} = \tan^{-1}\left.\frac{\frac{\varsigma\omega^{\beta-1}\sin\frac{\beta\pi}{2}}{\sqrt{1-\frac{c}{m}\omega^{\beta-2}\cos\frac{\beta\pi}{2}}}}{\sqrt{1-\frac{\varsigma^2\omega^{2(\beta-1)}\sin^2\frac{\beta\pi}{2}}{1-\frac{c}{m}\omega^{\beta-2}\cos\frac{\beta\pi}{2}}}}\right|_{\beta=1} = \tan^{-1}\frac{\varsigma}{\sqrt{1-\varsigma^2}}. \tag{222}$$

8.4. Step Response to a Fractional Oscillator in Class III

Theorem 18 (Step response III). *Let $g_3(t)$ be the unit step response to a fractional oscillator in Class III. It is in the form, for $t \geq 0$, $1 < \alpha \leq 2$, and $0 < \beta \leq 1$,*

$$g_3(t) = \frac{1}{k}\left\{1 - \frac{e^{-\frac{m\omega^{\alpha-1}\sin\frac{\alpha\pi}{2}+c\omega^{\beta-1}\sin\frac{\beta\pi}{2}}{2(m\omega^{\alpha-2}|\cos\frac{\alpha\pi}{2}|-c\omega^{\beta-2}\cos\frac{\beta\pi}{2})}t}\cos\left(\omega_n\frac{\sqrt{1-\frac{\left(m\omega^{\alpha-1}\sin\frac{\alpha\pi}{2}+c\omega^{\beta-1}\sin\frac{\beta\pi}{2}\right)^2}{4\left[-\left(m\omega^{\alpha-2}\cos\frac{\alpha\pi}{2}+c\omega^{\beta-2}\cos\frac{\beta\pi}{2}\right)k\right]}}}{\sqrt{-\left(\omega^{\alpha-2}\cos\frac{\alpha\pi}{2}+\frac{c}{m}\omega^{\beta-2}\cos\frac{\beta\pi}{2}\right)}}t - \phi_3\right)}{\sqrt{1 - \left[\frac{m\omega^{\alpha-1}\sin\frac{\alpha\pi}{2}+c\omega^{\beta-1}\sin\frac{\beta\pi}{2}}{2\sqrt{-\left(m\omega^{\alpha-2}\cos\frac{\alpha\pi}{2}+c\omega^{\beta-2}\cos\frac{\beta\pi}{2}\right)k}}\right]^2}}\right\}, \tag{223}$$

where

$$\phi_3 = \tan^{-1}\frac{\varsigma_{eq3}}{\sqrt{1-\varsigma_{eq3}^2}} = \tan^{-1}\frac{\frac{c\omega^{\beta-1}\sin\frac{\beta\pi}{2}}{2\sqrt{\left(m-c\omega^{\beta-2}\cos\frac{\beta\pi}{2}\right)k}}}{\sqrt{1-\left(\frac{c^2\omega^{2(\beta-1)}\sin^2\frac{\beta\pi}{2}}{4\left(m-c\omega^{\beta-2}\cos\frac{\beta\pi}{2}\right)k}\right)}}. \tag{224}$$

Proof. Replacing ς_{eq3} by that in (147) on the left side of the following produces the right side in the form

$$g_3(t) = \frac{1}{k}\left[1 - \frac{e^{-\varsigma_{eq3}\omega_{eqn,3}t}}{\sqrt{1-\varsigma_{eq3}^2}}\cos\left(\omega_{eqd,3}t - \phi_3\right)\right]$$

$$= \frac{1}{k}\left[1 - \frac{e^{-\frac{m\omega^{\alpha-1}\sin\frac{\alpha\pi}{2}+c\omega^{\beta-1}\sin\frac{\beta\pi}{2}}{2\sqrt{-(m\omega^{\alpha-2}\cos\frac{\alpha\pi}{2}+c\omega^{\beta-2}\cos\frac{\beta\pi}{2})k}}\omega_{eqn,3}t}\cos\left(\omega_{eqd,3}t-\phi_3\right)}{\sqrt{1 - \left[\frac{m\omega^{\alpha-1}\sin\frac{\alpha\pi}{2}+c\omega^{\beta-1}\sin\frac{\beta\pi}{2}}{2\sqrt{-\left(m\omega^{\alpha-2}\cos\frac{\alpha\pi}{2}+c\omega^{\beta-2}\cos\frac{\beta\pi}{2}\right)k}}\right]^2}}\right]. \tag{225}$$

Further, replacing $\omega_{eqn,3}$ with the one in (137) in the above yields

$$g_3(t) = \frac{1}{k}\left[1 - \frac{e^{-\frac{\omega_n\left(m\omega^{\alpha-1}\sin\frac{\alpha\pi}{2}+c\omega^{\beta-1}\sin\frac{\beta\pi}{2}\right)}{2\sqrt{-\left(m\omega^{\alpha-2}\cos\frac{\alpha\pi}{2}+c\omega^{\beta-2}\cos\frac{\beta\pi}{2}\right)k}}\sqrt{-\left(\omega^{\alpha-2}\cos\frac{\alpha\pi}{2}+\frac{c}{m}\omega^{\beta-2}\cos\frac{\beta\pi}{2}\right)}\,t}\cos\left(\omega_{eqd,3}t-\phi_3\right)}{\sqrt{1-\left[\frac{m\omega^{\alpha-1}\sin\frac{\alpha\pi}{2}+c\omega^{\beta-1}\sin\frac{\beta\pi}{2}}{2\sqrt{-\left(m\omega^{\alpha-2}\cos\frac{\alpha\pi}{2}+c\omega^{\beta-2}\cos\frac{\beta\pi}{2}\right)k}}\right]^2}}\right] \tag{226}$$

$$= \frac{1}{k}\left[1 - \frac{e^{-\frac{m\omega^{\alpha-1}\sin\frac{\alpha\pi}{2}+c\omega^{\beta-1}\sin\frac{\beta\pi}{2}}{2\left(m\omega^{\alpha-2}|\cos\frac{\alpha\pi}{2}|-c\omega^{\beta-2}\cos\frac{\beta\pi}{2}\right)}t}\cos\left(\omega_{eqd,3}t-\phi_3\right)}{\sqrt{1-\left[\frac{m\omega^{\alpha-1}\sin\frac{\alpha\pi}{2}+c\omega^{\beta-1}\sin\frac{\beta\pi}{2}}{2\sqrt{-\left(m\omega^{\alpha-2}\cos\frac{\alpha\pi}{2}+c\omega^{\beta-2}\cos\frac{\beta\pi}{2}\right)k}}\right]^2}}\right].$$

Finally, considering $\omega_{eqd,3}$ expressed by (160), we have (223) and (224). Hence, the proof finishes. $\square$

Figure 48 illustrates $g_3(t)$ in time with fixed ω while Figure 49 is with variable ω. Its illustrations in t-ω plane are shown in Figure 50.

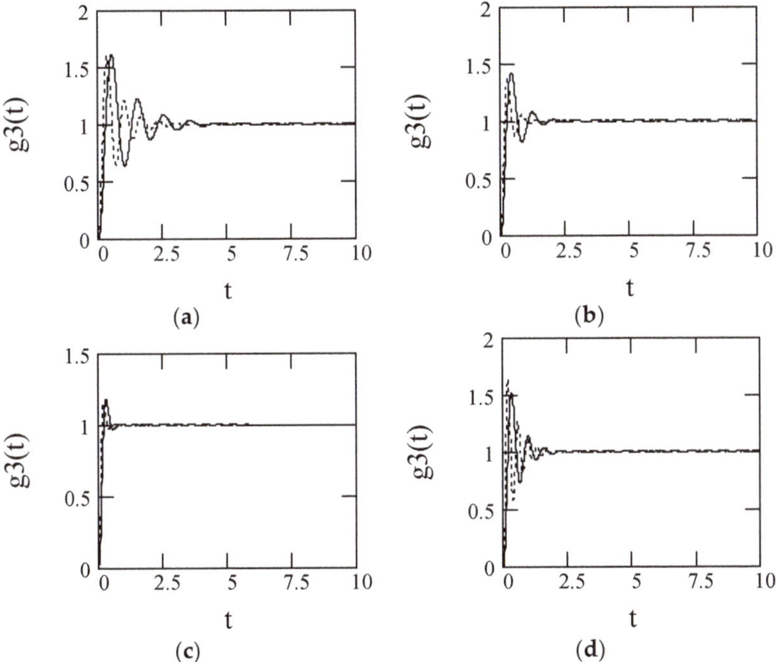

Figure 48. *Cont.*

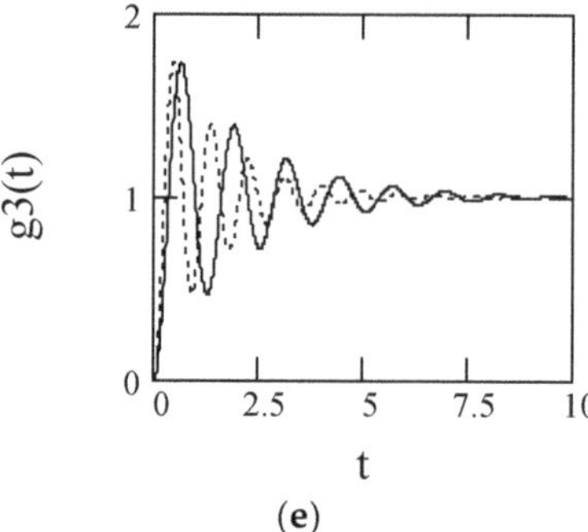

(e)

Figure 48. Illustrating step response $g_3(t)$ with fixed ω for $m = c = 1, k = 25$ ($\omega_n = 5$). (**a**) (α, β) = (1.8, 0.8), solid line: $\omega = 1$ ($\varsigma_{eq3} = 0.13$), dot line: $\omega = 2$ ($\varsigma_{eq3} = 0.05$). (**b**) (α, β) = (1.5, 0.8), solid line: $\omega = 1$ ($\varsigma_{eq3} = 0.33$), dot line: $\omega = 2$ ($\varsigma_{eq3} = 0.15$). (**c**) (α, β) = (1.3, 0.8), solid line: $\omega = 1$ ($\varsigma_{eq3} = 0.49$), dot line: $\omega = 2$ ($\varsigma_{eq3} = 0.24$). (**d**) (α, β) = (1.8, 0.5), solid line: $\omega = 1$ ($\varsigma_{eq3} = 0.09$), dot line: $\omega = 2$ ($\varsigma_{eq3} = 0.03$). (**e**) (α, β) = (2, 1), solid line: $\omega = 1$ ($\varsigma_{eq3} = 0$), dot line: $\omega = 2$ ($\varsigma_{eq3} = 0$).

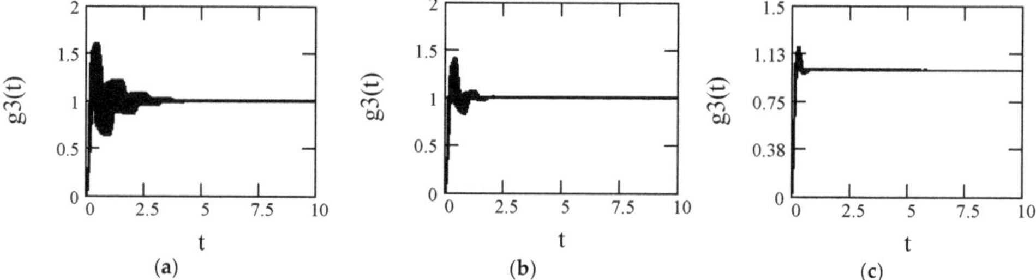

(a) **(b)** **(c)**

Figure 49. Demonstrating step response $g_3(t)$ with variable ω (= 1, 2, ..., 5) for $m = c = 1, k = 25$ ($\omega_n = 5$). (**a**) (α, β) = (1.8, 0.8) ($0.13 \leq \varsigma_{eq3} \leq 1.22$). (**b**) ($\alpha, \beta$) = (1.5, 0.8) ($0.34 \leq \varsigma_{eq3} \leq 0.91$). (**c**) ($\alpha, \beta$) = (1.3, 0.8) ($0.49 \leq \varsigma_{eq3} \leq 1.14$).

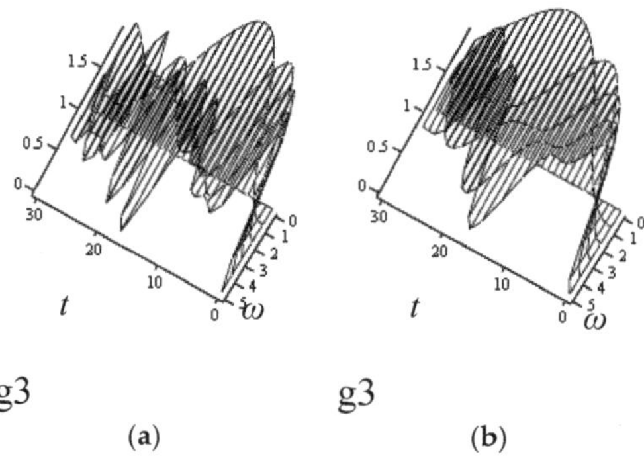

(a) **(b)**

Figure 50. *Cont.*

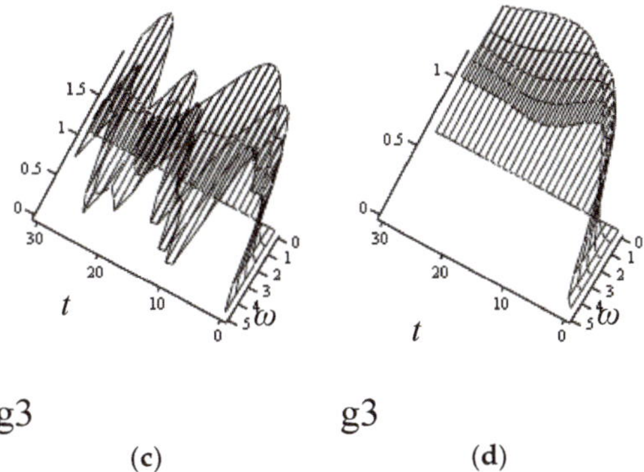

Figure 50. Illustrating step response $g_3(t)$ in t-ω plane for $m = c = k = 1$, with $t = 0, 1, ..., 30$, $\omega = 1, 2, ..., 5$. (**a**) $(\alpha, \beta) = (1.8, 0.3)$ $(0.05 \leq \varsigma_{eq3} \leq 0.10)$. (**b**) $(\alpha, \beta) = (1.8, 0.5)$ $(0.09 \leq \varsigma_{eq3} \leq 0.20)$. (**c**) $(\alpha, \beta) = (1.5, 0.6)$ $(0.25 \leq \varsigma_{eq3} \leq 0.55)$. (**d**) $(\alpha, \beta) = (2, 1)$ $(0.49 \leq \varsigma_{eq3} \leq 0.96)$.

Note 8.3: For $(\alpha, \beta) = (2, 1)$, $g_3(t)$ reduces to the conventional step response. Indeed,

$$g_3(t)\big|_{\alpha=2,\beta=1} = \frac{1}{k}\left[1 - \frac{e^{-\varsigma\omega_n t}}{\sqrt{1-\varsigma^2}}\cos\left(\omega_n\sqrt{1-\varsigma^2}t - \phi_3\big|_{\alpha=2,\beta=1}\right)\right], \tag{227}$$

where

$$\phi_3\big|_{\alpha=2,\beta=1} = \tan^{-1}\left.\frac{\dfrac{m\omega^{\alpha-1}\sin\frac{\alpha\pi}{2}+c\omega^{\beta-1}\sin\frac{\beta\pi}{2}}{2\sqrt{-\left(m\omega^{\alpha-2}\cos\frac{\alpha\pi}{2}+c\omega^{\beta-2}\cos\frac{\beta\pi}{2}\right)k}}}{\sqrt{1-\left[\dfrac{m\omega^{\alpha-1}\sin\frac{\alpha\pi}{2}+c\omega^{\beta-1}\sin\frac{\beta\pi}{2}}{2\sqrt{-\left(m\omega^{\alpha-2}\cos\frac{\alpha\pi}{2}+c\omega^{\beta-2}\cos\frac{\beta\pi}{2}\right)k}}\right]^2}}\right|_{\alpha=2,\beta=1} = \tan^{-1}\frac{\varsigma}{\sqrt{1-\varsigma^2}}. \tag{228}$$

8.5. Application to Represetenting Mittag-Leffler Function (3)

The step response to fractional oscillators in Class I by using the generalized Mittag-Leffler function is in the form (Uchaikin ([38], Chapter 7))

$$g_1(t) = t^\alpha E_{\alpha,\alpha+1}\left[-(\omega_n t)^\alpha\right], 1 < \alpha \leq 2, t \geq 0. \tag{229}$$

In the following corollary, we propose the representation of (229) by elementary functions.

Corollary 17. *The generalized Mittag-Leffler function expressed by (229) can be represented by using the elementary functions described in Theorem 16. Precisely, for $t \geq 0$ and $1 < \alpha \leq 2$, we have*

$$t^\alpha E_{\alpha,\alpha+1}\left[-(\omega_n t)^\alpha\right] = \frac{1}{k}\left[1 - \frac{e^{-\frac{\omega\sin\frac{\alpha\pi}{2}}{2|\cos\frac{\alpha\pi}{2}|}t}\cos\left(\omega_n\dfrac{\sqrt{1-\dfrac{\omega^\alpha\sin^2\frac{\alpha\pi}{2}}{4\omega_n^2\left|\cos\frac{\alpha\pi}{2}\right|}}}{\sqrt{-\omega^{\alpha-2}\cos\frac{\alpha\pi}{2}}}t - \phi_1\right)}{\sqrt{1-\left(\dfrac{\omega^{\frac{\alpha}{2}}\sin\frac{\alpha\pi}{2}}{2\omega_n\sqrt{-\cos\frac{\alpha\pi}{2}}}\right)^2}}\right], \tag{230}$$

where ϕ_1 is given by (212).

Proof. The left side of (230) equals to $g_1(t)$ following Theorem 16. According to (229), therefore, (230) holds. This completes the proof. $\square$

9. Frequency Responses to Three Classes of Fractional Oscillators

We put forward frequency responses to three classes of fractional oscillators in this section. They are expressed by elementary functions based on the theory of three equivalent oscillators addressed in Section 4.

9.1. General Form of Frequency Responses to Three Classes of Fractional Oscillators

Denote by $H_j(\omega)$ the Fourier transform of the impulse response $h_j(t)$ to a fractional oscillator in Class j ($j = 1, 2, 3$), where $h_j(t)$ is given by (196). Then, it is the frequency response function to a fractional oscillator in Class j ($j = 1, 2, 3$).

In fact, doing the Fourier transform on the both sides of (195) produces

$$\left(-\omega^2 + i2\varsigma_{eqj}\omega_{eqn,j}\omega + \omega_{eqn,j}^2\right)H_j(\omega) = \frac{1}{m_{eqj}}. \tag{231}$$

Thus, we have

$$H_j(\omega) = \frac{1}{m_{eqj}\left(\omega_{eqn,j}^2 - \omega^2 + i2\varsigma_{eqj}\omega_{eqn,j}\omega\right)} = \frac{1}{m_{eqj}\omega_{eqn,j}^2\left(1 - \frac{\omega^2}{\omega_{eqn,j}^2} + i2\varsigma_{eqj}\frac{\omega}{\omega_{eqn,j}}\right)}. \tag{232}$$

Note that

$$m_{eqj}\omega_{eqn,j}^2 = m_{eqj}\frac{k}{m_{eqj}} = k. \tag{233}$$

Therefore, by letting γ_{eqj} be the equivalent frequency ratio of a fractional oscillator in Class j, $H_j(\omega)$ may be expressed by

$$H_j(\omega) = \frac{1}{k\left(1 - \gamma_{eqj}^2 + i2\varsigma_{eqj}\gamma_{eqj}\right)}, j = 1, 2, 3. \tag{234}$$

The amplitude of $H_j(\omega)$ is

$$\left|H_j(\omega)\right| = \frac{1}{k}\frac{1}{\sqrt{\left(1 - \gamma_{eqj}^2\right)^2 + \left(2\varsigma_{eqj}\gamma_{eqj}\right)^2}}, j = 1, 2, 3. \tag{235}$$

Its phase frequency response is given by

$$\varphi_j(\omega) = \tan^{-1}\frac{2\varsigma_{eqj}\gamma_{eqj}}{1 - \gamma_{eqj}^2}, j = 1, 2, 3. \tag{236}$$

9.2. Frequency Response to a Fractional Oscillator in Class I

Theorem 19 (Frequency response I)**.** *Let $H_1(\omega)$ be the frequency response to a fractional oscillator in Class I. Then, for $1 < \alpha \le 2$, it is in the form*

$$H_1(\omega) = \frac{1}{k\left(1 - \frac{\omega^\alpha\left|\cos\frac{\alpha\pi}{2}\right|}{\omega_n^2} + i\frac{\omega^\alpha\sin\frac{\alpha\pi}{2}}{\omega_n^2}\right)}. \tag{237}$$

Proof. In the equation below,

$$H_1(\omega) = \frac{1}{k\left(1 - \gamma_{eq1}^2 + i2\varsigma_{eq1}\gamma_{eq1}\right)}, \tag{238}$$

when replacing γ_{eq1} by

$$\gamma_{eq1} = \gamma_{eq1}(\omega, \alpha) = \frac{\omega}{\omega_{eqn,1}} = \frac{\omega\sqrt{\omega^{\alpha-2}\left|\cos\frac{\alpha\pi}{2}\right|}}{\omega_n}, \tag{239}$$

and $2\varsigma_{eq1}\gamma_{eq1}$ by

$$2\varsigma_{eq1}\gamma_{eq1} = 2\frac{\omega^{\frac{\alpha}{2}}\sin\frac{\alpha\pi}{2}}{2\omega_n\sqrt{\left|\cos\frac{\alpha\pi}{2}\right|}}\frac{\omega\sqrt{\omega^{\alpha-2}\left|\cos\frac{\alpha\pi}{2}\right|}}{\omega_n} = \frac{\omega^{\alpha}\sin\frac{\alpha\pi}{2}}{\omega_n^2}, \tag{240}$$

we have (237). This completes the proof. $\square$

From Theorem 19, we have the amplitude of $H_1(\omega)$ given by

$$|H_1(\omega)| = \frac{1/k}{\sqrt{\left(1 - \frac{\omega^{\alpha}\left|\cos\frac{\alpha\pi}{2}\right|}{\omega_n^2}\right)^2 + \left(\frac{\omega^{\alpha}\sin\frac{\alpha\pi}{2}}{\omega_n^2}\right)^2}}, \tag{241}$$

and the phase in the form

$$\varphi_1(\omega) = \tan^{-1}\frac{\dfrac{\omega^{\alpha}\sin\frac{\alpha\pi}{2}}{\omega_n^2}}{1 - \dfrac{\omega^{\alpha}\left|\cos\frac{\alpha\pi}{2}\right|}{\omega_n^2}} = \tan^{-1}\frac{\omega^{\alpha}\sin\frac{\alpha\pi}{2}}{\omega_n^2 - \omega^{\alpha}\left|\cos\frac{\alpha\pi}{2}\right|}. \tag{242}$$

Note 9.1 (Equivalent frequency ratio I): The equivalent frequency ratio γ_{eq1} is a function of oscillation frequency ω and the fractional order α. It may be denoted by $\gamma_{eq1}(\omega, \alpha)$.

Figure 51 shows the plot of γ_{eq1}. Figure 52 indicates the illustrations of $H_1(\omega)$.

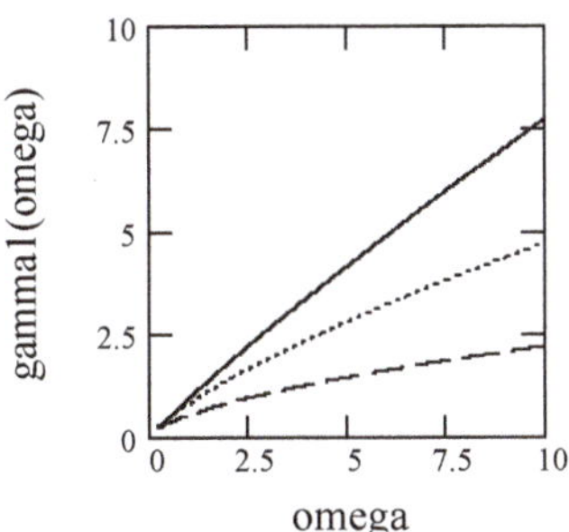

Figure 51. Equivalent frequency ratio $\gamma_{eq1}(\omega, \alpha)$ for fractional oscillators in Class I with $m = k = 1$. Solid line: $\alpha = 1.8$. Dot line: $\alpha = 1.5$. Dash line: $\alpha = 1.2$.

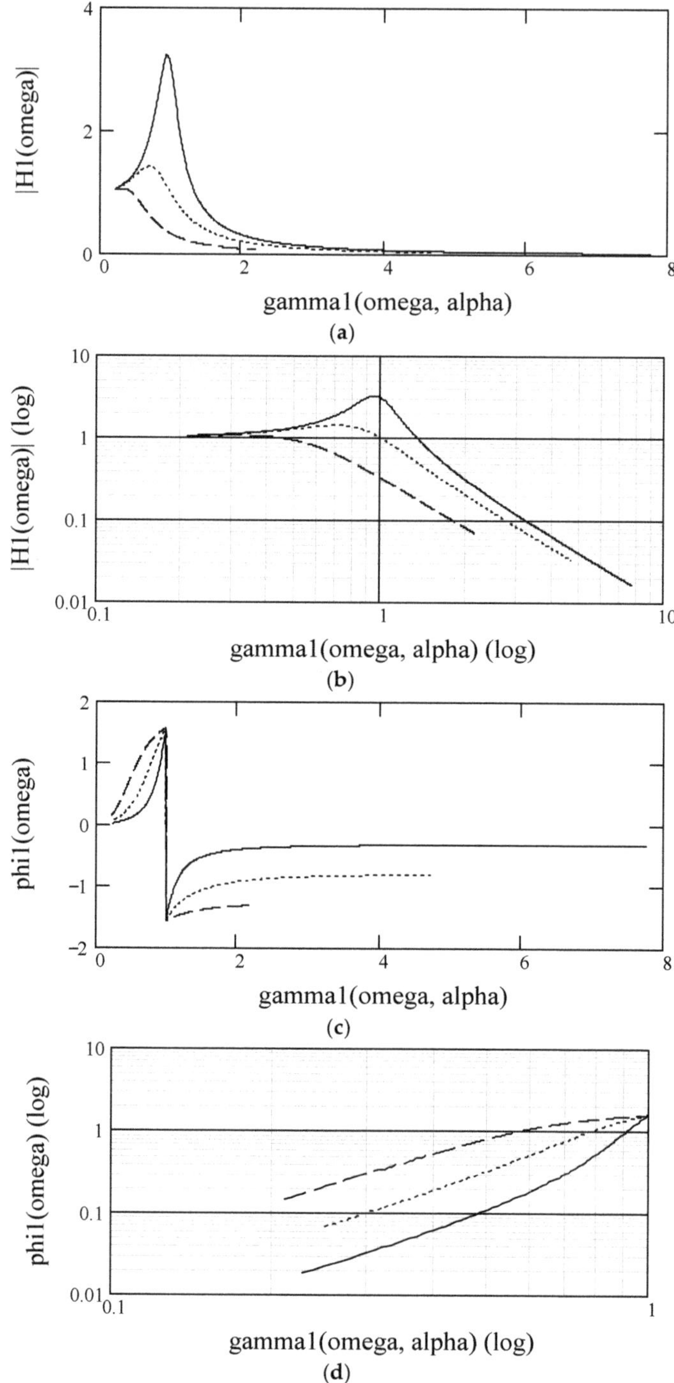

Figure 52. Frequency response $H_1(\omega)$ to fractional oscillators in Class I with $m = k = 1$. Solid line: $\alpha = 1.8$ ($0.04 \leq \varsigma_{eq1} \leq 0.06$). Dot line: $\alpha = 1.5$ ($0.13 \leq \varsigma_{eq1} \leq 0.19$). Dash line: $\alpha = 1.2$ ($0.33 \leq \varsigma_{eq1} \leq 0.46$). (**a**) Amplitude $|H_1(\omega)|$ in ordinary coordinate. (**b**) $|H_1(\omega)|$ in log-log. (**c**) Phase $\varphi_1(\omega)$ in ordinary coordinate. (**d**) $\varphi_1(\omega)$ in log-log.

Note 9.2: If $\alpha = 2$, $H_1(\omega) \to \infty$ at $\omega = \omega_n$. In that case, $H_1(\omega)$ turns to be the ordinary frequency response with damping free in the form

$$H_1(\omega)|_{\alpha=2} = \left.\frac{1/k}{1 - \frac{\omega^\alpha \left|\cos\frac{\alpha\pi}{2}\right|}{\omega_n^2} + i\frac{\omega^\alpha \sin\frac{\alpha\pi}{2}}{\omega_n^2}}\right|_{\alpha=2} = \frac{1/k}{1 - \frac{\omega^2}{\omega_n^2}} = \frac{1/k}{1 - \gamma^2}. \tag{243}$$

9.3. Frequency Response to a Fractional Oscillator in Class II

Theorem 20 (Frequency response II). *Denote by $H_2(\omega)$ the frequency response to a fractional oscillator in Class II. Then, for $0 < \beta \leq 1$, it is given by*

$$H_2(\omega) = \frac{1/k}{1 - \gamma^2\left(1 - \frac{c}{m}\omega^{\beta-2}\cos\frac{\beta\pi}{2}\right) + i\frac{2\varsigma\omega^\beta \sin\frac{\beta\pi}{2}}{\omega_n}}, \tag{244}$$

where $\gamma = \frac{\omega}{\omega_n}$ is the ordinary frequency ratio.

Proof. Consider

$$H_2(\omega) = \frac{1}{k\left(1 - \gamma_{eq2}^2 + i2\varsigma_{eq2}\gamma_{eq2}\right)}. \tag{245}$$

Note that

$$\gamma_{eq2} = \gamma_{eq2}(\omega, \beta) = \frac{\omega}{\omega_{eqn,2}} = \frac{\omega}{\omega_n}\sqrt{1 - \frac{c}{m}\omega^{\beta-2}\cos\frac{\beta\pi}{2}} = \gamma\sqrt{1 - \frac{c}{m}\omega^{\beta-2}\cos\frac{\beta\pi}{2}}. \tag{246}$$

Besides,

$$2\varsigma_{eq}\gamma_{eq2} = \frac{2\varsigma\omega^{\beta-1}\sin\frac{\beta\pi}{2}}{\sqrt{1 - \frac{c}{m}\omega^{\beta-2}\cos\frac{\beta\pi}{2}}}\left(\frac{\omega}{\omega_n}\sqrt{1 - \frac{c}{m}\omega^{\beta-2}\cos\frac{\beta\pi}{2}}\right) = \frac{2\varsigma\omega^\beta \sin\frac{\beta\pi}{2}}{\omega_n}. \tag{247}$$

Therefore, (245) becomes

$$\begin{aligned}
H_2(\omega) &= \frac{1}{k\left(1 - \gamma_{eq2}^2 + i2\varsigma_{eq2}\gamma_{eq2}\right)} \\[4pt]
&= \frac{1/k}{1 - \left(\frac{\omega}{\omega_n}\sqrt{1 - \frac{c}{m}\omega^{\beta-2}\cos\frac{\beta\pi}{2}}\right)^2 + i\frac{2\varsigma\omega^\beta \sin\frac{\beta\pi}{2}}{\omega_n}} \\[4pt]
&= \frac{1/k}{1 - \gamma^2\left(1 - \frac{c}{m}\omega^{\beta-2}\cos\frac{\beta\pi}{2}\right) + i\frac{2\varsigma\omega^\beta \sin\frac{\beta\pi}{2}}{\omega_n}}.
\end{aligned}$$

This finishes the proof. $\square$

From Theorem 20, we have the amplitude of $H_2(\omega)$ in the form

$$|H_2(\omega)| = \frac{1/k}{\sqrt{\left[1 - \gamma^2\left(1 - \frac{c}{m}\omega^{\beta-2}\cos\frac{\beta\pi}{2}\right)\right]^2 + \left(\frac{2\varsigma\omega^\beta \sin\frac{\beta\pi}{2}}{\omega_n}\right)^2}}, \tag{248}$$

and its phase given by

$$\varphi_2(\omega) = \tan^{-1}\frac{\frac{2\varsigma\omega^\beta \sin\frac{\beta\pi}{2}}{\omega_n}}{1 - \gamma^2\left(1 - \frac{c}{m}\omega^{\beta-2}\cos\frac{\beta\pi}{2}\right)}. \tag{249}$$

Note 9.3 (Equivalent frequency ratio II): The equivalent frequency ratio γ_{eq2} is dependent on oscillation frequency ω and the fractional order β as can be seen from (9.16). We denote it by $\gamma_{eq2}(\omega, \beta)$. Figure 53 indicates the plot of $\gamma_{eq2}(\omega, \beta)$. $H_2(\omega)$ is shown in Figure 54.

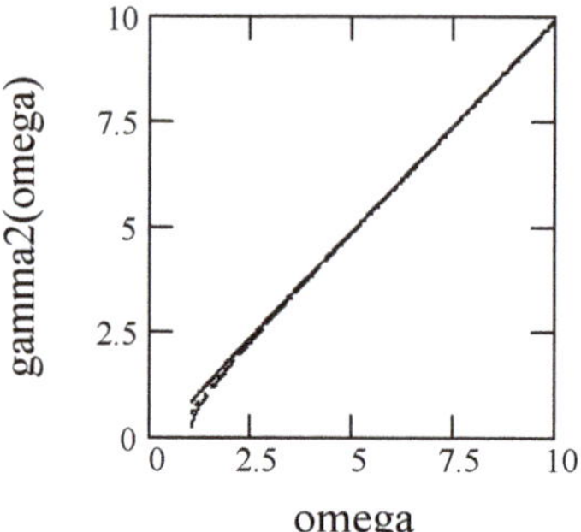

Figure 53. Equivalent frequency ratio $\gamma_{eq2}(\omega, \beta)$ of fractional oscillators in Class II with $m = c = k = 1$. Solid line: $\beta = 0.8$. Dot line: $\beta = 0.5$. Dash line: $\beta = 0.2$.

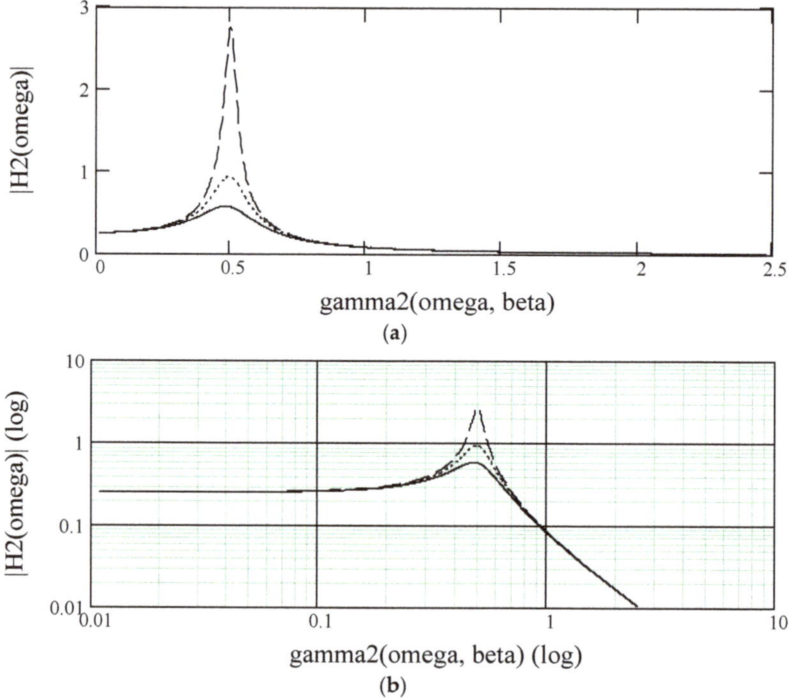

Figure 54. *Cont.*

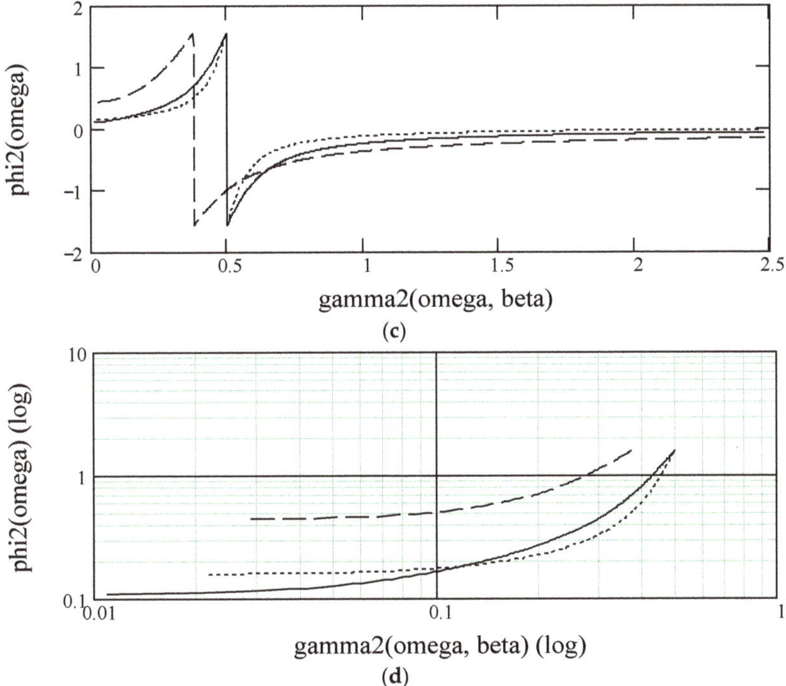

Figure 54. Frequency response $H_2(\omega)$ to fractional oscillators of Class II type with $m = c = 1$ and $k = 4$. Solid line: $\beta = 0.8$ ($0.15 \leq \varsigma_{eq2} \leq 0.29$). Dot line: $\beta = 0.5$ ($0.06 \leq \varsigma_{eq2} \leq 0.33$). Dash line: $\beta = 0.2$ ($0.01 \leq \varsigma_{eq2} \leq 0.35$). (**a**) Amplitude $|H_2(\omega)|$ in ordinary coordinate. (**b**) $|H_2(\omega)|$ in log-log. (**c**) Phase $\varphi_2(\omega)$ in ordinary coordinate. (**d**) $\varphi_2(\omega)$ in log-log.

Note 9.4: When $\beta = 1$, $H_2(\omega)$ reduces to that of an ordinary oscillator's in the form (also see Figure 55).

$$H_2(\omega)|_{\beta=1} = \left. \frac{1/k}{1-\gamma^2\left(1-\frac{c}{m}\omega^{\beta-2}\cos\frac{\beta\pi}{2}\right)+i\frac{2\varsigma\omega^\beta\sin\frac{\beta\pi}{2}}{\omega_n}} \right|_{\beta=1}$$

$$= \frac{1/k}{1-\gamma^2+i2\varsigma\frac{\omega}{\omega_n}} = \frac{1/k}{1-\gamma^2+i2\varsigma\gamma}. \tag{250}$$

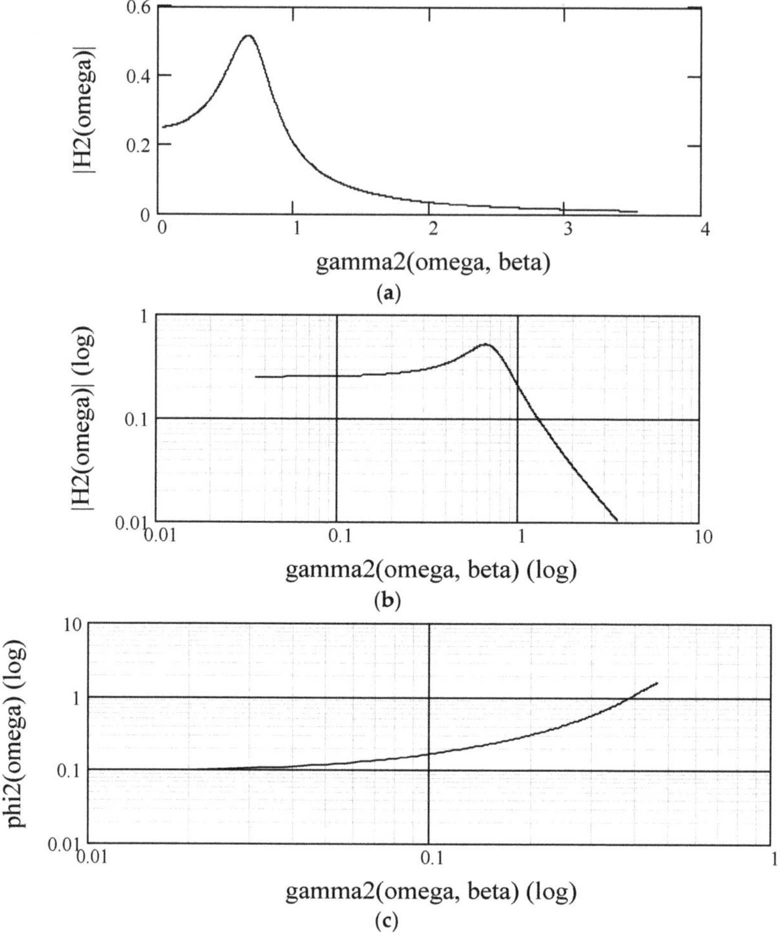

Figure 55. $H_2(\omega)$ for $\beta = 1$ with $m = c = 1$ and $k = 4$ ($\zeta_{eq2} = 0.25$). (**a**) $|H_2(\omega)|$ in ordinary coordinate. (**b**) $|H_2(\omega)|$ in log-log. (**c**) Phase $\varphi_2(\omega)$ in log-log.

9.4. Frequency Response to a Fractional Oscillator in Class III

Theorem 21 (Frequency response III). *Let $H_3(\omega)$ be the frequency response to a fractional oscillator of Class III type. Then, for $1 < \alpha \leq 2$ and $0 < \beta \leq 1$, $H_3(\omega)$ is in the form*

$$H_3(\omega) = \frac{1/k}{1 - \gamma^2 \left(\omega^{\alpha-2} \left| \cos \frac{\alpha\pi}{2} \right| - \frac{c\omega^{\beta-2} \cos \frac{\beta\pi}{2}}{m} \right) + i \dfrac{\gamma \left(\omega^{\alpha-1} \sin \frac{\alpha\pi}{2} + 2\varsigma\omega_n\omega^{\beta-1} \sin \frac{\beta\pi}{2} \right)}{\omega_n \left(\omega^{\alpha-2} \left| \cos \frac{\alpha\pi}{2} \right| - 2\varsigma\omega_n\omega^{\beta-2} \cos \frac{\beta\pi}{2} \right)}}. \tag{251}$$

Proof. In the equation below

$$H_3(\omega) = \frac{1}{k \left(1 - \gamma_{eq3}^2 + i2\varsigma_{eq3}\gamma_{eq3} \right)}, \tag{252}$$

we notice

$$\gamma_{eq3} = \gamma_{eq3}(\omega, \alpha, \beta) = \frac{\omega}{\omega_{eqn,3}} = \frac{\omega}{\omega_n}\sqrt{-\left(\omega^{\alpha-2}\cos\frac{\alpha\pi}{2} + \frac{c}{m}\omega^{\beta-2}\cos\frac{\beta\pi}{2}\right)}$$

$$= \gamma\sqrt{-\left(\omega^{\alpha-2}\cos\frac{\alpha\pi}{2} + \frac{c}{m}\omega^{\beta-2}\cos\frac{\beta\pi}{2}\right)}. \tag{253}$$

In addition,

$$2\varsigma_{eq3}\gamma_{eq3} = 2\frac{\left[\dfrac{\left(\omega^{\alpha-1}\sin\frac{\alpha\pi}{2} + 2\varsigma\omega_n\omega^{\beta-1}\sin\frac{\beta\pi}{2}\right)}{\gamma\sqrt{-\left(\omega^{\alpha-2}\cos\frac{\alpha\pi}{2} + \frac{c}{m}\omega^{\beta-2}\cos\frac{\beta\pi}{2}\right)}}\right]}{2\omega_n\sqrt{-\left(\omega^{\alpha-2}\cos\frac{\alpha\pi}{2} + 2\varsigma\omega_n\omega^{\beta-2}\cos\frac{\beta\pi}{2}\right)}} \tag{254}$$

$$= \frac{\gamma\left(\omega^{\alpha-1}\sin\frac{\alpha\pi}{2} + 2\varsigma\omega_n\omega^{\beta-1}\sin\frac{\beta\pi}{2}\right)}{\omega_n\left(\omega^{\alpha-2}\left|\cos\frac{\alpha\pi}{2}\right| - 2\varsigma\omega_n\omega^{\beta-2}\cos\frac{\beta\pi}{2}\right)}.$$

Thus, (252) becomes

$$H_3(\omega) = \frac{1/k}{1 - \gamma^2\left(\omega^{\alpha-2}\left|\cos\frac{\alpha\pi}{2}\right| - \dfrac{c\omega^{\beta-2}\cos\frac{\beta\pi}{2}}{m}\right) + i\dfrac{\gamma\left(\omega^{\alpha-1}\sin\frac{\alpha\pi}{2} + 2\varsigma\omega_n\omega^{\beta-1}\sin\frac{\beta\pi}{2}\right)}{\omega_n\left(\omega^{\alpha-2}\left|\cos\frac{\alpha\pi}{2}\right| - 2\varsigma\omega_n\omega^{\beta-2}\cos\frac{\beta\pi}{2}\right)}}.$$

Therefore, the proof completes. $\square$

From Theorem 21, we obtain $|H_3(\omega)|$ in the form

$$|H_3(\omega)| = \frac{1/k}{\sqrt{\left\{\left[1 - \gamma^2\left(\omega^{\alpha-2}\left|\cos\frac{\alpha\pi}{2}\right| - \frac{c}{m}\omega^{\beta-2}\cos\frac{\beta\pi}{2}\right)\right]^2 + \left[\dfrac{\gamma\left(\omega^{\alpha-1}\sin\frac{\alpha\pi}{2} + 2\varsigma\omega_n\omega^{\beta-1}\sin\frac{\beta\pi}{2}\right)}{\omega_n\left(\omega^{\alpha-2}\left|\cos\frac{\alpha\pi}{2}\right| - 2\varsigma\omega_n\omega^{\beta-2}\cos\frac{\beta\pi}{2}\right)}\right]^2\right\}}}. \tag{255}$$

The phase $\varphi_3(\omega)$ is given by

$$\varphi_3(\omega) = \tan^{-1}\frac{\dfrac{\gamma\left(\omega^{\alpha-1}\sin\frac{\alpha\pi}{2} + 2\varsigma\omega_n\omega^{\beta-1}\sin\frac{\beta\pi}{2}\right)}{\omega_n\left(\omega^{\alpha-2}\left|\cos\frac{\alpha\pi}{2}\right| - 2\varsigma\omega_n\omega^{\beta-2}\cos\frac{\beta\pi}{2}\right)}}{1 - \gamma^2\left(\omega^{\alpha-2}\left|\cos\frac{\alpha\pi}{2}\right| - \frac{c}{m}\omega^{\beta-2}\cos\frac{\beta\pi}{2}\right)}. \tag{256}$$

Note 9.5 (Equivalent frequency ratio III): γ_{eq3} relates to ω and a pair of fractional orders (α, β). Figure 56 indicates its plots. Figure 57 demonstrates $H_3(\omega)$.

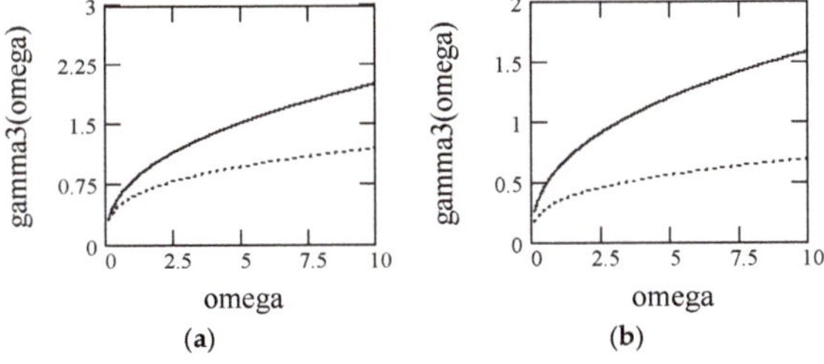

Figure 56. Plots of equivalent frequency ratio $\gamma_{eq3}(\omega, \alpha, \beta)$ for $m = c = k = 1$. (**a**) Solid line: $(\alpha, \beta) = (1.8, 0.8)$. Dot line: $(\alpha, \beta) = (1.5, 0.8)$. (**b**) Solid line: $(\alpha, \beta) = (1.5, 0.8)$. Dot line: $(\alpha, \beta) = (1.5, 0.6)$.

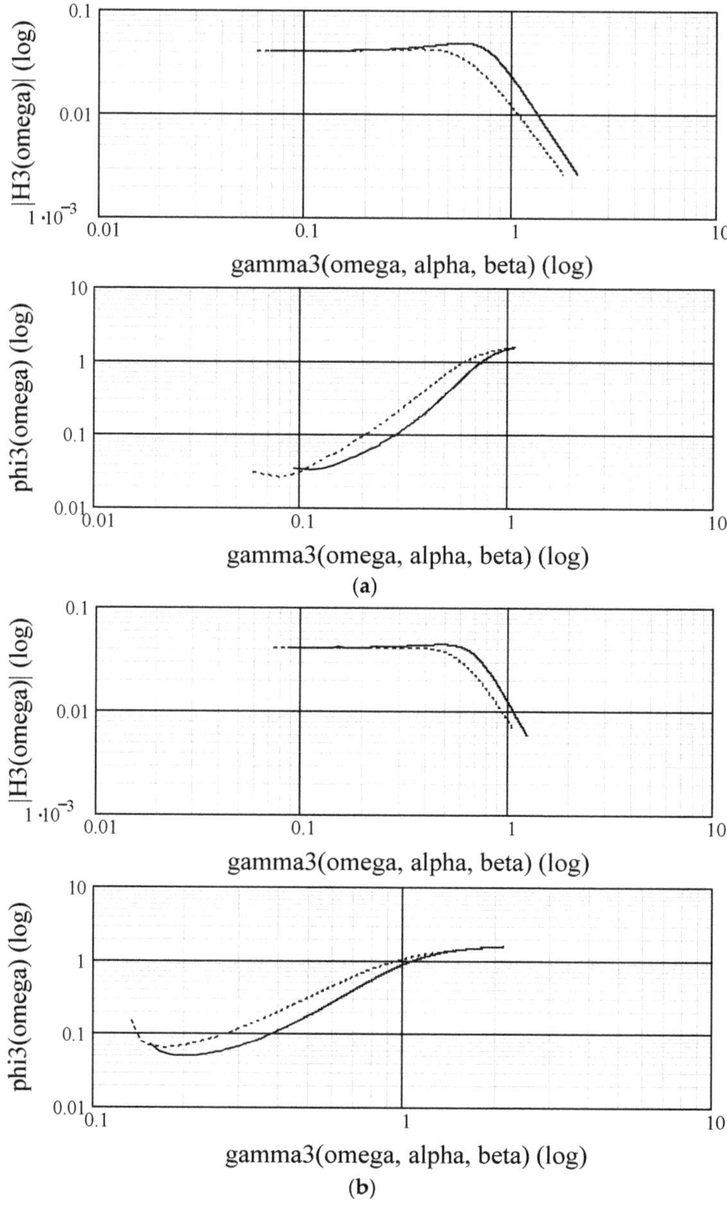

Figure 57. *Cont.*

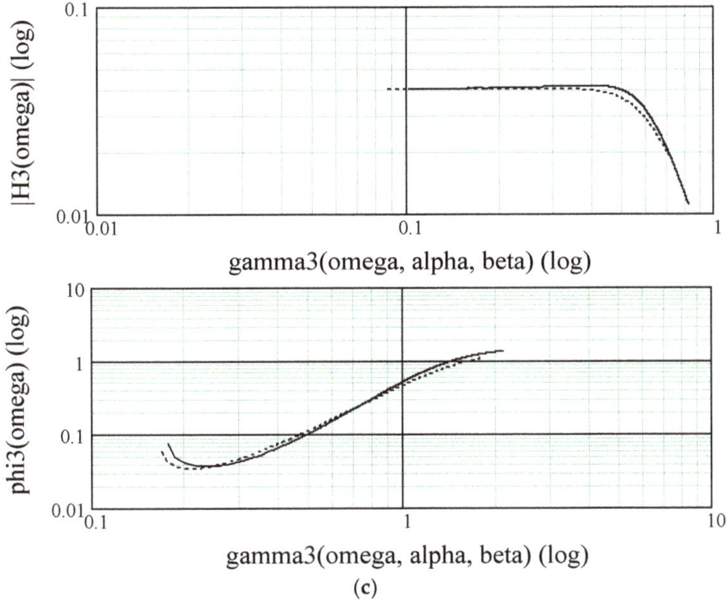

Figure 57. Illustrations of frequency response $H_3(\omega)$ to fractional oscillators of Class III with $m = c = 1$, $k = 25$. (**a**) $|H_3(\omega)|$ and $\varphi_3(\omega)$. Solid line: $(\alpha, \beta) = (1.8, 0.9)$ $(0.23 \leq \varsigma_{eq3} \leq 0.54)$. Dot line: $(\alpha, \beta) = (1.5, 0.9)$ $(0.36 \leq \varsigma_{eq3} \leq 1.04)$. (**b**) $|H_3(\omega)|$ and $\varphi_3(\omega)$. Solid line: $(\alpha, \beta) = (1.8, 0.7)$ $(0.27 \leq \varsigma_{eq3} \leq 0.50)$. Dot line: $(\alpha, \beta) = (1.5, 0.7)$ $(0.50 \leq \varsigma_{eq3} \leq 0.95)$. (**c**) $|H_3(\omega)|$ and $\varphi_3(\omega)$. Solid line: $(\alpha, \beta) = (1.8, 0.55)$ $(0.31 \leq \varsigma_{eq3} \leq 0.46)$. Dot line: $(\alpha, \beta) = (1.5, 0.55)$ $(0.86 \leq \varsigma_{eq3} \leq 0.97)$.

Note 9.6: If $(\alpha, \beta) = (2, 1)$, $H_3(\omega)$ reduces to the ordinary one given by

$$H_3(\omega)\big|_{(\alpha,\beta)=(2,1)} = \frac{1/k}{1 - \gamma^2 + i2\varsigma\gamma}. \tag{257}$$

10. Sinusoidal Responses of Three Classes of Fractional Oscillators

When the excitation force takes the sinusoidal one in the form of $\cos\omega t$ or $\sin\omega t$, the response is termed sinusoidal response, which plays a role in the field of oscillations.

10.1. Stating Problem

Note that the sinusoidal response to fractional oscillators attracts research interests but it is yet a problem that has not been solved satisfactorily. In fact, the existence of the sinusoidal response to fractional oscillators remains a problem. In mathematics, it is regarded as a problem of periodic solution to fractional oscillators. Kaslik and Sivasundaram stated that the exact periodic solution does not exist ([81], p. 1495, Remark 5). The view of Kaslik and Sivasundaram's in [81] is also implied in other works of researchers. Taking fractional oscillators in Class I as an example, Mainardi noticed that the solution to fractional oscillators for $1 < \alpha < 2$, when driven by sinusoidal function, does not exhibit permanent oscillations but asymptotically algebraic decayed ([25], p. 1469), also see Achar et al. ([33], lines above Equation (14)), Duan et al. ([39], p. 49).

As a matter of fact, when considering a fractional oscillator of Class I type for $1 < \alpha < 2$ without the case of $\alpha = 2$ in the form

$$m\frac{d^\alpha y_1(t)}{dt^\alpha} + ky_1(t) = \cos\omega t, 1 < \alpha < 2, \tag{258}$$

it is obvious that $y_1(t)$ must contain steady-state component that is not equal to 0 for $t \to \infty$ no matter what value of $\alpha \in (1, 2)$ is. Otherwise, the conservation law of energy would be violated. The problem is what the complete solution of $y_1(t)$ should be.

The actual solution $y_1(t)$ should, in reality, consist of two parts. One is the steady-state part, denoted by $y_{1s}(t)$, where the subscript s stands for steady-state, which is not equal to 0 for $t \to \infty$ and for any value of $\alpha \in (1, 2)$. The other is the transient part, denoted by $y_{1tr}(t)$, where the subscript tr means transient. Thus, the complete solution should, qualitatively, be in the form

$$y_1(t) = y_{1tr}(t) + y_{1s}(t). \tag{259}$$

We contribute the complete solutions to three classes of fractional oscillators regarding their sinusoidal responses in this section. Our results will show that there exist steady-state components for fractional oscillators in either class with any value of $\alpha \in (1, 2)$ for those in Class I, or $\beta \in (0, 1)$ in Class II, or any combination of $\alpha \in (1, 2)$ with $\beta \in (0, 1)$ for those in Class III.

10.2. Stating Research Thought

Consider the sinusoidal responses to three classes fractional oscillators based on the equivalent oscillation equation in the form

$$\begin{cases} m_{eqj}\dfrac{d^2 x_j(t)}{dt^2} + c_{eqj}\dfrac{dx_j(t)}{dt} + kx_j(t) = A\cos\omega t \\ x_j(0) = x_{j0}, \ \dfrac{dx_j(t)}{dt}\Big|_{t=0} = v_{j0} \end{cases}, j = 1, 2, 3. \tag{260}$$

The complete response $x_j(t)$ consists of the zero state response, denoted by $x_{jzs}(t)$, and zero input response denoted by $x_{jzi}(t)$, according to the theory of differential equations. Therefore,

$$x_j(t) = x_{jzs}(t) + x_{jzi}(t), \tag{261}$$

where $x_{jzi}(t)$ is solved from

$$\begin{cases} m_{eqj}\dfrac{d^2 x_{jzi}(t)}{dt^2} + c_{eqj}\dfrac{dx_{jzi}(t)}{dt} + kx_{jzi}(t) = 0 \\ x_j(0) = x_{j0}, \ \dfrac{dx_j(t)}{dt}\Big|_{t=0} = v_{j0} \end{cases}, j = 1, 2, 3. \tag{262}$$

On the other hand, $x_{jzs}(t)$ is the solution to

$$\begin{cases} m_{eqj}\dfrac{d^2 x_{jzs}(t)}{dt^2} + c_{eqj}\dfrac{dx_{jzs}(t)}{dt} + kx_{jzs}(t) = A\cos\omega t \\ x_j(0) = 0, \ \dfrac{dx_j(t)}{dt}\Big|_{t=0} = 0 \end{cases}, j = 1, 2, 3. \tag{263}$$

Note that $x_{jzi}(t)$ is actually the free response to the fractional oscillators in Class j. It has been solved in Section 6. Thus, the focus of this section is on (263).

10.3. General Form of Sinusoidal Responses to Three Classes of Fractional Oscillators

The solution to (263) in the general form, for $t > 0$, $j = 1, 2, 3$, is given by

$$x_{jzs}(t) = \frac{1}{m_{eqj}\omega_{eqd,j}} \cdot \frac{A\left\{ \begin{array}{l} \left(\omega_{eqn,j}^2 - \omega^2\right)\cos\omega t + 2\varsigma_{eqj}\omega_{eqn,j}\omega\sin\omega t \\[2ex] + e^{-\varsigma_{eqj}\omega_{eqn,j}t}\left[\left(\omega_{eqn,j}^2 - \omega^2\right)\cos\omega_{eqd,j}t - \dfrac{\varsigma_{eqj}\left(\omega_{eqn,j}^2 + \omega^2\right)\sin\omega_{eqd,j}t}{\sqrt{1-\varsigma_{eqj}^2}}\right] \end{array} \right\}}{\left(\omega_{eqn,j}^2 - \omega^2\right)^2 + \left(2\varsigma_{eqj}\omega_{eqn,j}\omega\right)^2}. \tag{264}$$

10.4. Sinusoidal Response to Fractional Oscillators in Class I

Theorem 22 (Sinusoidal response I). *Let $x_{1zs}(t)$ be the zero state sinusoidal response to a fractional oscillator in Class I. Then, for $t > 0$ and $1 < \alpha \le 2$, it is in the form*

$$
x_{1zs}(t) = \frac{1}{\left(m\omega_n \sqrt{-\omega^{\alpha-2}\cos\frac{\alpha\pi}{2}} \, \sqrt{1-\frac{\omega^\alpha \sin^2\frac{\alpha\pi}{2}}{4\omega_n^2\left|\cos\frac{\alpha\pi}{2}\right|}} \right)} \frac{A}{\left[\omega^4\left(\frac{\omega_n^2}{-\omega^\alpha\cos\frac{\alpha\pi}{2}}-1\right)^2 + \frac{\omega^4\sin^2\frac{\alpha\pi}{2}}{\left|\cos\frac{\alpha\pi}{2}\right|^2}\right]}
$$

$$
\left\{
\omega^4\left(\frac{\omega_n^2}{-\omega^\alpha\cos\frac{\alpha\pi}{2}}-1\right)^2 \cos\omega t + \frac{\omega^2\sin\frac{\alpha\pi}{2}}{\left|\cos\frac{\alpha\pi}{2}\right|}\sin\omega t \right.
$$
$$
\left. + e^{-\frac{\omega\sin\frac{\alpha\pi}{2}}{2\left|\cos\frac{\alpha\pi}{2}\right|}t}
\left[\omega^4\left(\frac{\omega_n^2}{-\omega^\alpha\cos\frac{\alpha\pi}{2}}-1\right)^2 \cos\frac{\omega_n\sqrt{1-\frac{\omega^\alpha\sin^2\frac{\alpha\pi}{2}}{4\omega_n^2\left|\cos\frac{\alpha\pi}{2}\right|}}}{\sqrt{-\omega^{\alpha-2}\cos\frac{\alpha\pi}{2}}}t \right.\right.
$$
$$
\left.\left. - \frac{\frac{\omega^{\frac{\alpha}{2}}\sin\frac{\alpha\pi}{2}}{2\omega_n\sqrt{-\cos\frac{\alpha\pi}{2}}}\omega^4\left(\frac{\omega_n^2}{-\omega^\alpha\cos\frac{\alpha\pi}{2}}+1\right)^2}{\sqrt{1-\left(\frac{\omega^{\frac{\alpha}{2}}\sin\frac{\alpha\pi}{2}}{2\omega_n\sqrt{-\cos\frac{\alpha\pi}{2}}}\right)^2}} \sin\frac{\omega_n\sqrt{1-\frac{\omega^\alpha\sin^2\frac{\alpha\pi}{2}}{4\omega_n^2\left|\cos\frac{\alpha\pi}{2}\right|}}}{\sqrt{-\omega^{\alpha-2}\cos\frac{\alpha\pi}{2}}}t \right]\right\}.
$$
(265)

Proof. Consider the expression below

$$
x_{1zs}(t) = \frac{1}{m_{eq1}\omega_{eqd,1}}\frac{A\left\{\begin{array}{l}\left(\omega_{eqn,1}^2-\omega^2\right)\cos\omega t + 2\varsigma_{eq1}\omega_{eqn,1}\omega\sin\omega t \\[4pt] + e^{-\varsigma_{eq1}\omega_{eqn,1}t}\left[\left(\omega_{eqn,1}^2-\omega^2\right)\cos\omega_{eqd,1}t - \frac{\varsigma_{eq1}}{\sqrt{1-\varsigma_{eq1}^2}}\left(\omega_{eqn,1}^2+\omega^2\right)\sin\omega_{eqd,1}t\right]\end{array}\right\}}{\left(\omega_{eqn,1}^2-\omega^2\right)^2+\left(2\varsigma_{eq1}\omega_{eqn,1}\omega\right)^2}.
$$
(266)

In the above, replacing m_{eq1} by the one in Section 4, ς_{eq1}, $\omega_{eqd,1}$ and $\omega_{eqn,1}$ by those in Section 5, respectively, produces (265). This finishes the proof. $\square$

Denote by $x_{1zs,s}(t)$ and $x_{1zs,tr}(t)$ the steady component and the instantaneous one, respectively. Then, we have

$$
x_{1zs,s}(t) = \left[\frac{1}{m\omega_n\sqrt{-\omega^{\alpha-2}\cos\frac{\alpha\pi}{2}}\sqrt{1-\frac{\omega^\alpha\sin^2\frac{\alpha\pi}{2}}{4\omega_n^2\left|\cos\frac{\alpha\pi}{2}\right|}}}\frac{A}{\omega^4\left(\frac{\omega_n^2}{-\omega^\alpha\cos\frac{\alpha\pi}{2}}-1\right)^2+\frac{\omega^4\sin^2\frac{\alpha\pi}{2}}{\left|\cos\frac{\alpha\pi}{2}\right|^2}}\right]
$$
$$
\left[\omega^4\left(\frac{\omega_n^2}{-\omega^\alpha\cos\frac{\alpha\pi}{2}}-1\right)^2\cos\omega t + \frac{\omega^2\sin\frac{\alpha\pi}{2}}{\left|\cos\frac{\alpha\pi}{2}\right|}\sin\omega t\right].
$$
(267)

and

$$
x_{1zs,tr}(t) = \left[\frac{1}{m\omega_n\sqrt{-\omega^{\alpha-2}\cos\frac{\alpha\pi}{2}}\sqrt{1-\frac{\omega^\alpha\sin^2\frac{\alpha\pi}{2}}{4\omega_n^2\left|\cos\frac{\alpha\pi}{2}\right|}}}\frac{A}{\omega^4\left(\frac{\omega_n^2}{-\omega^\alpha\cos\frac{\alpha\pi}{2}}-1\right)^2+\frac{\omega^4\sin^2\frac{\alpha\pi}{2}}{\left|\cos\frac{\alpha\pi}{2}\right|^2}}\right]
$$
$$
\left\{ e^{-\frac{\omega\sin\frac{\alpha\pi}{2}}{2\left|\cos\frac{\alpha\pi}{2}\right|}t}
\left[\omega^4\left(\frac{\omega_n^2}{-\omega^\alpha\cos\frac{\alpha\pi}{2}}-1\right)^2 \cos\frac{\omega_n\sqrt{1-\frac{\omega^\alpha\sin^2\frac{\alpha\pi}{2}}{4\omega_n^2\left|\cos\frac{\alpha\pi}{2}\right|}}}{\sqrt{-\omega^{\alpha-2}\cos\frac{\alpha\pi}{2}}}t \right.\right.
$$
$$
\left.\left. - \frac{\frac{\omega^{\frac{\alpha}{2}}\sin\frac{\alpha\pi}{2}}{2\omega_n\sqrt{-\cos\frac{\alpha\pi}{2}}}}{\sqrt{1-\left(\frac{\omega^{\frac{\alpha}{2}}\sin\frac{\alpha\pi}{2}}{2\omega_n\sqrt{-\cos\frac{\alpha\pi}{2}}}\right)^2}}\omega^4\left(\frac{\omega_n^2}{-\omega^\alpha\cos\frac{\alpha\pi}{2}}+1\right)^2\sin\frac{\omega_n\sqrt{1-\frac{\omega^\alpha\sin^2\frac{\alpha\pi}{2}}{4\omega_n^2\left|\cos\frac{\alpha\pi}{2}\right|}}}{\sqrt{-\omega^{\alpha-2}\cos\frac{\alpha\pi}{2}}}t \right]\right\}.
$$
(268)

Figure 58 shows the plots of $x_{1zs}(t)$.

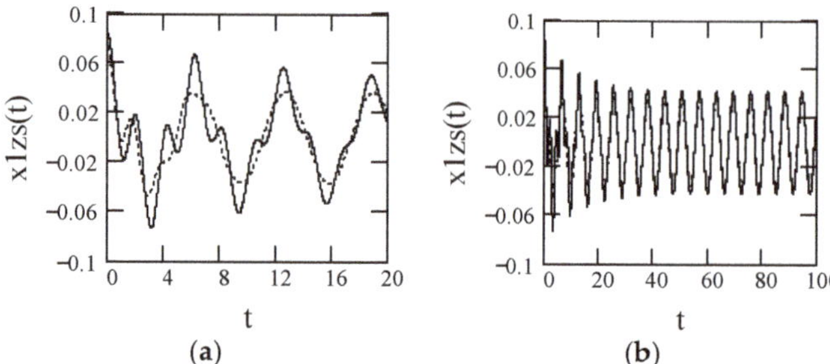

(a)   **(b)**

Figure 58. Illustrating $x_{1zs}(t)$ for $m = 1$, $k = 9$ ($\omega_n = 3$), $\omega = 1$. Solid line: $\alpha = 1.9$ ($\varsigma_{eq1} = 0.03$). Dot line: $\alpha = 1.6$ ($\varsigma_{eq1} = 0.11$). (**a**) $t = 0, 1, ..., 20$. (**b**) $t = 0, 1, ..., 100$.

Note 10.1: $x_{1zs}(t)$ is not a pure harmonic function as can be seen from Figure 58.

Remark 29. *We found that the sinusoidal response to fractional oscillators in Class I for any value of $\alpha \in (1, 2)$ does have steady-state component $x_{1zs,s}(t)$ expressed by (267), also see Figure 59.*

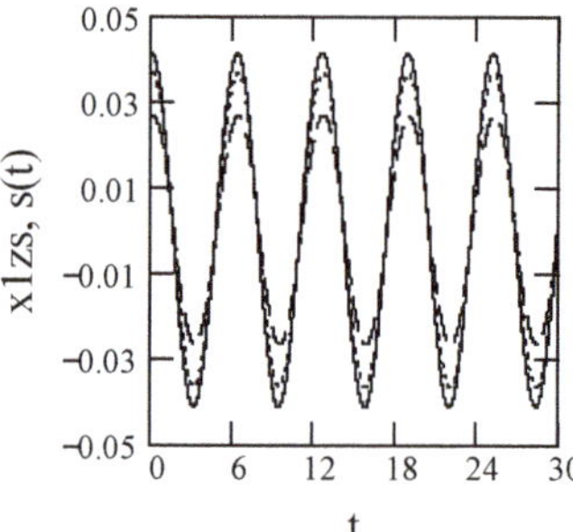

Figure 59. Steady-state component, $x_{1zs,s}(t)$, of sinusoidal response to a fractional oscillator in Class I for $m = 1$, $k = 9$ ($\omega_n = 3$), $\omega = 1$. Solid line: $\alpha = 1.9$ ($\varsigma_{eq1} = 0.03$). Dot line: $\alpha = 1.6$ ($\varsigma_{eq1} = 0.11$). Dash dot line: $\alpha = 1.3$ ($\varsigma_{eq1} = 0.22$).

The illustration of $x_{1zs,tr}(t)$ is indicated in Figure 60.

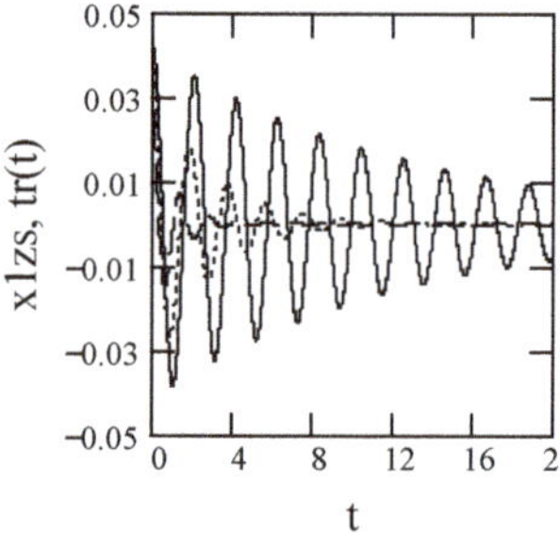

Figure 60. Instantaneous component $x_{1zs,tr}(t)$ for $m = 1$, $k = 9$ ($\omega_n = 3$), $\omega = 1$. Solid line: $\alpha = 1.9$ ($\varsigma_{eq1} = 0.03$). Dot line: $\alpha = 1.6$ ($\varsigma_{eq1} = 0.11$). Dash dot line: $\alpha = 1.3$ ($\varsigma_{eq1} = 0.22$).

Note 10.2: For $\alpha = 2$, we have

$$x_{1zs}(t) = \frac{1}{m\omega_n} \frac{A}{\omega^4\left(\frac{\omega_n^2}{\omega^2}-1\right)^2}\left[\omega^4\left(\frac{\omega_n^2}{\omega^2}-1\right)^2\cos\omega\, t + \omega^4\left(\frac{\omega_n^2}{\omega^2}-1\right)^2\cos\omega_n t\right]$$

$$= \frac{A}{m\omega_n}(\cos\omega\, t + \cos\omega_n t).$$

(269)

10.5. Sinusoidal Response to Fractional Oscillators in Class II

Theorem 23 (Sinusoidal response II). *Denote by $x_{2zs}(t)$ the zero state sinusoidal response to a fractional oscillator in Class II. Then, for $t > 0$ and $0 < \beta \leq 1$, it is expressed by*

$$x_{2zs}(t) = \frac{1}{m\omega_n\sqrt{1-\frac{c}{m}\omega^{\beta-2}\cos\frac{\beta\pi}{2}}\sqrt{1-\left(\frac{\varsigma^2\omega^{2(\beta-1)}\sin^2\frac{\beta\pi}{2}}{1-\frac{c}{m}\omega^{\beta-2}\cos\frac{\beta\pi}{2}}\right)^2}}$$

$$\frac{A}{\left(\frac{\omega_n^2}{1-\frac{c}{m}\omega^{\beta-2}\cos\frac{\beta\pi}{2}}-\omega^2\right)^2 + \left(\frac{2\varsigma\omega_n\omega^\beta\sin\frac{\beta\pi}{2}}{1-\frac{c}{m}\omega^{\beta-2}\cos\frac{\beta\pi}{2}}\right)^2}$$

$$\left\{\left(\frac{\omega_n^2}{1-\frac{c}{m}\omega^{\beta-2}\cos\frac{\beta\pi}{2}}-\omega^2\right)\cos\omega t + \frac{2\varsigma\omega_n\omega^\beta\sin\frac{\beta\pi}{2}}{1-\frac{c}{m}\omega^{\beta-2}\cos\frac{\beta\pi}{2}}\sin\omega t\right.$$

$$+ e^{-\frac{\varsigma\omega_n\omega^{\beta-1}\sin\frac{\beta\pi}{2}}{1-\frac{c}{m}\omega^{\beta-2}\cos\frac{\beta\pi}{2}}t}\left[\left[\left(\frac{\omega_n^2}{1-\frac{c}{m}\omega^{\beta-2}\cos\frac{\beta\pi}{2}}-\omega^2\right)\cos\frac{\omega_n\sqrt{1-\left(\frac{\varsigma^2\omega^{2(\beta-1)}\sin^2\frac{\beta\pi}{2}}{1-\frac{c}{m}\omega^{\beta-2}\cos\frac{\beta\pi}{2}}\right)^2}\,t}{\sqrt{1-\frac{c}{m}\omega^{\beta-2}\cos\frac{\beta\pi}{2}}}\right.\right.$$

$$\left.\left.- \frac{\frac{\varsigma\omega^{\beta-1}\sin\frac{\beta\pi}{2}}{\sqrt{1-\frac{c}{m}\omega^{\beta-2}\cos\frac{\beta\pi}{2}}}\left(\frac{\omega_n^2}{1-\frac{c}{m}\omega^{\beta-2}\cos\frac{\beta\pi}{2}}+\omega^2\right)}{\sqrt{1-\left(\frac{\varsigma\omega^{\beta-1}\sin\frac{\beta\pi}{2}}{\sqrt{1-\frac{c}{m}\omega^{\beta-2}\cos\frac{\beta\pi}{2}}}\right)^2}}\right.\right.$$

$$\left.\left.\sin\frac{\omega_n\sqrt{1-\left(\frac{\varsigma^2\omega^{2(\beta-1)}\sin^2\frac{\beta\pi}{2}}{1-\frac{c}{m}\omega^{\beta-2}\cos\frac{\beta\pi}{2}}\right)^2}\,t}{\sqrt{1-\frac{c}{m}\omega^{\beta-2}\cos\frac{\beta\pi}{2}}}\right]\right\}.$$

(270)

Proof. In the following expression,

$$x_{2zs}(t) = \frac{1}{m_{eq2}\omega_{eqd,2}}\frac{A}{\left(\omega_{eqn,2}^2-\omega^2\right)^2+\left(2\varsigma_{eq2}\omega_{eqn,2}\omega\right)^2}$$

$$\left\{\left(\omega_{eqn,2}^2-\omega^2\right)\cos\omega t + 2\varsigma_{eq2}\omega_{eqn,2}\omega\sin\omega t\right.$$

$$\left.+ e^{-\varsigma_{eq2}\omega_{eqn,2}t}\left[\left(\omega_{eqn,2}^2-\omega^2\right)\cos\omega_{eqd,2}t - \frac{\varsigma_{eq2}}{\sqrt{1-\varsigma_{eq2}^2}}\left(\omega_{eqn,2}^2+\omega^2\right)\sin\omega_{eqd,2}t\right]\right\},$$

(271)

replacing m_{eq2} with the one in Section 4, ς_{eq2}, $\omega_{eqd,2}$ and $\omega_{eqn,2}$ by those in Section 5, results in (270). The proof completes. $\square$

The stead-state part of $x_{2zs,s}(t)$ is represented by

$$x_{2zs,s}(t) = \cfrac{\cfrac{1}{m\omega_n\sqrt{1-\frac{c}{m}\omega^{\beta-2}\cos\frac{\beta\pi}{2}}\sqrt{1-\frac{\varsigma^2\omega^{2(\beta-1)}\sin^2\frac{\beta\pi}{2}}{1-\frac{c}{m}\omega^{\beta-2}\cos\frac{\beta\pi}{2}}}}{\left(\cfrac{\omega_n^2}{1-\frac{c}{m}\omega^{\beta-2}\cos\frac{\beta\pi}{2}}-\omega^2\right)^2+\left(\cfrac{2\varsigma\omega_n\omega^\beta\sin\frac{\beta\pi}{2}}{1-\frac{c}{m}\omega^{\beta-2}\cos\frac{\beta\pi}{2}}\right)^2}}{}$$

$$\left[\left(\cfrac{\omega_n^2}{1-\frac{c}{m}\omega^{\beta-2}\cos\frac{\beta\pi}{2}}-\omega^2\right)\cos\omega t+\cfrac{2\varsigma\omega_n\omega^\beta\sin\frac{\beta\pi}{2}}{1-\frac{c}{m}\omega^{\beta-2}\cos\frac{\beta\pi}{2}}\sin\omega t\right]. \tag{272}$$

On the other side, the transient part $x_{2zs,tr}(t)$ is given by

$$x_{2zs,tr}(t) = \cfrac{\cfrac{1}{m\omega_n\sqrt{1-\frac{c}{m}\omega^{\beta-2}\cos\frac{\beta\pi}{2}}\sqrt{1-\left(\cfrac{\varsigma^2\omega^{2(\beta-1)}\sin^2\frac{\beta\pi}{2}}{1-\frac{c}{m}\omega^{\beta-2}\cos\frac{\beta\pi}{2}}\right)}}{\left(\cfrac{\omega_n^2}{1-\frac{c}{m}\omega^{\beta-2}\cos\frac{\beta\pi}{2}}-\omega^2\right)^2+\left(\cfrac{2\varsigma\omega_n\omega^\beta\sin\frac{\beta\pi}{2}}{1-\frac{c}{m}\omega^{\beta-2}\cos\frac{\beta\pi}{2}}\right)^2}}{}$$

$$\left\{e^{-\frac{\varsigma\omega_n\omega^{\beta-1}\sin\frac{\beta\pi}{2}}{1-\frac{c}{m}\omega^{\beta-2}\cos\frac{\beta\pi}{2}}t}\left[\left[\left(\cfrac{\omega_n^2}{1-\frac{c}{m}\omega^{\beta-2}\cos\frac{\beta\pi}{2}}-\omega^2\right)\cos\cfrac{\omega_n\sqrt{1-\left(\frac{\varsigma^2\omega^{2(\beta-1)}\sin^2\frac{\beta\pi}{2}}{1-\frac{c}{m}\omega^{\beta-2}\cos\frac{\beta\pi}{2}}\right)}}{\sqrt{1-\frac{c}{m}\omega^{\beta-2}\cos\frac{\beta\pi}{2}}}t\right]\right.\right.$$

$$-\cfrac{\cfrac{\varsigma\omega^{\beta-1}\sin\frac{\beta\pi}{2}}{\sqrt{1-\frac{c}{m}\omega^{\beta-2}\cos\frac{\beta\pi}{2}}}\left(\cfrac{\omega_n^2}{1-\frac{c}{m}\omega^{\beta-2}\cos\frac{\beta\pi}{2}}+\omega^2\right)}{\sqrt{1-\left(\cfrac{\varsigma\omega^{\beta-1}\sin\frac{\beta\pi}{2}}{\sqrt{1-\frac{c}{m}\omega^{\beta-2}\cos\frac{\beta\pi}{2}}}\right)^2}}$$

$$\left.\left.\sin\cfrac{\omega_n\sqrt{1-\left(\frac{\varsigma^2\omega^{2(\beta-1)}\sin^2\frac{\beta\pi}{2}}{1-\frac{c}{m}\omega^{\beta-2}\cos\frac{\beta\pi}{2}}\right)}}{\sqrt{1-\frac{c}{m}\omega^{\beta-2}\cos\frac{\beta\pi}{2}}}t\right]\right\}. \tag{273}$$

Figures 61–63 show the plots of $x_{2zs}(t)$, $x_{2zs,s}(t)$, and $x_{2zs,tr}(t)$.

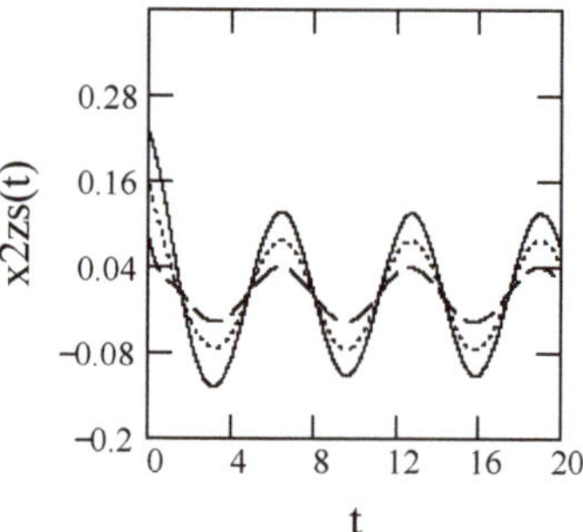

Figure 61. Sinusoidal response $x_{2zs}(t)$ to a fractional oscillator in Class II with $\beta = 0.9$ (solid line) ($\varsigma_{eq2} = 0.14$), $\beta = 0.6$ (dot line) ($\varsigma_{eq2} = 0.07$), $\beta = 0.3$ (dash dot line) ($\varsigma_{eq2} = 0.03$) with $m = c = 1$, $\omega_n = 3$ and $\omega = 1$.

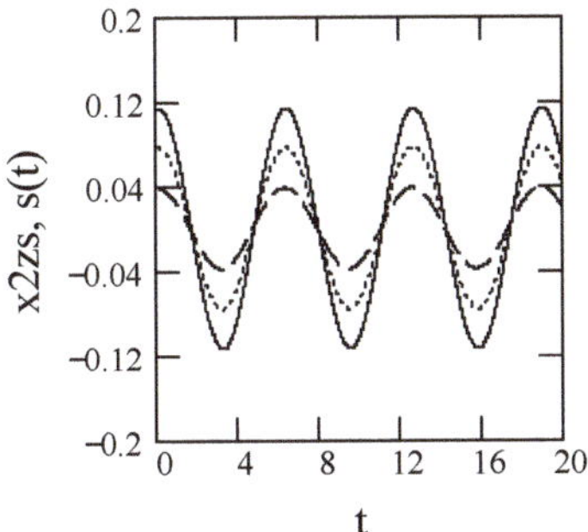

Figure 62. Steady-state sinusoidal part of $x_{2zs}(t)$ with $\beta = 0.9$ (solid line) ($\varsigma_{eq2} = 0.14$), $\beta = 0.6$ (dot line) ($\varsigma_{eq2} = 0.07$), $\beta = 0.3$ (dash dot line) ($\varsigma_{eq2} = 0.03$) with $m = c = 1$, $\omega_n = 3$ and $\omega = 1$.

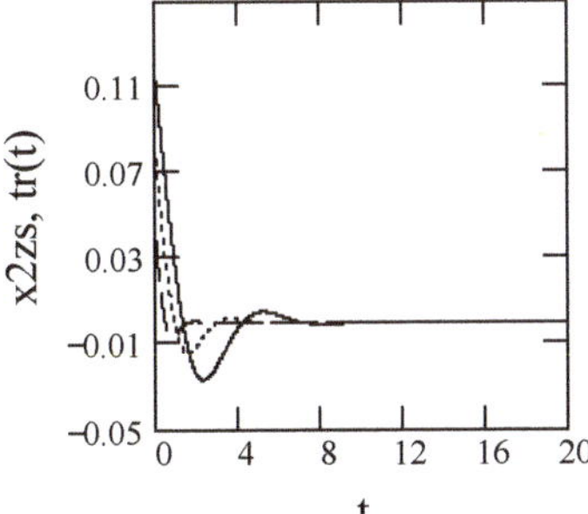

Figure 63. Transient part of $x_{2zs}(t)$ with $\beta = 0.9$ (solid line) ($\varsigma_{eq2} = 0.14$), $\beta = 0.6$ (dot line) ($\varsigma_{eq2} = 0.07$), $\beta = 0.3$ (dash dot line) ($\varsigma_{eq2} = 0.03$) with $m = c = 1$, $\omega_n = 3$ and $\omega = 1$.

Note 10.3: If $\beta = 1$, we obtain the zero-state response of the conventional sinusoidal response to a 2-order oscillator in the form

$$x_{2zs}(t)|_{\beta=1} = \frac{1}{m\omega_n\sqrt{1-\varsigma^2}} \frac{Ae^{-\varsigma\omega_n t}\left[(\omega_n^2 - \omega^2)\cos\omega_n\sqrt{1-\varsigma^2}t - \dfrac{\varsigma(\omega_n^2+\omega^2)\sin\omega_n\sqrt{1-\varsigma^2}t}{\sqrt{1-\varsigma^2}} \right]}{(\omega_n^2 - \omega^2)^2 + (2\varsigma\omega_n\omega)^2}. \tag{274}$$

Remark 30. *We discovered that the sinusoidal response to fractional oscillators in Class II for any value of $\beta \in (0, 1)$ does have steady-state component $x_{2zs,s}(t)$ described by (272), also see Figure 62.*

10.6. Sinusoidal Response to Fractional Oscillators in Class III

Theorem 24 (Sinusoidal response III). *Let $x_{3zs}(t)$ be the zero state sinusoidal response to a fractional oscillator of Class III type. Then, for $t > 0$, $1 < \alpha \leq 2$, and $0 < \beta \leq 1$, it is written in the form*

$$x_{3zs}(t) = \left\{ \frac{\omega_n m_{eq3}\sqrt{\dfrac{-\left(\omega^{\alpha-2}\cos\frac{\alpha\pi}{2}+\frac{c}{m}\omega^{\beta-2}\cos\frac{\beta\pi}{2}\right)}{1-\dfrac{\left(m\omega^{\alpha-1}\sin\frac{\alpha\pi}{2}+c\omega^{\beta-1}\sin\frac{\beta\pi}{2}\right)^2}{4\left[-\left(m\omega^{\alpha-2}\cos\frac{\alpha\pi}{2}+c\omega^{\beta-2}\cos\frac{\beta\pi}{2}\right)k\right]}}}}{A\left[\left[\dfrac{\omega_n^2}{-\left(\omega^{\alpha-2}\cos\frac{\alpha\pi}{2}+\frac{c}{m}\omega^{\beta-2}\cos\frac{\beta\pi}{2}\right)}-\omega^2\right]^2+\left(\dfrac{\omega^\alpha\sin\frac{\alpha\pi}{2}+2\varsigma\omega_n\omega^{\beta-1}\sin\frac{\beta\pi}{2}}{\omega^{\alpha-2}\left|\cos\frac{\alpha\pi}{2}\right|-2\varsigma\omega_n\omega^{\beta-2}\cos\frac{\beta\pi}{2}}\right)^2\right]} \right\}$$

$$\left\{\left\{\left[\dfrac{\omega_n^2}{-\left(\omega^{\alpha-2}\cos\frac{\alpha\pi}{2}+\frac{c}{m}\omega^{\beta-2}\cos\frac{\beta\pi}{2}\right)}-\omega^2\right]\cos\omega t + \dfrac{\omega^\alpha\sin\frac{\alpha\pi}{2}+2\varsigma\omega_n\omega^{\beta-1}\sin\frac{\beta\pi}{2}}{\omega^{\alpha-2}\left|\cos\frac{\alpha\pi}{2}\right|-2\varsigma\omega_n\omega^{\beta-2}\cos\frac{\beta\pi}{2}}\sin\omega t\right\}\right.$$

$$+e^{-\frac{\omega^{\alpha-1}\sin\frac{\alpha\pi}{2}+2\varsigma\omega_n\omega^{\beta-1}\sin\frac{\beta\pi}{2}}{2\left(\omega^{\alpha-2}\left|\cos\frac{\alpha\pi}{2}\right|-2\varsigma\omega_n\omega^{\beta-2}\cos\frac{\beta\pi}{2}\right)}t}\left[\left[\dfrac{\omega_n^2}{-\left(\omega^{\alpha-2}\cos\frac{\alpha\pi}{2}+\frac{c}{m}\omega^{\beta-2}\cos\frac{\beta\pi}{2}\right)}-\omega^2\right]\cos\omega_{eqd,3}t\right.$$

$$-\left\{\dfrac{\dfrac{\omega^{\alpha-1}\sin\frac{\alpha\pi}{2}+2\varsigma\omega_n\omega^{\beta-1}\sin\frac{\beta\pi}{2}}{2\omega_n\sqrt{-\left(\omega^{\alpha-2}\cos\frac{\alpha\pi}{2}+2\varsigma\omega_n\omega^{\beta-2}\cos\frac{\beta\pi}{2}\right)}}}{\sqrt{1-\left[\dfrac{\omega^{\alpha-1}\sin\frac{\alpha\pi}{2}+2\varsigma\omega_n\omega^{\beta-1}\sin\frac{\beta\pi}{2}}{2\omega_n\sqrt{-\left(\omega^{\alpha-2}\cos\frac{\alpha\pi}{2}+2\varsigma\omega_n\omega^{\beta-2}\cos\frac{\beta\pi}{2}\right)}}\right]^2}}\right\}$$

$$\left.\left.\left(\dfrac{\omega_n^2}{-\left(\omega^{\alpha-2}\cos\frac{\alpha\pi}{2}+\frac{c}{m}\omega^{\beta-2}\cos\frac{\beta\pi}{2}\right)}+\omega^2\right)\sin\omega_{eqd,3}t\right]\right\}, \tag{275}$$

where m_{eq3} and $\omega_{eqd,\,3}$ are given by (119) and (160), respectively.

Proof. In the following expression,

$$x_{3zs}(t) = \frac{1}{m_{eq3}\omega_{eqd,3}}\frac{A\left\{\left(\omega_{eqn,3}^2-\omega^2\right)\cos\omega t + 2\varsigma_{eq3}\omega_{eqn,3}\omega\sin\omega t + e^{-\varsigma_{eq3}\omega_{eqn,3}t}\left[\left(\omega_{eqn,3}^2-\omega^2\right)\cos\omega_{eqd,3}t - \dfrac{\varsigma_{eq3}}{\sqrt{1-\varsigma_{eq3}^2}}\left(\omega_{eqn,3}^2+\omega^2\right)\sin\omega_{eqd,3}t\right]\right\}}{\left(\omega_{eqn,3}^2-\omega^2\right)^2+\left(2\varsigma_{eq3}\omega_{eqn,3}\omega\right)^2}, \tag{276}$$

Substituting ς_{eq3}, $\omega_{eqn,3}$ and $\omega_{eqd,3}$ with those in Section 5 yields the Theorem 24. That completes the proof. $\square$

Figure 64 illustrates $x_{3zs}(t)$.

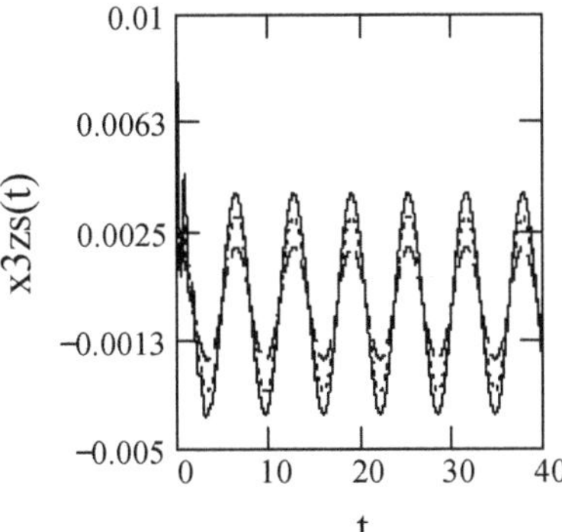

Figure 64. Indicating the sinusoidal response $x_{3zs}(t)$ to a fractional oscillator in Class III with $(\alpha, \beta) = (1.8, 0.8)$ (solid line) $(\varsigma_{eq3} = 0.13)$, $(\alpha, \beta) = (1.5, 0.8)$ (dot line) $(\varsigma_{eq3} = 0.22)$, $(\alpha, \beta) = (1.3, 0.8)$ (dash dot line) $(\varsigma_{eq3} = 0.40)$ with $m = c = 1, k = 36$ $(\omega_n = 6)$ and $\omega = 1$.

The steady-state part of $x_{3zs}(t)$ is in the form

$$
x_{3zs,s}(t) = \left\{ \frac{\dfrac{\sqrt{-\left(\omega^{\alpha-2}\cos\frac{\alpha\pi}{2} + \frac{c}{m}\omega^{\beta-2}\cos\frac{\beta\pi}{2}\right)}}{\omega_n m_{eq3}\sqrt{1 - \dfrac{\left(m\omega^{\alpha-1}\sin\frac{\alpha\pi}{2} + c\omega^{\beta-1}\sin\frac{\beta\pi}{2}\right)^2}{4\left[-\left(m\omega^{\alpha-2}\cos\frac{\alpha\pi}{2} + c\omega^{\beta-2}\cos\frac{\beta\pi}{2}\right)k\right]}}}}{\left\{\left[\dfrac{\omega_n^2}{-\left(\omega^{\alpha-2}\cos\frac{\alpha\pi}{2} + \frac{c}{m}\omega^{\beta-2}\cos\frac{\beta\pi}{2}\right)} - \omega^2\right]^2 + \left(\dfrac{\omega^{\alpha}\sin\frac{\alpha\pi}{2} + 2\varsigma\omega_n\omega^{\beta-1}\sin\frac{\beta\pi}{2}}{\omega^{\alpha-2}\left|\cos\frac{\alpha\pi}{2}\right| - 2\varsigma\omega_n\omega^{\beta-2}\cos\frac{\beta\pi}{2}}\right)^2\right\}} \times A \right.
$$

$$
\left\{\left[\frac{\omega_n^2}{-\left(\omega^{\alpha-2}\cos\frac{\alpha\pi}{2} + \frac{c}{m}\omega^{\beta-2}\cos\frac{\beta\pi}{2}\right)} - \omega^2\right]\cos\omega t + \frac{\left(\omega^{\alpha}\sin\frac{\alpha\pi}{2} + 2\varsigma\omega_n\omega^{\beta-1}\sin\frac{\beta\pi}{2}\right)\sin\omega t}{\omega^{\alpha-2}\left|\cos\frac{\alpha\pi}{2}\right| - 2\varsigma\omega_n\omega^{\beta-2}\cos\frac{\beta\pi}{2}}\right\}. \tag{277}
$$

Its transient part, taking into account (160), is given by

$$
\begin{aligned}
x_{3zs,tr}(t) = {} & \left\{ \frac{\sqrt{-\left(\omega^{\alpha-2}\cos\frac{\alpha\pi}{2}+\frac{c}{m}\omega^{\beta-2}\cos\frac{\beta\pi}{2}\right)}}{\omega_n m_{eq3}\sqrt{1-\frac{\left(m\omega^{\alpha-1}\sin\frac{\alpha\pi}{2}+c\omega^{\beta-1}\sin\frac{\beta\pi}{2}\right)^2}{4\left[-\left(m\omega^{\alpha-2}\cos\frac{\alpha\pi}{2}+c\omega^{\beta-2}\cos\frac{\beta\pi}{2}\right)k\right]}}} \right. \\[2mm]
& \times \frac{A}{\left\{\left[\frac{\omega_n^2}{-\left(\omega^{\alpha-2}\cos\frac{\alpha\pi}{2}+\frac{c}{m}\omega^{\beta-2}\cos\frac{\beta\pi}{2}\right)}-\omega^2\right]^2+\left(\frac{\omega^\alpha\sin\frac{\alpha\pi}{2}+2\varsigma\omega_n\omega^{\beta-1}\sin\frac{\beta\pi}{2}}{\omega^{\alpha-2}\left|\cos\frac{\alpha\pi}{2}\right|-2\varsigma\omega_n\omega^{\beta-2}\cos\frac{\beta\pi}{2}}\right)^2\right\}} \\[3mm]
& \times e^{-\frac{\omega^{\alpha-1}\sin\frac{\alpha\pi}{2}+2\varsigma\omega_n\omega^{\beta-1}\sin\frac{\beta\pi}{2}}{2\left(\omega^{\alpha-2}\left|\cos\frac{\alpha\pi}{2}\right|-2\varsigma\omega_n\omega^{\beta-2}\cos\frac{\beta\pi}{2}\right)}t} \\[3mm]
& \times \left[\left\{\left[\frac{\omega_n^2}{-\left(\omega^{\alpha-2}\cos\frac{\alpha\pi}{2}+\frac{c}{m}\omega^{\beta-2}\cos\frac{\beta\pi}{2}\right)}-\omega^2\right]\right.\right. \\[2mm]
& \times \cos\left\{\frac{\omega_n}{\sqrt{-\left(\omega^{\alpha-2}\cos\frac{\alpha\pi}{2}+\frac{c}{m}\omega^{\beta-2}\cos\frac{\beta\pi}{2}\right)}}\sqrt{1-\frac{\left(m\omega^{\alpha-1}\sin\frac{\alpha\pi}{2}+c\omega^{\beta-1}\sin\frac{\beta\pi}{2}\right)^2}{4\left[-\left(m\omega^{\alpha-2}\cos\frac{\alpha\pi}{2}+c\omega^{\beta-2}\cos\frac{\beta\pi}{2}\right)k\right]}}\,t\right\} \\[3mm]
& -\left\{2\omega_n\sqrt{-\left(\omega^{\alpha-2}\cos\frac{\alpha\pi}{2}+2\varsigma\omega_n\omega^{\beta-2}\cos\frac{\beta\pi}{2}\right)}\sqrt{1-\left[\frac{\omega^{\alpha-1}\sin\frac{\alpha\pi}{2}+2\varsigma\omega_n\omega^{\beta-1}\sin\frac{\beta\pi}{2}}{2\omega_n\sqrt{-\left(\omega^{\alpha-2}\cos\frac{\alpha\pi}{2}+2\varsigma\omega_n\omega^{\beta-2}\cos\frac{\beta\pi}{2}\right)}}\right]^2}\right\} \\[3mm]
& \times \left(\frac{\omega_n^2}{-\left(\omega^{\alpha-2}\cos\frac{\alpha\pi}{2}+\frac{c}{m}\omega^{\beta-2}\cos\frac{\beta\pi}{2}\right)}+\omega^2\right) \\[3mm]
& \left.\left. \times \sin\left\{\frac{\omega_n\sqrt{1-\frac{\left(m\omega^{\alpha-1}\sin\frac{\alpha\pi}{2}+c\omega^{\beta-1}\sin\frac{\beta\pi}{2}\right)^2}{4\left[-\left(m\omega^{\alpha-2}\cos\frac{\alpha\pi}{2}+c\omega^{\beta-2}\cos\frac{\beta\pi}{2}\right)k\right]}}\,t}{\sqrt{-\left(\omega^{\alpha-2}\cos\frac{\alpha\pi}{2}+\frac{c}{m}\omega^{\beta-2}\cos\frac{\beta\pi}{2}\right)}}\right\}\right]\right\}
\end{aligned}
\tag{278}
$$

The steady-state component and the transient one of $x_{3zs}(t)$ are shown in Figures 65 and 66, respectively.

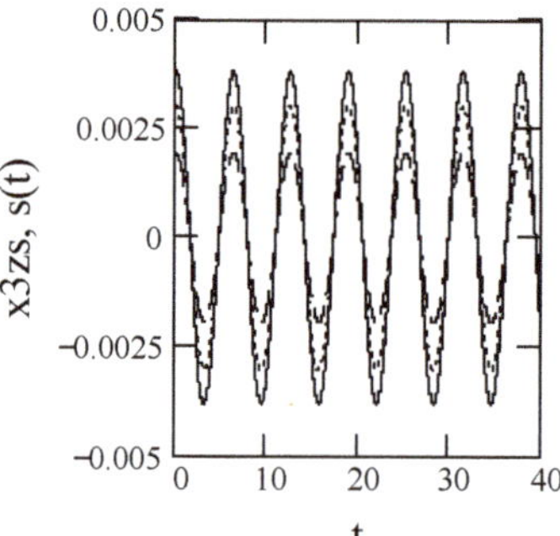

Figure 65. Indicating the steady-state component of $x_{3zs}(t)$ with $(\alpha, \beta) = (1.8, 0.8)$ (solid line) ($\varsigma_{eq3} = 0.13$), $(\alpha, \beta) = (1.5, 0.8)$ (dot line) ($\varsigma_{eq3} = 0.22$), $(\alpha, \beta) = (1.3, 0.8)$ (dash dot line) ($\varsigma_{eq3} = 0.40$) with $m = c = 1, k = 36$ ($\omega_n = 6$) and $\omega = 1$.

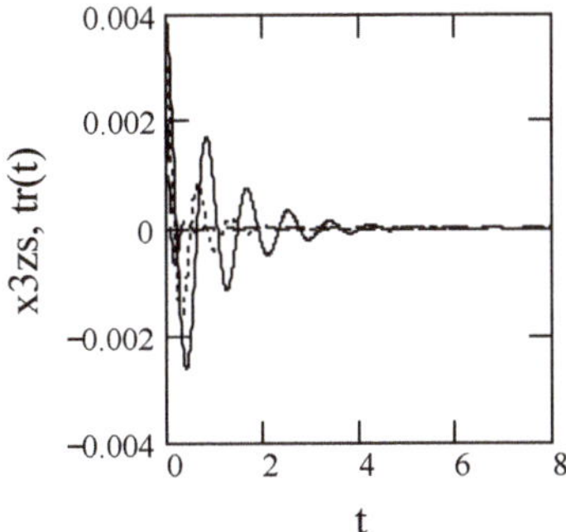

Figure 66. Transient component of $x_{3zs}(t)$ with $(\alpha, \beta) = (1.8, 0.8)$ (solid line) ($\varsigma_{eq3} = 0.13$), $(\alpha, \beta) = (1.5, 0.8)$ (dot line) ($\varsigma_{eq3} = 0.22$), $(\alpha, \beta) = (1.3, 0.8)$ (dash dot line) ($\varsigma_{eq3} = 0.40$) with $m = c = 1$, $k = 36$ ($\omega_n = 6$) and $\omega = 1$.

Note 10.4: When $(\alpha, \beta) = (2, 1)$, $x_{3zs}(t)$ reduces to the ordinary zero-state sinusoidal response to a 2-order oscillator in the form

$$
x_{3zs}(t)\big|_{\alpha=2,\beta=1} = \frac{\dfrac{A}{m\omega_n\sqrt{1-\varsigma^2}}e^{-\varsigma\omega_n t}\left[\left(\omega_n^2-\omega^2\right)\cos\omega_n\sqrt{1-\varsigma^2}t - \dfrac{\varsigma\left(\omega_n^2+\omega^2\right)\sin\omega_n\sqrt{1-\varsigma^2}t}{\sqrt{1-\varsigma^2}}\right]}{\left(\omega_n^2-\omega^2\right)^2+(2\varsigma\omega_n\omega)^2}
$$

$$
= \frac{1}{m\omega_d}\frac{Ae^{-\varsigma\omega_n t}\left[\left(\omega_n^2-\omega^2\right)\cos\omega_d t - \dfrac{\varsigma\left(\omega_n^2+\omega^2\right)\sin\omega_d t}{\sqrt{1-\varsigma^2}}\right]}{\left(\omega_n^2-\omega^2\right)^2+(2\varsigma\omega_n\omega)^2}.
\tag{279}
$$

Remark 31. *We revealed that the sinusoidal response to fractional oscillators in Class III for any value of $\alpha \in (1, 2)$ and $\beta \in (0, 1)$ does have steady-state component $x_{3zs,s}(t)$ described by (277), also see Figure 65.*

Remark 32. *The results presented above show that the exact periodic solutions to three classes of fractional oscillators exist.*

11. Discussion

Three classes of fractional oscillators previously studied are usually characterized by constant-coefficient fractional differential equations. The basic theory and key point I presented in Section 4 is to equivalently represent them by the second-order differential equations with variable-coefficients. In this way, three classes of fractional oscillators, which are nonlinear in nature, all reduce to linear oscillators with variable-coefficients. In methodology, that may open a new way of the linearization to describe and research fractional oscillators.

In addition to keep fractional properties of fractional oscillators with its equivalences, for instance, the characteristic roots of a fractional oscillator being infinitely large as explained by Li et al. [18] and Duan et al. [39], based on the proposed equivalent oscillators, we also reveal other properties of fractional oscillators, which may be very difficult, if not impossible, to be described directly from the point of view of fractional differential equations, such as the equivalent, i.e., intrinsic, masses m_{eqj}, equivalent dampings c_{eqj}, equivalent natural frequencies $\omega_{eqn,j}$ and $\omega_{eqd,j}$ ($j = 1, 2, 3$) of fractional oscillators, which are nonlinear with the power laws in terms of oscillation frequency ω as stated in Sections 4 and 5.

The significance of the presented theory with respect to three classes of fractional oscillators in both theory and practice is about the closed form analytic formulas of the responses to fractional oscillators explained in Sections 6–10 by using elementary functions, making the matters much better in engineering.

Note that power laws plays a role in understanding the nature in general, see, e.g., Gabaix et al. [82], Stanley [83]. As a matter of fact, the fractional order α relates to the fractal dimension, see Lim et al. [20–22]. Thus, my study of the power laws previously stated is quite beginning in the aspect of fractional oscillations. Further research is needed in future. In addition to that, our future work will consider the applications of the present equivalent theory of the fractional oscillators to fractional noise in communication systems (Levy and Pinchas [84], Pinchas [85]), partial differential equations, such as transient phenomena of complex systems or fractional diffusion equations (Toma [86], Bakhoum and Toma [87], Cattani [88], Mardani et al. [89]).

12. Conclusions

We have established a theory of equivalent oscillators with respect to three classes of fractional oscillation systems. Its principle is to represent a fractional oscillator with constant coefficients (mass and damping) by a 2-order oscillator equivalently with variable mass and damping. The analytic expressions of equivalent masses, equivalent dampings, equivalent damping ratios, equivalent natural frequencies, and equivalent frequency ratios have been presented. We have revealed that the equivalent masses and dampings of three classes of fractional oscillators follow power laws in terms of oscillation frequency. By using elementary functions, we have put forward the closed form representations of responses (free, impulse, step, frequency, sinusoidal) to three classes of fractional oscillators. Additionally, analytic expressions of the logarithmic decrements of three classes of fractional oscillators have been proposed. As by products, we have stated the representations of four types of the generalized Mittag-Leffler functions in the closed form with elementary functions.

Acknowledgments: This work was supported in part by the National Natural Science Foundation of China under the project grant numbers 61672238 and 61272402. The views and conclusions contained in this document are those of the author and should not be interpreted as representing the official policies, either expressed or implied, of NSFC or the Chinese government. The author shows his appreciation for the valuable comments from reviewers on the manuscript.

Conflicts of Interest: The authors declare no conflict of interest.

References

1. Timoshenko, S.P. *Vibration Problems in Engineering*, 2nd ed.; Fifth Printing; D. Van Nostrand Company, Inc.: New York, NY, USA, 1955.
2. Timoshenko, S.P. *Mechanical Vibrations*; D. Van Nostrand Company, Inc.: New York, NY, USA, 1955.
3. Den Hartog, J.P. *Mechanical Vibrations*; McGraw-Hill: New York, NY, USA, 1956.
4. Harris, C.M. *Shock and Vibration Handbook*, 5th ed.; McGraw-Hill: New York, NY, USA, 2002.
5. Palley, O.M.; Bahizov, Г.Б.; Voroneysk, E.Я. *Handbook of Ship Structural Mechanics*; Translated from Russian by Xu, B.H., Xu, X., Xu, M.Q.; National Defense Industry Publishing House: Beijing, China, 2002. (In Chinese)
6. Rayleigh, J.W.S.B., III. *The Theory of Sound*; Macmillan & Co., Ltd.: London, UK, 1877; Volume 1.
7. Lalanne, C. *Mechanical Vibration and Shock*, 2nd ed.; Wiley: New York, NY, USA, 2013; Volumes 1–5.
8. Nilsson, J.W.; Riedel, S.A. *Electronic Circuits*, 9th ed.; Prentice Hall: Upper Saddle River, NJ, USA, 2011.
9. Diaz, H.F.; Markgraf, V. *El Niño and the Southern Oscillation Multiscale Variability and Global and Regional Impacts*; Cambridge University Press: Cambridge, UK, 2000.
10. Hurrell, J.W.; Kushnir, Y.; Ottersen, G.; Visbeck, M. *The North Atlantic Oscillation: Climatic Significance and Environmental Impact*; John Wiley & Sons: New York, NY, USA, 2003; Volume 134.
11. Xu, L.; Chen, Z.; Hu, K.; Stanley, H.E.; Ivanov, P.C. Spurious detection of phase synchronization in coupled nonlinear oscillators. *Phys. Rev. E* **2006**, *73*, 065201. [CrossRef] [PubMed]
12. Fourier, J. *Thkorie Analytique de la Chaleur*; Didot: Paris, France, 1822.
13. Bernoulli, D. *Hydrodynamica, Sive de Viribus et Motibus Fluidorum Commentarii. Opus Academicum ab Auctore, dum Petropoli Ageret, Congestum*; Argentorati, Sumptibus Johannis Reinholdi Dulseckeri, Typis Joh. Deckeri, Typographi Basiliensis: ETH-Bibliothek Zürich, Zürich, Switzerland, 1738. [CrossRef]
14. Timoshenko, S.P. On the transverse vibrations of bars of uniform cross-section. *Philos. Mag.* **1922**, *43*, 125–131. [CrossRef]

15. Searle, J.H.C. The Effect of rotatory inertia on the vibration of bars. *Philos. Mag.* **1907**, *14*, 35–60. [CrossRef]
16. Steinmetz, C.P. *Theory and Calculation of Alternating Current Phenomena*; McGraw-Hill: New York, NY, USA, 1897.
17. Steinmetz, C.P. *Engineering Mathematics: A Series of Lectures Delivered at Union College*; McGraw-Hill: New York, NY, USA, 1917.
18. Li, M.; Lim, S.C.; Cattani, C.; Scalia, M. Characteristic roots of a class of fractional oscillators. *Adv. High Energy Phys.* **2013**, *2013*, 853925. [CrossRef]
19. Li, M.; Lim, S.C.; Chen, S.Y. Exact solution of impulse response to a class of fractional oscillators and its stability. *Math. Probl. Eng.* **2011**, *2011*, 657839. [CrossRef]
20. Lim, S.C.; Li, M.; Teo, L.P. Langevin equation with two fractional orders. *Phys. Lett. A* **2008**, *372*, 6309–6320. [CrossRef]
21. Lim, S.C.; Li, M.; Teo, L.P. Locally self-similar fractional oscillator processes. *Fluct. Noise Lett.* **2007**, *7*, L169–L179. [CrossRef]
22. Lim, S.C.; Teo, L.P. The fractional oscillator process with two indices. *J. Phys. A Stat. Mech. Appl.* **2009**, *42*, 065208. [CrossRef]
23. West, B.J.; Geneston, E.L.; Grigolini, P. Maximizing information exchange between complex networks. *Phys. Rep.* **2008**, *468*, 1–99. [CrossRef]
24. Duan, J.-S. The periodic solution of fractional oscillation equation with periodic input. *Adv. Math. Phys.* **2013**, *2013*, 869484. [CrossRef]
25. Mainardi, F. Fractional relaxation-oscillation and fractional diffusion-wave phenomena. *Chaos Solitons Fractals* **1996**, *7*, 1461–1477. [CrossRef]
26. Zurigat, M. Solving fractional oscillators using Laplace homotopy analysis method. *Ann. Univ. Craiova Math. Comput. Sci. Ser.* **2011**, *38*, 1–11.
27. Blaszczyk, T.; Ciesielski, M. Fractional oscillator equation—Transformation into integral equation and numerical solution. *Appl. Math. Comput.* **2015**, *257*, 428–435. [CrossRef]
28. Blaszczyk, T.; Ciesielski, M.; Klimek, M.; Leszczynski, J. Numerical solution of fractional oscillator equation. *Appl. Math. Comput.* **2011**, *218*, 2480–2488. [CrossRef]
29. Al-rabtah, A.; Ertürk, V.S.; Momani, S. Solutions of a fractional oscillator by using differential transform method. *Comput. Math. Appl.* **2010**, *59*, 1356–1362. [CrossRef]
30. Drozdov, A.D. Fractional oscillator driven by a Gaussian noise. *Phys. A Stat. Mech. Appl.* **2007**, *376*, 237–245. [CrossRef]
31. Stanislavsky, A.A. Fractional oscillator. *Phys. Rev. E* **2004**, *70*, 051103. [CrossRef] [PubMed]
32. Achar, B.N.N.; Hanneken, J.W.; Clarke, T. Damping characteristics of a fractional oscillator. *Phys. A Stat. Mech. Appl.* **2004**, *339*, 311–319. [CrossRef]
33. Achar, B.N.N.; Hanneken, J.W.; Clarke, T. Response characteristics of a fractional oscillator. *Phys. A Stat. Mech. Appl.* **2002**, *309*, 275–288. [CrossRef]
34. Achar, B.N.N.; Hanneken, J.W.; Enck, T.; Clarke, T. Dynamics of the fractional oscillator. *Phys. A Stat. Mech. Appl.* **2001**, *297*, 361–367. [CrossRef]
35. Tofighi, A. The intrinsic damping of the fractional oscillator. *Phys. A Stat. Mech. Appl.* **2003**, *329*, 29–34. [CrossRef]
36. Ryabov, Y.E.; Puzenko, A. Damped oscillations in view of the fractional oscillator equation. *Phys. Rev. B* **2002**, *66*, 184201. [CrossRef]
37. Ahmad, W.E.; Elwakil, A.S.R. Fractional-order Wien-bridge oscillator. *Electron. Lett.* **2001**, *37*, 1110–1112. [CrossRef]
38. Uchaikin, V.V. *Fractional Derivatives for Physicists and Engineers*; Springe: Berlin Heidelberg, 2013; Volumes II.
39. Duan, J.-S.; Wang, Z.; Liu, Y.-L.; Qiu, X. Eigenvalue problems for fractional ordinary differential equations. *Chaos Solitons Fractals* **2013**, *46*, 46–53. [CrossRef]
40. Lin, L.-F.; Chen, C.; Zhong, S.-C.; Wang, H.-Q. Stochastic resonance in a fractional oscillator with random mass and random frequency. *J. Stat. Phys.* **2015**, *160*, 497–511. [CrossRef]
41. Duan, J.-S. A modified fractional derivative and its application to fractional vibration equation. *Appl. Math. Inf. Sci.* **2016**, *10*, 1863–1869. [CrossRef]

42. Alkhaldi, H.S.; Abu-Alshaikh, I.M.; Al-Rabadi, A.N. Vibration control of fractionally-damped beam subjected to a moving vehicle and attached to fractionally-damped multi-absorbers. *Adv. Math. Phys.* **2013**, *2013*, 232160. [CrossRef]

43. Dai, H.; Zheng, Z.; Wang, W. On generalized fractional vibration equation. *Chaos Solitons Fractals* **2017**, *95*, 48–51. [CrossRef]

44. Ren, R.; Luo, M.; Deng, K. Stochastic resonance in a fractional oscillator driven by multiplicative quadratic noise. *J. Stat. Mech. Theory Exp.* **2017**, *2017*, 023210. [CrossRef]

45. Xu, Y.; Li, Y.; Liu, D.; Jia, W.; Huang, H. Responses of Duffing oscillator with fractional damping and random phase. *Nonlinear Dyn.* **2013**, *74*, 745–753. [CrossRef]

46. He, G.; Tian, Y.; Wang, Y. Stochastic resonance in a fractional oscillator with random damping strength and random spring stiffness. *J. Stat. Mech. Theory Exp.* **2013**, *2013*, P09026. [CrossRef]

47. Leung, A.Y.T.; Guo, Z.; Yang, H.X. Fractional derivative and time delay damper characteristics in Duffing-van der Pol oscillators. *Commun. Nonlinear Sci. Numer. Simul.* **2013**, *18*, 2900–2915. [CrossRef]

48. Chen, L.C.; Zhuang, Q.Q.; Zhu, W.Q. Response of SDOF nonlinear oscillators with lightly fractional derivative damping under real noise excitations. *Eur. Phys. J. Spec. Top.* **2011**, *193*, 81–92. [CrossRef]

49. Deü, J.-F.; Matignon, D. Simulation of fractionally damped mechanical systems by means of a Newmark-diffusive scheme. *Comput. Math. Appl.* **2010**, *59*, 1745–1753. [CrossRef]

50. Drăgănescu, G.E.; Bereteu, L.; Ercuţa, A.; Luca, G. Anharmonic vibrations of a nano-sized oscillator with fractional damping. *Commun. Nonlinear Sci. Numer. Simul.* **2010**, *15*, 922–926. [CrossRef]

51. Rossikhin, Y.A.; Shitikova, M.V. New approach for the analysis of damped vibrations of fractional oscillators. *Shock Vib.* **2009**, *16*, 365–387. [CrossRef]

52. Xie, F.; Lin, X. Asymptotic solution of the van der Pol oscillator with small fractional damping. *Phys. Scr.* **2009**, *2009*, 014033. [CrossRef]

53. Gomez-Aguilar, J.F.; Rosales-Garcia, J.J.; Bernal-Alvarado, J.J.; Cordova-Fraga, T.; Guzman-Cabrera, R. Fractional mechanical oscillators. *Rev. Mex. Fis.* **2012**, *58*, 348–352.

54. Liu, Q.X.; Liu, J.K.; Chen, Y.M. An analytical criterion for jump phenomena in fractional Duffing oscillators. *Chaos Solitons Fractals* **2017**, *98*, 216–219. [CrossRef]

55. Leung, A.Y.T.; Yang, H.X.; Guo, Z.J. The residue harmonic balance for fractional order van der Pol like oscillators. *J. Sound Vib.* **2012**, *331*, 1115–1126. [CrossRef]

56. Kavyanpoor, M.; Shokrollahi, S. Challenge on solutions of fractional Van Der Pol oscillator by using the differential transform method. *Chaos Solitons Fractals* **2017**, *98*, 44–45. [CrossRef]

57. Xiao, M.; Zheng, W.X.; Cao, J. Approximate expressions of a fractional order Van der Pol oscillator by the residue harmonic balance method. *Math. Comput. Simul.* **2013**, *89*, 1–12. [CrossRef]

58. Chen, L.; Wang, W.; Li, Z.; Zhu, W. Stationary response of Duffing oscillator with hardening stiffness and fractional derivative. *Int. J. Non-Linear Mech.* **2013**, *48*, 44–50. [CrossRef]

59. Wen, S.-F.; Shen, Y.-J.; Yang, S.-P.; Wang, J. Dynamical response of Mathieu-Duffing oscillator with fractional-order delayed feedback. *Chaos Solitons Fractals* **2017**, *94*, 54–62. [CrossRef]

60. Liao, H. Optimization analysis of Duffing oscillator with fractional derivatives. *Nonlinear Dyn.* **2015**, *79*, 1311–1328. [CrossRef]

61. Abu-Gurra, S.; Ertürk, V.S.; Momani, S. Application of the modified differential transform method to fractional oscillators. *Kybernetes* **2011**, *40*, 751–761. [CrossRef]

62. Ikeda, T.; Harata, Y.; Hiraoka, R. Intrinsic localized modes of 1/2-order subharmonic oscillations in nonlinear oscillator arrays. *Nonlinear Dyn.* **2015**, *81*, 1759–1777. [CrossRef]

63. Department of Mathematics, Fudan University. *Vibration Theory*; Shanghai Scientific & Technical Publishing House: Shanghai, China, 1960. (In Chinese)

64. Andronov, A.A.; Victor, A.A.; Hayijin, C.Э. *Oscillation Theory*; Translated from Russian by Gao, W.B., Yang, R.W., Xiao, Z.Y.; Science Press: Beijing, China, 1974. (In Chinese)

65. Gabel, R.A.; Roberts, R.A. *Signals and Linear Systems*; John Wiley & Sons: New York, NY, USA, 1973.

66. Zheng, J.L.; Ying, Q.Y.; Yang, W.L. *Signals and Systems*, 2nd ed.; Higher Education Press: Beijing, China, 2000; Volume 1. (In Chinese)

67. Gelfand, I.M.; Vilenkin, K. *Generalized Functions*; Academic Press: New York, NY, USA, 1964; Volume 1.

68. Griffel, D.H. *Applied Functional Analysis*; John Wiley & Sons: New York, NY, USA, 1981.

69. Miller, K.S.; Ross, B. *An Introduction to the Fractional Calculus and Fractional Differential Equations*; John Wiley: New York, NY, USA, 1993.
70. Klafter, J.; Lim, S.C.; Metzler, R. *Fractional Dynamics: Recent Advances*; World Scientific: Singapore, 2012.
71. Lavoie, J.L.; Osler, T.J.; Tremblay, R. Fractional derivatives and special functions. *SIAM Rev.* **1976**, *18*, 240–268. [CrossRef]
72. Uchaikin, V.V. *Fractional Derivatives for Physicists and Engineers*; Springer: Berlin Heidelberg, 2013; Volume I.
73. Mathai, A.M.; Haubold, H.J. *Special Functions for Applied Scientists*; Springer: New York, NY, USA, 2008.
74. Gorenflo, R.; Kilbas, A.A.; Mainardi, F.; Rogosin, S.V. *Mittag-Leffler Functions: Related Topics and Applications, Theory and Applications*; Monographs in Mathematics; Springer: Berlin Heidelberg, 2014.
75. Erdelyi, A.; Magnus, W.; Oberhettinger, F.; Tricomi, F.G. *Higher Transcendental Functions*; McGraw Hill: New York, NY, USA, 1955; Volume I.
76. Papoulis, A. *The Fourier Integral and Its Applications*; McGraw-Hill Inc.: New York, NY, USA, 1962.
77. Chung, W.S.; Jung, M. Fractional damped oscillators and fractional forced oscillators. *J. Korean Phys. Soc.* **2014**, *64*, 186–191. [CrossRef]
78. Korotkin, A.I. *Added Masses of Ship Structures, Fluid Mechanics and Its Applications*; Springer: Berlin Heidelberg, 2009; Volume 88.
79. Jin, X.D.; Xia, L.J. *Ship Hull Vibration*; The Press of Shanghai Jiaotong University: Shanghai, China, 2011. (In Chinese)
80. Nakagawa, K.; Ringo, M. *Engineering Vibrations*; Translated from Japanese by Xia, S.R.; Shanghai Science and Technology Publishing House: Shanghai, China, 1981. (In Chinese)
81. Kaslik, E.; Sivasundaram, S. Non-existence of periodic solutions in fractional-order dynamical systems and a remarkable difference between integer and fractional-order derivatives of periodic functions. *Nonlinear Anal. Real World Appl.* **2012**, *13*, 1489–1497. [CrossRef]
82. Gabaix, X.; Gopikrishnan, P.; Plerou, V.; Stanley, H.E. A theory of power-law distributions in financial market fluctuations. *Nature* **2003**, *423*, 267–270. [CrossRef] [PubMed]
83. Stanley, H.E. Phase transitions: Power laws and universality. *Nature* **1995**, *378*, 554. [CrossRef]
84. Levy, C.; Pinchas, M. Maximum likelihood estimation of clock skew in IEEE 1588 with fractional Gaussian noise. *Math. Probl. Eng.* **2015**, *2015*, 174289. [CrossRef]
85. Pinchas, M. Symbol error rate for non-blind adaptive equalizers applicable for the SIMO and fGn case. *Math. Probl. Eng.* **2014**, *2014*, 606843. [CrossRef]
86. Toma, C. Wavelets-computational aspects of Sterian-realistic approach to uncertainty principle in high energy physics: A transient approach. *Adv. High Energy Phys.* **2013**, *2013*, 735452. [CrossRef]
87. Bakhoum, E.G.; Toma, C. Transient aspects of wave propagation connected with spatial coherence. *Math. Probl. Eng.* **2013**, *2013*, 691257. [CrossRef]
88. Cattani, C.; Ciancio, A.; Laserra, E.; Bochicchio, I. On the fractal distribution of primes and prime-indexed primes by the binary image analysis. *Phys. A Stat. Mech. Appl.* **2016**, *460*, 222–229. [CrossRef]
89. Mardani, A.; Heydari, M.H.; Hooshmandasl, M.R.; Cattani, C. A meshless method for solving time fractional advection-diffusion equation with variable coefficients. *Comput. Math. Appl.* **2017**, in press. [CrossRef]

Article

Computing Zagreb Indices and Zagreb Polynomials for Symmetrical Nanotubes

Zehui Shao [1], Muhammad Kamran Siddiqui [2,*] and Mehwish Hussain Muhammad [3]

[1] Institute of Computing Science and Technology, Guangzhou University, Guangzhou 510006, China; zshao@gzhu.edu.cn
[2] Department of Mathematics, COMSATS University Islamabad, Sahiwal Campus, Sahiwal 57000, Pakistan
[3] College of Chemistry and Molecular Engineering, Zhengzhou University, Zhengzhou 450001, China; mehwish@foxmail.com
* Correspondence: kamransiddiqui75@gmail.com

Received: 26 May 2018; Accepted: 24 June 2018; 28 June 2018

Abstract: Topological indices are numbers related to sub-atomic graphs to allow quantitative structure-movement/property/danger connections. These topological indices correspond to some specific physico-concoction properties such as breaking point, security, strain vitality of chemical compounds. The idea of topological indices were set up in compound graph hypothesis in view of vertex degrees. These indices are valuable in the investigation of mitigating exercises of specific Nanotubes and compound systems. In this paper, we discuss Zagreb types of indices and Zagreb polynomials for a few Nanotubes covered by cycles.

Keywords: first multiple Zagreb index; second multiple Zagreb index, hyper-Zagreb index; Zagreb polynomials; Nanotubes

1. Introduction

Mathematical chemistry becomes an interesting branch of science in which we talk about and foresee the concoction structure by utilizing numerical apparatuses and does not really allude to the quantum mechanics. As a branch of numerical science where we apply devices of graph hypothesis, chemical graph theory was introduced and extensively studied to show the compound wonder scientifically. This is more imperative to state that the hydrogen particles are regularly overlooked in any sub-atomic graph. Topological indices are really a numeric measures related to the constitution synthetic material implying for relationship of concoction structure with numerous physio-substance features, compound responsiveness or biological activity. Motivated by the wide applications of topological indices, the topological indices of graphs are studied extensively [1–3].

A nano structure is a question of middle size among both molecular and microscopic structures. Such a material is determined through designing at atomic scale, which is something that has a physical measurement littler than one hundred nanometers, running from bunches of particles to many dimensional layers. Carbon Nanotubes (CNTs) with allotropes of carbon whose shapes are usually hollow and round possess some kinds of nanostructure.

For a graph G, the degree of a vertex w is the cardinality of edges incident to w and denoted by $dgr(s)$. A molecular graph is a basic limited graph in which vertices mean the atoms and edges indicate the compound bonds in fundamental substance structures.

For a graph G, a topological index $Tp(G)$ is a value which can be obtained by a computing method from G. Moreover, if graphs G and F are isomorphic, then the result $Tp(F) = Tp(G)$ holds. Wiener [4] initially figured out an idea for a topological index in the early years, and at that time, he took a shot at breaking point of paraffin. He defined this record to be the way number. Afterwards, such a concept was renamed the Wiener index. As we know, the Wiener record is the first posed index and it is one of the most attractive indices, from not only a hypothetical perspective but applications, and characterized as the total of separations among vertices in G, see for subtle elements [5].

The first Zagreb index, a very old topological index, was initiated in 1972 [6] and later many variations of Zagreb index were proposed, e.g., Shirdel et al. [7] in 2013 described a novel index under the name of "hyper-Zagreb index" and it was defined to be

$$HM(G) = \sum_{sr \in E(G)} \left[dgr(s) + dgr(r) \right]^2 \tag{1}$$

in [8], two new versions of Zagreb indices were put forward, which are the first multiple Zagreb index $PM_1(G)$ and the second multiple Zagreb index $PM_2(G)$. More precisely, they are formulated as follows.

$$PM_1(G) = \prod_{sr \in E(G)} \left[dgr(s) + dgr(r) \right] \tag{2}$$

$$PM_2(G) = \prod_{sr \in E(G)} \left[dgr(s) \times dgr(r) \right] \tag{3}$$

Some properties of the indices $PM_1(G), PM_2(G)$ of specific chemical structures were investigated in [9].

To investigate more interesting properties of $PM_1(G), PM_2(G)$ of a graph G, the first Zagreb Polynomial $M_1(G, x)$ and the second Zagreb Polynomial $M_2(G, x)$ are proposed [10,11] and put forward as

$$M_1(G, x) = \sum_{sr \in E(G)} x^{[dgr(s)+dgr(r)]} \tag{4}$$

$$M_2(G, x) = \sum_{sr \in E(G)} x^{[dgr(s) \times dgr(r)]} \tag{5}$$

2. Applications of Nanostructure and Topological Indices

In the past few decades, graph theory was widely applied as a tool to study physical and chemical properties of materials. More and more people are interested in this field and as a result chemical graph theory was introduces, and later various topological indices were studied and defined. Moreover, as a combination of chemistry, mathematics and nano science, *nanotechnology* was also studied by means of chemical graph theory. Among these, quantitative structure–activity relationship (QSAR) and quantitative structure-property relationship (QSPR) are analyzed to predict the properties of nanostructure and biological activities. To study QSAR and QSPR, hyper-Zagreb index, first multiple Zagreb index, second multiple Zagreb index and Zagreb polynomials are applied to predict the bioactivity of nanostructures [12–15].

The Zagreb index is defined to be a topological descriptor which is related to substantial synthetic qualities of the atoms [16]. The particle bond network hyper Zagreb index gives a decent connection to the security of direct dendrimers and also the stretched pharmacies and for processing the strain vitality of cyclo alkanes [17–21]. To relate with some physico-concoction properties, multiple Zagreb index bears much preferred prescient control over the prescient energy of the dendrimers [22–24]. The first and second Zagreb indices were revealed to be used to research the π-electron energy of various microscopic particles [25–27].

3. $HAC_5C_7[p, q]$ Nanotube

The $HAC_5C_7[p, q]$ Nanotube can be studied as a C_5C_7 net and it consists of C_5s and C_7s with the trivalent decorations. An example is presented in Figure 1, which can be decorated in a cylindrical or toroidal manner. The 2-dimensional lattice of $HAC_5C_7[p, q]$ were ever been discussed in [28], in which p and q are the cardinalites of heptagons in one row and periods in whole lattice, respectively. As an example, such a Nanotube with three rows is presented in Figure 2.

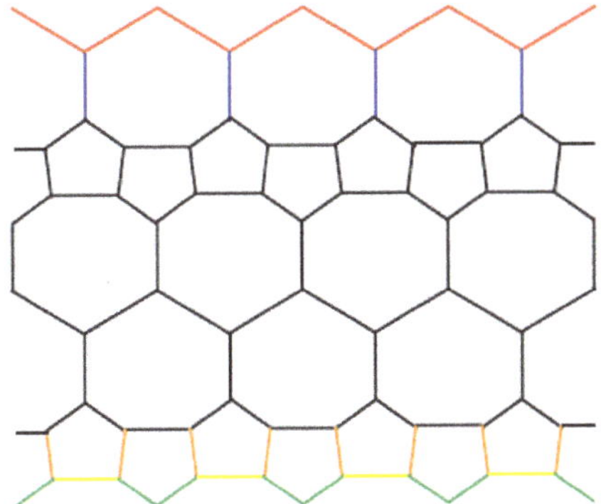

Figure 1. $HAC_5C_7[p,q]$ Nanotube with $p = 4$ and $q = 2$.

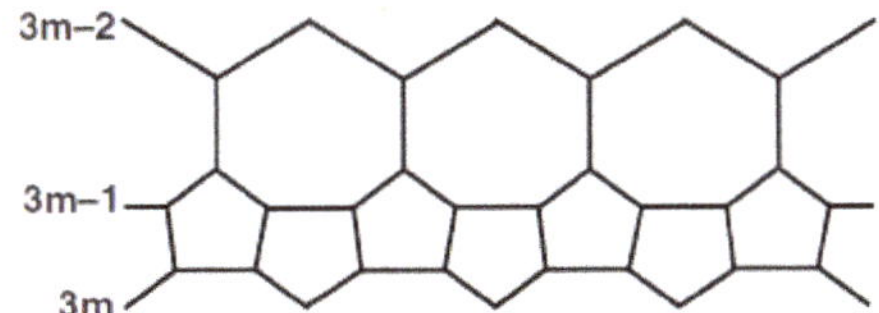

Figure 2. The mth period of $HAC_5C_7[p,q]$ Nanotube.

3.1. Methodology of $HAC_5C_7[p,q]$ Nanotube Formulas

In the Nanotube $HAC_5C_7[p,q], (p,q \geq 1)$, we have that $V(HAC_5C_7[p,q]) = 8pq + p$ and $E(HAC_5C_7[p,q]) = 12pq - p$. The cardinality of vertices of degree two and three are $2p + 2$ and $8pq - p - 2$, respectively. According to their sum the the degree over its neighbors of each vertex, the edge set can be partitioned to six disjoint sets as follows.

$$E_1(HAC_5C_7[p,q]) = \left\{ sr \in E(HAC_5C_7[p,q]) \mid dgr(s) = 6, dgr(r) = 7 \right\}$$
$$E_2(HAC_5C_7[p,q]) = \left\{ sr \in E(HAC_5C_7[p,q]) \mid dgr(s) = 6, dgr(r) = 8 \right\}$$
$$E_3(HAC_5C_7[p,q]) = \left\{ sr \in E(HAC_5C_7[p,q]) \mid dgr(s) = 7, dgr(r) = 9 \right\}$$
$$E_4(HAC_5C_7[p,q]) = \left\{ sr \in E(HAC_5C_7[p,q]) \mid dgr(s) = 8, dgr(r) = 8 \right\}$$
$$E_5(HAC_5C_7[p,q]) = \left\{ sr \in E(HAC_5C_7[p,q]) \mid dgr(s) = 8, dgr(r) = 9 \right\}$$
$$E_6(HAC_5C_7[p,q]) = \left\{ sr \in E(HAC_5C_7[p,q]) \mid dgr(s) = 9, dgr(r) = 9 \right\}$$

The cardinality of edges in $E_1(HAC_5C_7[p,q])$, $E_2(HAC_5C_7[p,q])$ and $E_5(HAC_5C_7[p,q])$ are $2p$. The cardinality of edges in $E_3(HAC_5C_7[p,q])$ and $E_4(HAC_5C_7[p,q])$ are p. The cardinality of edges in $E_6(HAC_5C_7[p,q])$ is $12pq - 9p$. Such a partition is shown in Figure 1 in which red, green, blue, yellow, brown and black edges are the edges belong to $E_1(HAC_5C_7[p,q])$, $E_2(HAC_5C_7[p,q])$, $E_3(HAC_5C_7[p,q])$, $E_4(HAC_5C_7[p,q])$, $E_5(HAC_5C_7[p,q])$ and $E_6(HAC_5C_7[p,q])$ respectively.

3.2. Main Results for $HAC_5C_7[p,q]$ Nanotube

In this section, we will obtain hyper-Zagreb index $HM(G)$, first multiple Zagreb index $PM_1(G)$, second multiple Zagreb index $PM_2(G)$, Zagreb polynomials $M_1(G,x)$, $M_2(G,x)$ for $HAC_5C_7[p,q], (p,q \geq 1)$ Nanotube.

- **Hyper Zagreb index of $HAC_5C_7[p,q]$ Nanotube**

  Let G be the $HAC_5C_7[p,q]$ Nanotube. Then by Equations (1), we have

$$
\begin{aligned}
HM(G) &= \sum_{sr \in E(G)} [dgr(s) + dgr(r)]^2 \\
HM(G) &= \sum_{sr \in E_1} [dgr(s) + dgr(r)]^2 + \sum_{sr \in E_2} [dgr(s) + dgr(r)]^2 + \sum_{sr \in E_3} [dgr(s) + dgr(r)]^2 \\
&\quad + \sum_{sr \in E_4} [dgr(s) + dgr(r)]^2 + \sum_{sr \in E_5} [dgr(s) + dgr(r)]^2 + \sum_{sr \in E_6} [dgr(s) + dgr(r)]^2 \\
&= 13^2|E_1| + 14^2|E_2| + 16^2|E_3| + 16^2|E_4| + 17^2|E_5| + 18^2|E_6| \\
&= 169(2p) + 196(2p) + 256p + 256p + 289(2p) + 324(12pq - 9p) = 3888pq - 1092p
\end{aligned}
$$

- **Multiple Zagreb indices of $HAC_5C_7[p,q]$ Nanotube**

  Let G be the $HAC_5C_7[p,q]$ Nanotube. Then by Equations (2) and (3), we have

$$
\begin{aligned}
PM_1(G) &= \prod_{sr \in E(G)} [dgr(s) + dgr(r)] \\
PM_1(G) &= \prod_{sr \in E_1} [dgr(s) + dgr(r)] \times \prod_{sr \in E_2} [dgr(s) + dgr(r)] \times \prod_{sr \in E_3} [dgr(s) + dgr(r)] \\
&\quad \times \prod_{sr \in E_4} [dgr(s) + dgr(r)] \times \prod_{sr \in E_5} [dgr(s) + dgr(r)] \times \prod_{sr \in E_6} [dgr(s) + dgr(r)] \\
&= 13^{|E_1|} \times 14^{|E_2|} \times 16^{|E_3|} \times 16^{|E_4|} \times 17^{|E_5|} \times 18^{|E_6|} \\
&= 13^{2p} \times 14^{2p} \times 16^p \times 16^p \times 17^{2p} \times 18^{(12pq - 9p)}
\end{aligned}
$$

$$
\begin{aligned}
PM_2(G) &= \prod_{sr \in E(G)} [dgr(s) \times dgr(r)] \\
PM_2(G) &= \prod_{sr \in E_1} [dgr(s) \times dgr(r)] \times \prod_{sr \in E_2} [dgr(s) \times dgr(r)] \times \prod_{sr \in E_3} [dgr(s) \times dgr(r)] \\
&\quad \times \prod_{sr \in E_4} [dgr(s) \times dgr(r)] \times \prod_{sr \in E_5} [dgr(s) \times dgr(r)] \times \prod_{sr \in E_6} [dgr(s) \times dgr(r)] \\
&= 42^{|E_1|} \times 48^{|E_2|} \times 63^{|E_3|} \times 64^{|E_4|} \times 72^{|E_5|} \times 81^{|E_6|} \\
&= 42^{2p} \times 48^{2p} \times 63^p \times 64^p \times 72^{2p} \times 81^{(12pq - 9p)}
\end{aligned}
$$

- **Zagreb Polynomials of $HAC_5C_7[p,q]$ Nanotube**

  Let G be the $HAC_5C_7[p,q]$ Nanotube. Then by Equations (4) and (5), we have

$$
\begin{aligned}
M_1(G,x) &= \sum_{sr \in E(G)} x^{[dgr(s) + dgr(r)]} \\
M_1(G,x) &= \sum_{sr \in E_1} x^{[dgr(s) + dgr(r)]} + \sum_{sr \in E_2} x^{[dgr(s) + dgr(r)]} + \sum_{sr \in E_3} x^{[dgr(s) + dgr(r)]} \\
&\quad + \sum_{sr \in E_4} x^{[dgr(s) + dgr(r)]} + \sum_{sr \in E_5} x^{[dgr(s) + dgr(r)]} + \sum_{sr \in E_6} x^{[dgr(s) + dgr(r)]} \\
&= \sum_{sr \in E_1} x^{13} + \sum_{sr \in E_2} x^{14} + \sum_{sr \in E_3} x^{16} + \sum_{sr \in E_4} x^{16} + \sum_{sr \in E_5} x^{17} + \sum_{sr \in E_6} x^{18} \\
&= |E_1|x^{13} + |E_2|x^{14} + |E_3|x^{16} + |E_4|x^{16} + |E_5|x^{17} + |E_6|x^{18} \\
&= 2px^{13} + 2px^{14} + px^{16} + px^{16} + 2px^{17} + (12pq - 9p)x^{18}
\end{aligned}
$$

$$M_2(G,x) = \sum_{sr \in E(G)} x^{[dgr(s) \times dgr(r)]}$$

$$M_2(G,x) = \sum_{sr \in E_1} x^{[dgr(s) \times dgr(r)]} + \sum_{sr \in E_2} x^{[dgr(s) \times dgr(r)]} + \sum_{sr \in E_3} x^{[dgr(s) \times dgr(r)]}$$

$$+ \sum_{sr \in E_4} x^{[dgr(s) \times dgr(r)]} + \sum_{sr \in E_5} x^{[dgr(s) \times dgr(r)]} + \sum_{sr \in E_6} x^{[dgr(s) \times dgr(r)]}$$

$$= \sum_{sr \in E_1} x^{42} + \sum_{sr \in E_2} x^{48} + \sum_{sr \in E_3} x^{63} + \sum_{sr \in E_4} x^{64} + \sum_{sr \in E_5} x^{72} + \sum_{sr \in E_6} x^{81}$$

$$= |E_1|x^{42} + |E_2|x^{48} + |E_3|x^{63} + |E_4|x^{64} + |E_5|x^{72} + |E_6|x^{81}$$

$$= 2px^{42} + 2px^{48} + px^{63} + px^{64} + 2px^{72} + (12pq - 9p)x^{81}$$

4. $HAC_5C_6C_7[p,q]$ Nanotube

The $HAC_5C_6C_7[p,q]$ Nanotube is a $C_5C_6C_7$ net and constructed by using C_5s, C_6s and C_7s alternately with the trivalent decorations as demonstrated in Figure 3. These tessellations of C_5s, C_6s and C_7s are usually decorated in a cylindrical or a torodial manner. The 2-dimensional lattice of $HAC_5C_6C_7[p,q]$ is obtained by repeating pentagons for q rows and p columns. The construction of this Nanotube can be found in [29]. As an example, a Nanotube with three rows is shown in Figure 4.

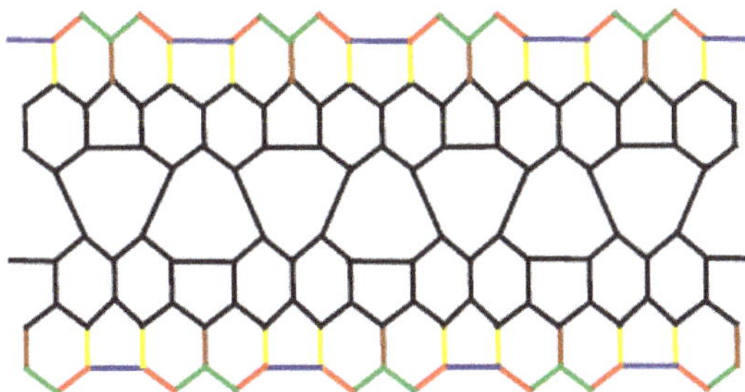

Figure 3. $HAC_5C_6C_7[p,q]$ Nanotube with $p = 4$ and $q = 2$.

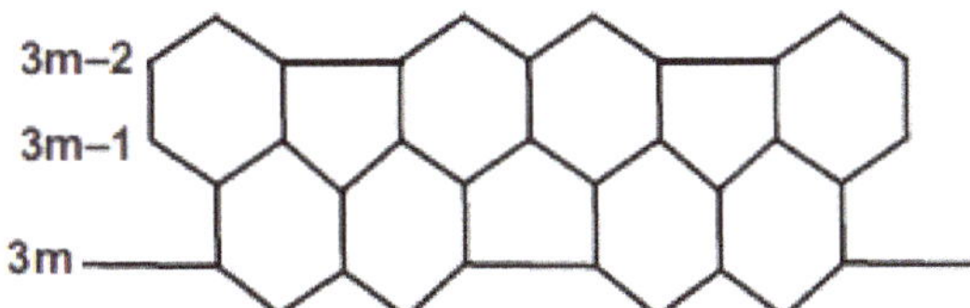

Figure 4. The mth period of $HAC_5C_6C_7[p,q]$ Nanotube.

4.1. Methodology of Carbon Graphite $HAC_5C_6C_7[p,q]$ Formulas

For the Nanotube $HAC_5C_6C_7[p,q], (p,q \geq 1)$ (see Figure 2), we have $V(HAC_5C_6C_7[p,q]) = 8pq + p$ and $E(HAC_5C_6C_7[p,q]) = 12pq - p$, and its edge set can be partitioned as follows.

$$E_1(HAC_5C_6C_7[p,q]) = \{sr \in E(HAC_5C_6C_7[p,q]) \mid dgr(s) = 6, dgr(r) = 7\}$$
$$E_2(HAC_5C_6C_7[p,q]) = \{sr \in E(HAC_5C_6C_7[p,q]) \mid dgr(s) = 6, dgr(r) = 8\}$$
$$E_3(HAC_5C_6C_7[p,q]) = \{sr \in E(HAC_5C_6C_7[p,q]) \mid dgr(s) = 7, dgr(r) = 8\}$$
$$E_4(HAC_5C_6C_7[p,q]) = \{sr \in E(HAC_5C_6C_7[p,q]) \mid dgr(s) = 8, dgr(r) = 8\}$$
$$E_5(HAC_5C_6C_7[p,q]) = \{sr \in E(HAC_5C_6C_7[p,q]) \mid dgr(s) = 8, dgr(r) = 9\}$$
$$E_6(HAC_5C_6C_7[p,q]) = \{sr \in E(HAC_5C_6C_7[p,q]) \mid dgr(s) = 9, dgr(r) = 9\}$$

The cardinality of edges in $E_1(HAC_5C_6C_7[p,q])$, $E_2(HAC_5C_6C_7[p,q])$ and $E_5(HAC_5C_6C_7[p,q])$ are $4p$, the cardinality of edges in $E_3(HAC_5C_6C_7[p,q])$ and $E_4(HAC_5C_6C_7[p,q])$ are $2p$ while the cardinality of edges in $E_6\big(HAC_5C_6C_7[p,q]\big)$ are $24pq - 18p$. The representatives of these partitioned edge set are demonstrated in Figure 3, in which the edge set with color green, red, brown, blue, yellow and black are $E_1\big(HAC_5C_6C_7[p,q]\big)$, $E_2\big(HAC_5C_6C_7[p,q]\big)$, $E_3\big(HAC_5C_6C_7[p,q]\big)$, $E_4\big(HAC_5C_6C_7[p,q]\big)$, $E_5\big(HAC_5C_6C_7[p,q]\big)$ and $E_6\big(HAC_5C_6C_7[p,q]\big)$ respectively.

4.2. Main Results for $HAC_5C_6C_7[p,q]$ Nanotube

In this section, we derive hyper-Zagreb index HM, first multiple Zagreb index PM_1, second multiple Zagreb index PM_2 and Zagreb polynomials for $HAC_5C_6C_7[p,q]$ Nanotube.

- **Hyper Zagreb index of $HAC_5C_6C_7[p,q]$ Nanotube**

Let G be the $HAC_5C_6C_7[p,q]$ Nanotube. Then by Equation (1), we have

$$
\begin{aligned}
HM(G) \;=\;& \sum_{sr\in E(G)} \big[dgr(s)+dgr(r)\big]^2 \\
HM(G) \;=\;& \sum_{sr\in E_1} \big[dgr(s)+dgr(r)\big]^2 + \sum_{sr\in E_2} \big[dgr(s)+dgr(r)\big]^2 + \sum_{sr\in E_3} \big[dgr(s)+dgr(r)\big]^2 \\
&+ \sum_{sr\in E_4} \big[dgr(s)+dgr(r)\big]^2 + \sum_{sr\in E_5} \big[dgr(s)+dgr(r)\big]^2 + \sum_{sr\in E_6} \big[dgr(s)+dgr(r)\big]^2 \\
=\;& 13^2|E_1| + 14^2|E_2| + 15^2|E_3| + 16^2|E_4| + 17^2|E_5| + 18^2|E_6| \\
=\;& 169(4p) + 196(4p) + 225(2p) + 256(2p) + 289(4p) + 324(24pq - 18p) \\
=\;& 7776pq - 2254p
\end{aligned}
$$

- **Multiple Zagreb indices of $HAC_5C_6C_7[p,q]$ Nanotube**

Let G be the $HAC_5C_6C_7[p,q]$ Nanotube. Then by Equations (2) and (3), we have

$$
\begin{aligned}
PM_1(G) \;=\;& \prod_{sr\in E(G)} \big[dgr(s)+dgr(r)\big] \\
PM_1(G) \;=\;& \prod_{sr\in E_1} \big[dgr(s)+dgr(r)\big] \times \prod_{sr\in E_2} \big[dgr(s)+dgr(r)\big] \times \prod_{sr\in E_3} \big[dgr(s)+dgr(r)\big] \\
&\times \prod_{sr\in E_4} \big[dgr(s)+dgr(r)\big] \times \prod_{sr\in E_5} \big[dgr(s)+dgr(r)\big] \times \prod_{sr\in E_6} \big[dgr(s)+dgr(r)\big] \\
=\;& 13^{|E_1|} \times 14^{|E_2|} \times 15^{|E_3|} \times 16^{|E_4|} \times 17^{|E_5|} \times 18^{|E_6|} \\
=\;& 13^{4p} \times 14^{4p} \times 15^{2p} \times 16^{2p} \times 17^{4p} \times 18^{(24pq-18p)}
\end{aligned}
$$

$$
\begin{aligned}
PM_2(G) \;=\;& \prod_{sr\in E(G)} \big[dgr(s)\times dgr(r)\big] \\
PM_2(G) \;=\;& \prod_{sr\in E_1} \big[dgr(s)\times dgr(r)\big] \times \prod_{sr\in E_2} \big[dgr(s)\times dgr(r)\big] \times \prod_{sr\in E_3} \big[dgr(s)\times dgr(r)\big] \\
&\times \prod_{sr\in E_4} \big[dgr(s)\times dgr(r)\big] \times \prod_{sr\in E_5} \big[dgr(s)\times dgr(r)\big] + \prod_{sr\in E_6} \big[dgr(s)\times dgr(r)\big] \\
=\;& 42^{|E_1|} \times 48^{|E_2|} \times 56^{|E_3|} \times 64^{|E_4|} \times 72^{|E_5|} \times 81^{|E_6|} \\
=\;& 42^{4p} \times 48^{4p} \times 56^{2p} \times 64^{2p} \times 72^{4p} \times 81^{(24pq-18p)}
\end{aligned}
$$

- **Zagreb polynomials of $HAC_5C_6C_7[p, q]$ Nanotube**

Let G be the graph of $HAC_5C_6C_7[p, q]$ Nanotube. Then by Equations (4) and (5), we have

$$
M_1(G, x) = \sum_{sr \in E(G)} x^{[dgr(s)+dgr(r)]}
$$

$$
M_1(G, x) = \sum_{sr \in E_1} x^{[dgr(s)+dgr(r)]} + \sum_{sr \in E_2} x^{[dgr(s)+dgr(r)]} + \sum_{sr \in E_3} x^{[dgr(s)+dgr(r)]}
$$

$$
+ \sum_{sr \in E_4} x^{[dgr(s)+dgr(r)]} + \sum_{sr \in E_5} x^{[dgr(s)+dgr(r)]} + \sum_{sr \in E_6} x^{[dgr(s)+dgr(r)]}
$$

$$
= \sum_{sr \in E_1} x^{13} + \sum_{sr \in E_2} x^{14} + \sum_{sr \in E_3} x^{16} + \sum_{sr \in E_4} x^{16} + \sum_{sr \in E_5} x^{17} + \sum_{sr \in E_6} x^{18}
$$

$$
= |E_1|x^{13} + |E_2|x^{14} + |E_3|x^{15} + |E_4|x^{16} + |E_5|x^{17} + |E_6|x^{18}
$$

$$
= 4px^{13} + 4px^{14} + 2px^{15} + 2px^{16} + 4px^{17} + (24pq - 18p)x^{18}
$$

$$
M_2(G, x) = \sum_{sr \in E(G)} x^{[dgr(s) \times dgr(r)]}
$$

$$
M_2(G, x) = \sum_{sr \in E_1} x^{[dgr(s) \times dgr(r)]} + \sum_{sr \in E_2} x^{[dgr(s) \times dgr(r)]} + \sum_{sr \in E_3} x^{[dgr(s) \times dgr(r)]}
$$

$$
+ \sum_{sr \in E_4} x^{[dgr(s) \times dgr(r)]} + \sum_{sr \in E_5} x^{[dgr(s) \times dgr(r)]} + \sum_{sr \in E_6} x^{[dgr(s) \times dgr(r)]}
$$

$$
= \sum_{sr \in E_1} x^{42} + \sum_{sr \in E_2} x^{48} + \sum_{sr \in E_3} x^{56} + \sum_{sr \in E_4} x^{64} + \sum_{sr \in E_5} x^{72} + \sum_{sr \in E_6} x^{81}
$$

$$
= |E_1|x^{42} + |E_2|x^{48} + |E_3|x^{56} + |E_4|x^{64} + |E_5|x^{72} + |E_6|x^{81}
$$

$$
= 4px^{42} + 4px^{48} + 2px^{56} + 2px^{64} + 4px^{72} + (24pq - 18p)x^{81}
$$

5. $TUC_4C_8[p, q]$ Nanotube and Nanotorus

We will use the notations and notions of Diudea and Graovac, and the 2D lattice of $TUC_4C_8[p, q]$ Nanotorus is denoted by $KTUC[p, q]$ (see Figure 5) and the $TUC_4C_8[p, q]$ Nanotube is denoted by $GTUC[p, q]$ (see Figure 6). A $TUC_4C_8[p, q]$ Nanotube is constructed in such a way that the total cardinality of octagons in each row equalsp p and the total cardinality of octagons in each column equals q. An example is presented in Figure 6. In $TUC_4C_8[p, q]$ Nanotube, the total cardinality of octagons and squares are the same as those in each row, and in $TUC_4C_8[p, q]$ Nanotorus the total cardinality of octagons and squares are the same as those in each row and column. In 2D lattice of $TUC_4C_8[p, q]$ Nanotorus, the total cardinality of squares in rows and columns are $(p + 1)$ and $(q + 1)$, respectively (cf. [30,31]).

The cardinalities of the vertex and edge set of $KTUC[p, q]$ and $GTUC[p, q]$ are presented in the following Table 1.

Table 1. Order and size of Nanotorus $KTUC[p, q]$ and Nanotube $GTUC[p, q]$.

$TUC_4C_8[p, q]$	$KTUC[p, q]$	$GTUC[p, q]$
$|V|$	$(4p^2 + 4p)(q + 1)$	$4pq + 4p$
$|E|$	$6pq + 5p + 5q + 4$	$6pq + 5p$

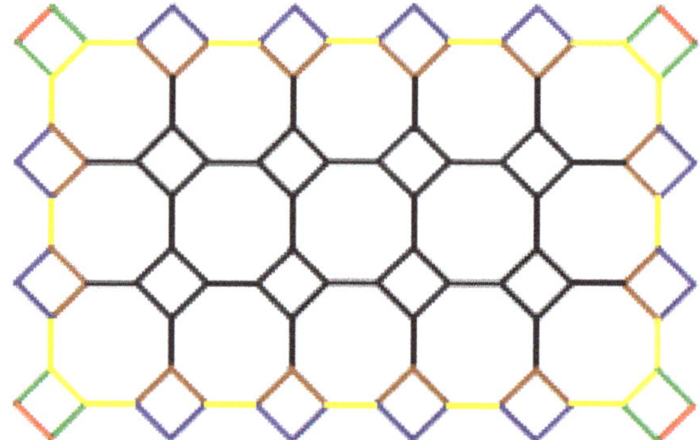

Figure 5. 2D-lattice of $TUC_4C_8(R)[p,q]$ Nanotorus with $p = 5$ and $q = 3$.

5.1. Methodology of $KTUC[p,q]$, $(p,q \geq 1)$ Nanotorus Formulas

For the Nanotorus $KTUC[p,q]$, $(p,q \geq 1)$, we have that the number of vertices in $KTUC[p,q]$ is $4p(p+1)(q+1)$ and the number of edges is $6pq + 5(p+q) + 4$. The edge set can be partitioned into the following six disjoint sets:

$$E_1(KTUC[p,q]) = \{sr \in E(KTUC[p,q]) \mid dgr(s) = 5, dgr(r) = 5\}$$
$$E_2(KTUC[p,q]) = \{sr \in E(KTUC[p,q]) \mid dgr(s) = 5, dgr(r) = 8\}$$
$$E_3(KTUC[p,q]) = \{sr \in E(KTUC[p,q]) \mid dgr(s) = 6, dgr(r) = 8\}$$
$$E_4(KTUC[p,q]) = \{sr \in E(KTUC[p,q]) \mid dgr(s) = 8, dgr(r) = 8\}$$
$$E_5(KTUC[p,q]) = \{sr \in E(KTUC[p,q]) \mid dgr(s) = 8, dgr(r) = 9\}$$
$$E_6(KTUC[p,q]) = \{sr \in E(KTUC[p,q]) \mid dgr(s) = 9, dgr(r) = 9\}$$

We can obtain that $|E_1(KTUC[p,q])| = 4$, $|E_2(KTUC[p,q])| = 8$, $|E_3(KTUC[p,q])| = 4(p + q - 2)$, $|E_4(KTUC[p,q])| = 2(p + q + 2)$, $|E_5(KTUC[p,q])| = 4(p + q - 2)$ and $|E_6(KTUC[p,q])| = 6pq - 5p - 5q + 4$, and the representatives of these partitioned edge set are demonstrated in Figure 5, in which the edge set with color red, green, blue, yellow, brown and black are $E_1(KTUC[p,q])$, $E_2(KTUC[p,q])$, $E_3(KTUC[p,q])$, $E_4(KTUC[p,q])$, $E_5(KTUC[p,q])$ and $E_6(KTUC[p,q])$ respectively.

5.2. Main Results for $KTUC[p,q]$, $(p,q \geq 1)$ Nanotorus

- **Hyper Zagreb index of $KTUC[p,q]$, $(p,q \geq 1)$ Nanotorus**

Let $G = KTUC[p,q]$. Now using Equation (1), we have

$$HM(G) = \sum_{sr \in E(G)} [dgr(s) + dgr(r)]^2$$

$$HM(K) = \sum_{sr \in E_1} [dgr(s) + dgr(r)]^2 + \sum_{sr \in E_2} [dgr(s) + dgr(r)]^2 + \sum_{sr \in E_3} [dgr(s) + dgr(r)]^2$$
$$+ \sum_{sr \in E_4} [dgr(s) + dgr(r)]^2 + \sum_{sr \in E_5} [dgr(s) + dgr(r)]^2 + \sum_{sr \in E_6} [dgr(s) + dgr(r)]^2$$
$$= 10^2|E_1| + 13^2|E_2| + 14^2|E_3| + 16^2|E_4| + 17^2|E_5| + 18^2|E_6|$$
$$= 100(4) + 169(8) + 196(4p + 4q - 8) + 256(2p + 2q + 4)$$
$$+ 289(4p + 4q - 8) + 324(6pq - 5p - 5q + 4)$$
$$= 1944pq + 832(p + q) + 192$$

- **Multiple Zagreb indices of $KTUC[p, q], (p, q \geq 1)$ Nanotorus**

  Let $G = KTUC[p, q]$. Now using Equations (2) and (3) we have

$$
\begin{aligned}
PM_1(G) &= \prod_{sr \in E(G)} [dgr(s) + dgr(r)] \\
PM_1(K) &= \prod_{sr \in E_1} [dgr(s) + dgr(r)] \times \prod_{sr \in E_2} [dgr(s) + dgr(r)] \times \prod_{sr \in E_3} [dgr(s) + dgr(r)] \\
&\quad \times \prod_{sr \in E_4} [dgr(s) + dgr(r)] \times \prod_{sr \in E_5} [dgr(s) + dgr(r)] \times \prod_{sr \in E_6} [dgr(s) + dgr(r)] \\
&= 10^{|E_1|} \times 13^{|E_2|} \times 14^{|E_3|} \times 16^{|E_4|} \times 17^{|E_5|} \times 18^{|E_6|} \\
&= 10^4 \times 13^8 \times 14^{(4p+4q-8)} \times 16^{(2p+2q+4)} \times 17^{(4p+4q-8)} \times 18^{(6pq-5p-5q+4)}
\end{aligned}
$$

$$
\begin{aligned}
PM_2(G) &= \prod_{sr \in E(G)} [dgr(s) \times dgr(r)] \\
PM_2(K) &= \prod_{sr \in E_1} [dgr(s) \times dgr(r)] \times \prod_{sr \in E_2} [dgr(s) \times dgr(r)] \times \prod_{sr \in E_3} [dgr(s) \times dgr(r)] \\
&\quad \times \prod_{sr \in E_4} [dgr(s) \times dgr(r)] \times \prod_{sr \in E_5} [dgr(s) \times dgr(r)] \times \prod_{sr \in E_6} [dgr(s) \times dgr(r)] \\
&= 25^{|E_1|} \times 40^{|E_2|} \times 48^{|E_3|} \times 64^{|E_4|} \times 72^{|E_5|} \times 81^{|E_6|} \\
&= 25^4 \times 40^8 \times 48^{(4p+4q-8)} \times 64^{(2p+2q+4)} \times 72^{(4p+4q-8)} \times 81^{(6pq-5p-5q+4)}
\end{aligned}
$$

- **Zagreb polynomials of $KTUC[p, q], (p, q \geq 1)$ Nanotorus**

  Let $G = KTUC[p, q]$. Now using Equations (4) and (5) we have

$$
\begin{aligned}
M_1(G, x) &= \sum_{sr \in E(G)} x^{[dgr(s) + dgr(r)]} \\
M_1(K, x) &= \sum_{sr \in E_1} x^{[dgr(s) + dgr(r)]} + \sum_{sr \in E_2} x^{[dgr(s) + dgr(r)]} + \sum_{sr \in E_3} x^{[dgr(s) + dgr(r)]} \\
&\quad + \sum_{sr \in E_4} x^{[dgr(s) + dgr(r)]} + \sum_{sr \in E_5} x^{[dgr(s) + dgr(r)]} + \sum_{sr \in E_6} x^{[dgr(s) + dgr(r)]}
\end{aligned}
$$

$$
\begin{aligned}
M_1(K, x) &= \sum_{sr \in E_1} x^{10} + \sum_{sr \in E_2} x^{13} + \sum_{sr \in E_3} x^{14} + \sum_{sr \in E_4} x^{16} + \sum_{sr \in E_5} x^{17} + \sum_{sr \in E_6} x^{18} \\
&= |E_1| x^{10} + |E_2| x^{13} + |E_3| x^{14} + |E_4| x^{16} + |E_5| x^{17} + |E_6| x^{18} \\
&= 4x^{10} + 8x^{13} + (4p + 4q - 8)x^{14} + (2p + 2q + 4)x^{16} \\
&\quad + (4p + 4q - 8)x^{17} + (6pq - 5p - 5q + 4)x^{18}
\end{aligned}
$$

$$
\begin{aligned}
M_2(G, x) &= \sum_{sr \in E(G)} x^{[dgr(s) \times dgr(r)]} \\
M_2(K, x) &= \sum_{sr \in E_1} x^{[dgr(s) \times dgr(r)]} + \sum_{sr \in E_2} x^{[dgr(s) \times dgr(r)]} + \sum_{sr \in E_3} x^{[dgr(s) \times dgr(r)]} \\
&\quad + \sum_{sr \in E_4} x^{[dgr(s) \times dgr(r)]} + \sum_{sr \in E_5} x^{[dgr(s) \times dgr(r)]} + \sum_{sr \in E_6} x^{[dgr(s) \times dgr(r)]} \\
&= \sum_{sr \in E_1} x^{25} + \sum_{sr \in E_2} x^{40} + \sum_{sr \in E_3} x^{48} + \sum_{sr \in E_4} x^{64} + \sum_{sr \in E_5} x^{72} + \sum_{sr \in E_6} x^{81} \\
&= |E_1| x^{25} + |E_2| x^{40} + |E_3| x^{48} + |E_4| x^{64} + |E_5| x^{72} + |E_6| x^{81} \\
&= 4x^{25} + 8x^{40} + (4p + 4q - 8)x^{48} + (2p + 2q + 4)x^{64} \\
&\quad + (4p + 4q - 8)x^{72} + (6pq - 5p - 5q + 4)x^{81}
\end{aligned}
$$

5.3. Methodology and Results of $GTUC[p,q], (p,q \geq 1)$ Nanotube Formulas

For the Nanotube $GTUC[p,q], (p,q \geq 1)$, we know that the number of vertices in $GTUC[p,q]$ are $4p(q+1)$ and the number of edges are $6pq+5p$. The edge set can be partitioned into the following four disjoint sets:

$$E_1(GTUC[p,q]) = \{sr \in E(GTUC[p,q]) \mid dgr(s) = 6, dgr(r) = 8\}$$
$$E_2(GTUC[p,q]) = \{sr \in E(GTUC[p,q]) \mid dgr(s) = 8, dgr(r) = 8\}$$
$$E_3(GTUC[p,q]) = \{sr \in E(GTUC[p,q]) \mid dgr(s) = 8, dgr(r) = 9\}$$
$$E_4(GTUC[p,q]) = \{sr \in E(GTUC[p,q]) \mid dgr(s) = 9, dgr(r) = 9\}$$

The cardinality of edges in $E_1(GTUC[p,q])$ are $4p$, in $E_2(GTUC[p,q])$ are $2p$, in $E_3(GTUC[p,q])$ are $4p$ and in $E_4(GTUC[p,q])$ are $6pq-5p$. The representatives of these edge set partitions are shown in Figure 6 in which red, green, blue and black edges are the edges belong to $E_1(GTUC[p,q])$, $E_2(GTUC[p,q])$, $E_3(GTUC[p,q])$ and $E_4(GTUC[p,q])$ respectively. Now using Equations (1)–(5), we have

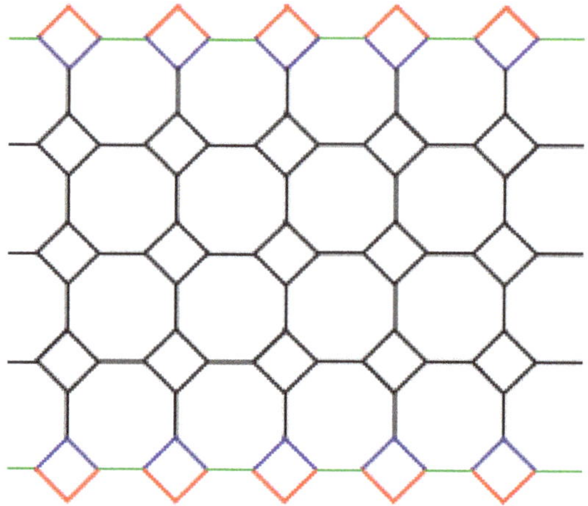

Figure 6. Nanotube $TUC_4C_8(R)[p,q]$ Nanotube with $p=5$ and $q=4$.

- **Hyper Zagreb index of $GTUC[p,q], (p,q \geq 1)$ Nanotube**

Let $G = GTUC[p,q]$. Now using Equation (1), we have

$$
\begin{aligned}
HM(G) &= \sum_{sr \in E(G)} [dgr(s)+dgr(r)]^2 \\
HM(G) &= \sum_{sr \in E_1} [dgr(s)+dgr(r)]^2 + \sum_{sr \in E_2} [dgr(s)+dgr(r)]^2 \\
&\quad + \sum_{sr \in E_3} [dgr(s)+dgr(r)]^2 + \sum_{sr \in E_4} [dgr(s)+dgr(r)]^2 \\
&= 14^2|E_1| + 16^2|E_2| + 17^2|E_3| + 18^2|E_4| \\
&= 196(4p) + 256(2p) + 289(4p) + 324(6pq-5p) \\
&= 1944pq + 832p
\end{aligned}
$$

- **Multiple Zagreb indices of $GTUC[p,q]$, $(p,q \geq 1)$ Nanotube**

Let $G = GTUC[p,q]$. Now using Equations (2) and (3) we have

$$
\begin{aligned}
PM_1(G) &= \prod_{sr \in E(G)} [dgr(s) + dgr(r)] \\
PM_1(G) &= \prod_{sr \in E_1} [dgr(s) + dgr(r)] \times \prod_{sr \in E_2} [dgr(s) + dgr(r)] \\
&\quad \times \prod_{sr \in E_3} [dgr(s) + dgr(r)] \times \prod_{sr \in E_4} [dgr(s) + dgr(r)] \\
&= 14^{|E_1|} \times 16^{|E_2|} \times 17^{|E_3|} \times 18^{|E_4|} \\
&= 14^{(4p)} \times 16^{(2p)} \times 17^{(4p)} \times 18^{(6pq-5p)} \\
PM_2(G) &= \prod_{sr \in E(G)} [dgr(s) \times dgr(r)] \\
PM_2(G) &= \prod_{sr \in E_1} [dgr(s) \times dgr(r)] \times \prod_{sr \in E_2} [dgr(s) \times dgr(r)] \\
&\quad \times \prod_{sr \in E_3} [dgr(s) \times dgr(r)] \times \prod_{sr \in E_4} [dgr(s) \times dgr(r)] \\
&= 48^{|E_1|} \times 64^{|E_2|} \times 72^{|E_3|} \times 81^{|E_4|} \\
&= 48^{(4p)} \times 64^{(2p)} \times 72^{(4p)} \times 81^{(6pq-5p)}
\end{aligned}
$$

- **Zagreb polynomials of $GTUC[p,q]$, $(p,q \geq 1)$ Nanotube**

Let $G = GTUC[p,q]$. Now using Equations (4) and (5) we have

$$
\begin{aligned}
M_1(G,x) &= \sum_{sr \in E(G)} x^{[dgr(s)+dgr(r)]} \\
M_1(G,x) &= \sum_{sr \in E_1} x^{[dgr(s)+dgr(r)]} + \sum_{sr \in E_2} x^{[dgr(s)+dgr(r)]} \\
&\quad + \sum_{sr \in E_3} x^{[dgr(s)+dgr(r)]} + \sum_{sr \in E_4} x^{[dgr(s)+dgr(r)]} \\
&= \sum_{sr \in E_1} x^{14} + \sum_{sr \in E_2} x^{16} + \sum_{sr \in E_3} x^{17} + \sum_{sr \in E_4} x^{18}
\end{aligned}
$$

$$
\begin{aligned}
M_1(G,x) &= |E_1|x^{14} + |E_2|x^{16} + |E_3|x^{17} + |E_4|x^{18} \\
&= (4p)x^{14} + (2p)x^{16} + (4p)x^{17} + (6pq-5p)x^{18} \\
M_2(G,x) &= \sum_{sr \in E(G)} x^{[dgr(s) \times dgr(r)]} \\
M_2(G,x) &= \sum_{sr \in E_1} x^{[dgr(s) \times dgr(r)]} + \sum_{sr \in E_2} x^{[dgr(s) \times dgr(r)]} \\
&\quad + \sum_{sr \in E_3} x^{[dgr(s) \times dgr(r)]} + \sum_{sr \in E_4} x^{[dgr(s) \times dgr(r)]} \\
&= \sum_{sr \in E_1} x^{48} + \sum_{sr \in E_2} x^{64} + \sum_{sr \in E_3} x^{72} + \sum_{sr \in E_4} x^{81} \\
&= |E_1|x^{48} + |E_2|x^{64} + |E_3|x^{72} + |E_4|x^{81} \\
&= (4p)x^{48} + (2p)x^{64} + (4p)x^{72} + (6pq-5p)x^{81}
\end{aligned}
$$

6. Comparisons and Discussion

- Firstly, we have obtained some indices of $HAC_5C_7[p,q]$ Nanotube for any p and q. Now from Table 2, it can be seen that all indices are in increasing order as the values of p,q increase. Finally, we depicted the the graphical representation of $HAC_5C_7[p,q]$ Nanotube for hyper Zagreb index, first and second multiple Zagreb index in Figure 7 and for first and second Zagreb polynomial in Figure 8.

Table 2. Comparison of all indices for $HAC_5C_7[p,q]$ Nanotube.

$[p,q]$	$HM(G)$	$PM_1(G)$	$PM_2(G)$
$[1,1]$	2796	2.11×10^{11}	3.4×10^{15}
$[2,2]$	13,368	3.31×10^{25}	4.5×10^{28}
$[3,3]$	31,716	4.21×10^{55}	6.61×10^{62}
$[4,4]$	57,840	6.57×10^{95}	8.72×10^{98}

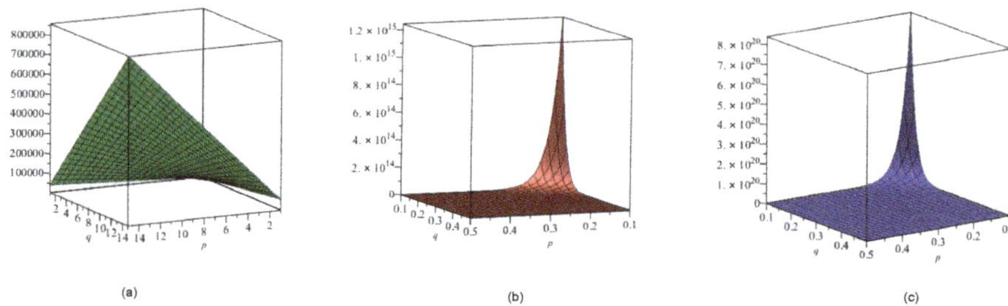

Figure 7. (**a**) Hyper Zagreb index; (**b**) First multiple Zagreb index; (**c**) Second multiple Zagreb index.

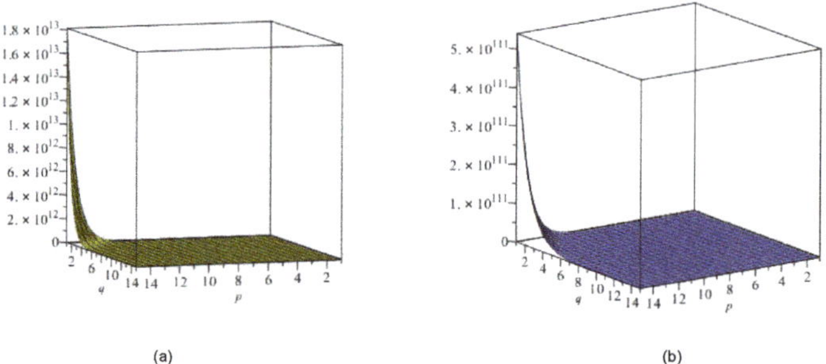

Figure 8. (**a**) First Zagreb polynomial; (**b**) Second multiple Zagreb polynomial.

- Secondly, we have worked out many indices of $HAC_5C_6C_7[p,q]$ Nanotube for each p and q. Now from Table 3, we can easily see that all indices are in increasing order as the values of p,q increase. Finally, we gave the the graphical representation of $HAC_5C_6C_7[p,q]$ Nanotube for hyper Zagreb index, first and second multiple Zagreb index in Figure 9 and for first and second Zagreb polynomial in Figure 10.

Table 3. Comparison of all indices for $HAC_5C_6C_7[p,q]$ Nanotube.

$[p,q]$	$HM(G)$	$PM_1(G)$	$PM_2(G)$
$[1,1]$	5522	3.4×10^{13}	4.5×10^{16}
$[2,2]$	26,596	5.3×10^{26}	6.5×10^{31}
$[3,3]$	63,222	6.31×10^{65}	7.62×10^{72}
$[4,4]$	115,400	7.57×10^{98}	9.82×10^{99}

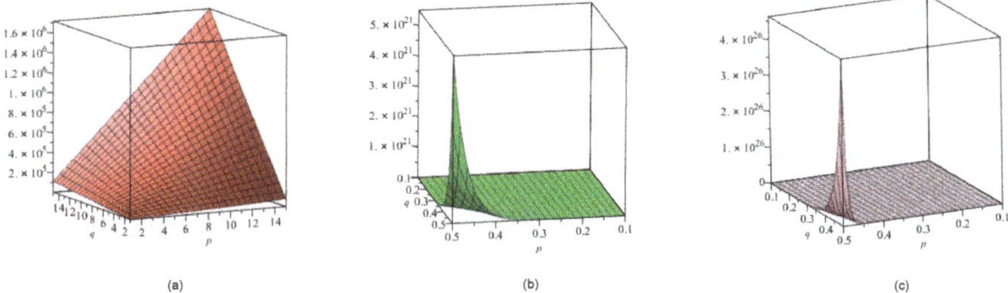

Figure 9. (**a**) Hyper Zagreb index; (**b**) First multiple Zagreb index; (**c**) Second multiple Zagreb index.

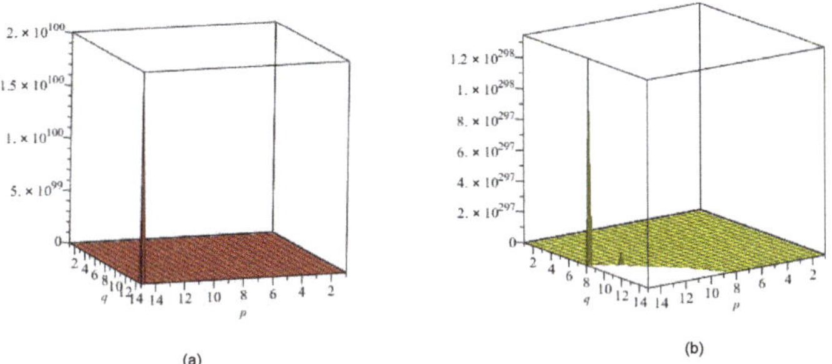

Figure 10. (**a**) First Zagreb polynomial; (**b**) Second multiple Zagreb polynomial.

- Now, we have worked out various indices of $KTUC[p,q], (p,q \geq 1)$ Nanotorus with different p and q. Now from Table 4, we have that each index increases with the values of p,q increasing. Finally, we depicted the the graphical representation of $KTUC[p,q], (p,q \geq 1)$ Nanotorus for hyper Zagreb index, first and second multiple Zagreb index in Figure 11 and for first and second Zagreb polynomial in Figure 12.

Table 4. Comparison of all indices for $KTUC[p,q], (p,q \geq 1)$ Nanotorus.

$[p,q]$	$HM(G)$	$PM_1(G)$	$PM_2(G)$
$[1,1]$	2996	3.2×10^{12}	4.5×10^{18}
$[2,2]$	11,296	4.6×10^{27}	5.7×10^{30}
$[3,3]$	22,680	6.8×10^{58}	6.21×10^{61}
$[4,4]$	37,952	8.7×10^{96}	7.8×10^{97}

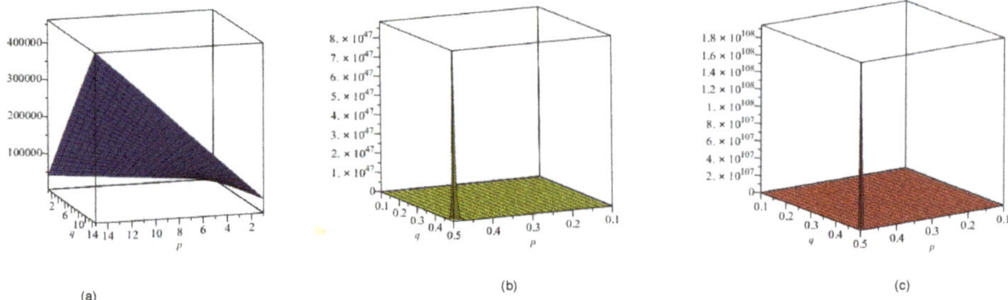

Figure 11. (**a**) Hyper Zagreb index; (**b**) First multiple Zagreb index; (**c**) Second multiple Zagreb index.

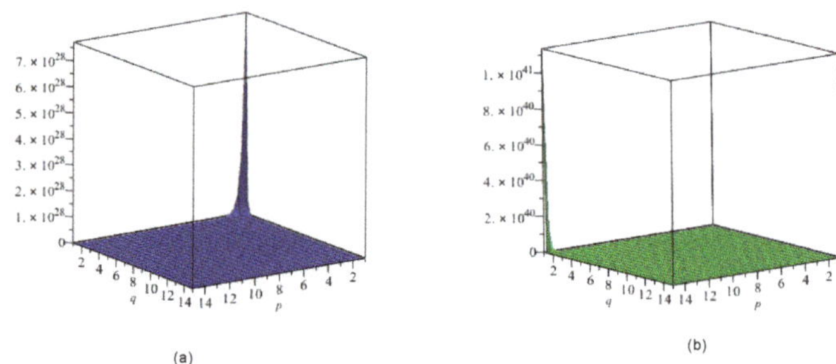

Figure 12. (**a**) First Zagreb polynomial; (**b**) Second multiple Zagreb polynomial.

- At the end of this section, we have computed substantial indices of $GTUC[p,q], (p,q \geq 1)$ Nanotube for different values of p, q. Now from Table 5, it can be seen that all indices are in increasing order as the values of p, q increase. We also provided the the graphical representation of $GTUC[p,q], (p,q \geq 1)$ Nanotube for hyper Zagreb index, first and second multiple Zagreb index in Figure 13 and for first and second Zagreb polynomial in Figure 14.

Table 5. Comparison of all indices for $GTUC[p,q], (p,q \geq 1)$ Nanotube.

$[p,q]$	$HM(G)$	$PM_1(G)$	$PM_2(G)$
$[1,1]$	2776	1.2×10^{14}	2.3×10^{15}
$[2,2]$	9440	3.6×10^{21}	4.5×10^{25}
$[3,3]$	19,992	5.8×10^{53}	5.23×10^{51}
$[4,4]$	34,432	7.7×10^{95}	6.7×10^{96}

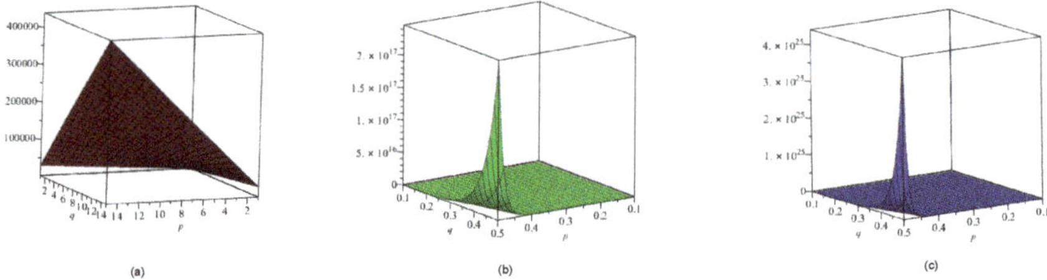

Figure 13. (**a**) Hyper Zagreb index; (**b**) First multiple Zagreb index; (**c**) Second multiple Zagreb index.

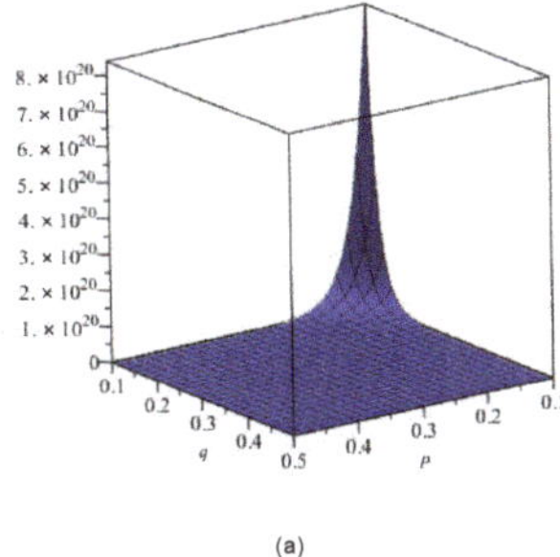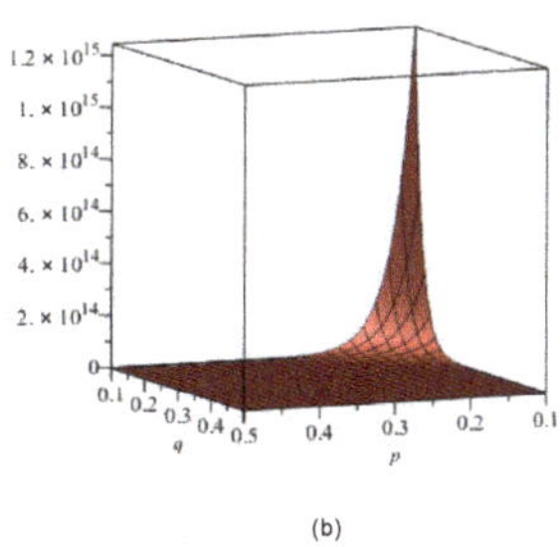

(a)

(b)

Figure 14. (**a**) First Zagreb polynomial; (**b**) Second multiple Zagreb polynomial.

7. Conclusions

In this paper, we computed various topological indices of Nanotubes. More precisely, we determined second multiple Zagreb index $PM_2(G)$, hyper-Zagreb index $HM(G)$, first multiple Zagreb index $PM_1(G)$, and Zagreb polynomials $M_1(G, x)$, $M_2(G, x)$ for certain Nanotubes. We conclude that the the the Zagreb indices are in increasing order as the values of p, q increase. In addition, the hyper Zagreb index gives a decent connection to the security of nonstructural objects and the stretched pharmacies, and for processing the strain vitality of Nanotubes. The first and second Zagreb polynomials are helpful to find the features of π-electron energy of the microscopic particles in the inner part of Nanostructural objects.

Author Contributions: Z. Shao contribute for conceptualization, designing the experiments, funding and analyzed the data curation. M.K. Siddiqui contribute for supervision, methodology, validation, project administration and formal analysing. M.H. Muhammad contribute for performed experiments, resources, software, some computations and wrote the initial draft of the paper which were investigated and approved by Z. Shao and M.K. Siddiqui and wrote the final draft. All authors read and approved the final version of the paper.

Funding: This research work is supported by Applied Basic Research (Key Project) of Sichuan Province under Grant No. 2017JY0095.

Conflicts of Interest: The authors declare no conflict of interest.

References

1. Naji, A.M.; Soner, N.D.; Gutman, I. On leap Zagreb indices of graphs. *Commun. Comb. Optim.* **2017**, *2*, 99–117.
2. Shao, Z.; Wu, P.; Gao, Y.; Gutman, I.; Zhang, X. On the maximum ABC index of graphs without pendent vertices. *Appl. Math. Comput.* **2017**, *315*, 298–312. [CrossRef]
3. Shao, Z.; Wu, P.; Zhang, X.; Dimitrov, D.; Liu, J. On the maximum ABC index of graphs with prescribed size and without pendent vertices. *IEEE Access* **2018**, *6*, 27604–27616. [CrossRef]
4. Wiener, H. Structural determination of paraffin boiling points. *J. Am. Chem. Soc.* **1947**, *69*, 17–20. [CrossRef] [PubMed]
5. Dobrynin, A.A.; Entringer, R.; Gutman, I. Wiener index of trees: Theory and applications. *Acta Appl. Math.* **2001**, *66*, 211–249. [CrossRef]
6. Gutman, I.; Trinajstic, N. Graph theory and molecular orbitals. Total ϕ-electron energy of alternant hydrocarbons. *Chem. Phys. Lett.* **1972**, *17*, 535–538. [CrossRef]
7. Shirdel, G.H.; RezaPour, H.; Sayadi, A.M. The hyper-Zagreb Index of Graph Operations. *Iran. J. Math. Chem.* **2013**, *4*, 213–220.
8. Ghorbani, M.; Azimi, N. Note on multiple Zagreb indices. *Iran. J. Math. Chem.* **2012**, *3*, 137–143.
9. Eliasi, M.; Iranmanesh, A.; Gutman, I. Multiplicative version of first Zagreb index. *MATCH Commun. Math. Comput. Chem.* **2012**, *68*, 217–230.
10. Furtula, B.; Gutman, I.; Dehmer, M. On structure-sensitivity of degree-based topological indices. *Appl. Math. Comput.* **2013**, *219*, 8973–8978. [CrossRef]

11. Gutman, I.; Das, K.C. Some Properties of the Second Zagreb Index. *MATCH Commun. Math. Comput. Chem.* **2004**, *50*, 103–112.

12. Farahani, M.R. The hyper zegreb index of $TUSC_4C_8(S)$ Nanotubes. *Int. J. Eng. Technol. Res.* **2015**, *3*, 1–6.

13. Hayat, S.; Imran, M. Computation of topological indices of certain networks. *Appl. Math. Comput.* **2014**, *240*, 213–228. [CrossRef]

14. Iranmanesh, A.; Alizadeh, Y. Computing wiener index of $HAC_5C_7[p,q]$ Nanotubes by gap program. *Iran. J. Math. Sci. Inform.* **2008**, *3*, 1–12.

15. Iranmanesh, A.; Zeraatkar, M. Computing ga index of $HAC_5C_7[p,q]$ and $HAC_5C_6C_7[p,q]$ Nanotubes. *Optoelectron. Adv. Mater.-Rapid Commun.* **2011**, *5*, 790–792.

16. Gao, W.; Siddiqui, M.K.; Naeem, M.; Rehman, N.A. Topological Characterization of Carbon Graphite and Crystal Cubic Carbon Structures. *Molecules* **2017**, *22*, 1496. [CrossRef] [PubMed]

17. Bača, M.; Horváthová, J.; Mokrišová, M.; Suhányiová, A. On topological indices of fullerenes. *Appl. Math. Comput.* **2015**, *251*, 154–161. [CrossRef]

18. Bača, M.; Horváthová, J.; Mokrišová, M.; Semanicová-Fenovcíková, A.; Suhányiová, A. On topological indices of carbon Nanotube network. *Can. J. Chem.* **2015**, *93*, 1–4. [CrossRef]

19. Gao, W.; Farahani, M.R.; Siddiqui, M.K.; Jamil, M.K. On the First and Second Zagreb and First and Second Hyper-Zagreb Indices of Carbon Nanocones $CNC_k[n]$. *J. Comput. Theor. Nanosci.* **2016**, *13*, 7475–7482. [CrossRef]

20. Gao, W.; Farahani, M.R.; Jamil, M.K.; Siddiqui, M.K. The Redefined First, Second and Third Zagreb Indices of Titania Nanotubes $TiO_2[m,n]$. *Open Biotechnol. J.* **2016**, *10*, 272–277. [CrossRef]

21. Gao, W.; Siddiqui, M.K.; Imran, M.; Jamil, M.K.; Farahani, M.R. Forgotten Topological Index of Chemical Structure in Drugs. *Saudi Pharm. J.* **2016**, *24*, 258–267. [CrossRef] [PubMed]

22. Imran, M.; Hayat, S.; Mailk, M.Y.H. On topological indices of certain interconnection networks. *Appl. Math. Comput.* **2014**, *244*, 936–951. [CrossRef]

23. Hayat, S.; Imran, M. Computation of certain topological indices of Nanotubes covered by C_5 and C_7. *J. Comput. Theor. Nanosci.* **2015**, *12*, 1–9.

24. Imran, M.; Siddiqui, M.K.; Naeem, M.; Iqbal, M.A. On Topological Properties of Symmetric Chemical Structures. *Symmetry* **2018**, *10*, 173. [CrossRef]

25. Siddiqui, M.K.; Gharibi, W. On Zagreb Indices, Zagreb Polynomials of Mesh Derived Networks. *J. Comput. Theor. Nanosci.* **2016**, *13*, 8683–8688. [CrossRef]

26. Siddiqui, M.K.; Imran, M.; Ahmad, A. On Zagreb indices, Zagreb polynomials of some nanostar dendrimers. *Appl. Math. Comput.* **2016**, *280*, 132–139. [CrossRef]

27. Siddiqui, M.K.; Naeem, M.; Rahman, N.A.; Imran, M. Computing topological indicesof certain networks. *J. Optoelectron. Adv. Mater.* **2016**, *18*, 884–892.

28. Mahmiani, A.; Iranmanesh, A. Edge-Szeged Index of $HAC_5C_7[r,p]$ Nanotube. *MATCH Commun. Math. Comput. Chem.* **2009**, *62*, 397–417.

29. Yazdani, J.; Bahrami, A. Padmakar-Ivan, Omega and Sadhana Polynomial of $HAC_5C_6C_7$ Nanotubes. *Dig. J. Nanomater. Biostruct.* **2009**, *4*, 507–517.

30. Ashrafi, A.R.; Dosli, T.; Saheli, M. The eccentric connectivity index of $TUC_4C_8(R)$ Nanotubes. *MATCH Commun. Math. Comput. Chem.* **2011**, *65*, 221–230.

31. Ghorbani, M. Computing ABC_4 index of nanostar dendrimers. *Optoelectron. Adv. Mater.-Rapid Commun.* **2010**, *4*, 1419–1422.

Article

Intelligent Prognostics of Degradation Trajectories for Rotating Machinery Based on Asymmetric Penalty Sparse Decomposition Model

Qing Li [1,*] and Steven Y. Liang [1,2]

1   College of Mechanical Engineering, Donghua University, Shanghai 201620, China;
    steven.liang@me.gatech.edu
2   George W. Woodruff School of Mechanical Engineering, Georgia Institute of Technology,
    Atlanta, GA 30332-0405, USA
*   Correspondence: suesliqing@163.com

Received: 22 May 2018; Accepted: 8 June 2018; Published: 12 June 2018

Abstract: The ability to accurately track the degradation trajectories of rotating machinery components is arguably one of the challenging problems in prognostics and health management (PHM). In this paper, an intelligent prediction approach based on asymmetric penalty sparse decomposition (APSD) algorithm combined with wavelet neural network (WNN) and autoregressive moving average-recursive least squares algorithm (ARMA-RLS) is proposed for degradation prognostics of rotating machinery, taking the accelerated life test of rolling bearings as an example. Specifically, the health indicators time series (e.g., peak-to-peak value and Kurtosis) is firstly decomposed into low frequency component (LFC) and high frequency component (HFC) using the APSD algorithm; meanwhile, the resulting non-convex regularization problem can be efficiently solved using the majorization-minimization (MM) method. In particular, the HFC part corresponds to the stable change around the zero line of health indicators which most extensively occurs; in contrast, the LFC part is essentially related to the evolutionary trend of health indicators. Furthermore, the nonparametric-based method, i.e., WNN, and parametric-based method, i.e., ARMA-RLS, are respectively introduced to predict the LFC and HFC that focus on abrupt degradation regions (e.g., last 100 points). Lastly, the final predicted data could be correspondingly obtained by integrating the predicted LFC and predicted HFC. The proposed methodology is tested using degradation health indicator time series from four rolling bearings. The proposed approach performed favorably when compared to some state-of-the-art benchmarks such as WNN and largest Lyapunov (LLyap) methods.

Keywords: degradation trajectories prognostic; asymmetric penalty sparse decomposition (APSD); rolling bearings; wavelet neural network (WNN); recursive least squares (RLS); health indicators

---

1. Introduction

Rotating machines are critical elements of almost all forms of mechanical assemblies, which play an important role in today's industrial applications. Unexpected failures and unscheduled system interruptions caused by harsh working environments may result in costly lapse in production or even catastrophic incidents. Accurate/timely prediction and health assessment of rotating machines are of great significance in the functionality and performance of these equipment, especially in the early stages of failure [1–3].

Accordingly, in recent years, a plethora of research works have proposed tracking the degradation trend of rotating machines and predicting their remaining useful life (RUL), wherein condition-based prognostics maintenance (CBPM) has become an efficient strategy for the PHM of rotating machines. The CBPM is based on the data information collected via embedded sensors to assess and judge the real-time state of the mechanical assemblies, and then predict their degradation trend to develop appropriate decisions on maintenance activities before failure occurs. More advanced prognostics techniques are focused on degradation assessment of the systems, but it is also a relatively complex problem due to the non-stationary and nonlinear characteristic of data information [4–6]. Generally, to fulfill the goal of prognostics, three crucial steps are needed:

(1) The historical data with long-terms, including normal and abnormal time series data, should be collected using sensors and other portable testing devices. Typically, good degradation data must capture the physical transitions that the rotating machinery such as rolling bearing undergo during different stages of its running life; fortunately, most of the historical data are easy to collect since the rotating machinery is frequently updated in real time.

(2) The health indicators (e.g., peak-to-peak value and Kurtosis) should be extracted to assess and continuously track the health condition of rotating machines or system. The health indicators are used to design a suitable prediction model that captures the evolution of the degradation trend of rotating machines.

(3) The reliable prediction models and possible failure models should be established and the remaining useful life (RUL) or future trajectory of rotating machines could be effectively predicted.

Among these three steps, model establishment and trajectory estimation are the most important and challenging steps because of the stochastic nature and nonlinear characteristics of the health indicators. Therefore, many researchers have laid their emphasis on model foundation in recent years. From the view of application, roughly, the existing prediction approaches can be divided into two categories: (a) parametric-based methods and (b) nonparametric-based methods. In the literature, parametric-based methods mainly include time-series methods such as autoregressive moving average (ARMA) [7,8], fractional autoregressive integrated moving average (FARIMA) [9,10], fractional Brownian motion (FBM) [11,12], hidden Markov model (HMM) [13,14] and grey theoretical model (GTM) [15,16], etc. Generally, the parametric-based methods overcome the hurdle of predictive availability during long-term prediction (according to needs) which assumes the model's parameters to be constants in the predicted region. However, due to the stochastic, complex and nonlinear properties of health indicators time series (HITS), the trade-off of parametric-based method is that only accurate prediction with real-time can help adjust the system model. Otherwise, the prediction performance may not be ensured. In addition, the parametric-based methods are often developed case by case, thus model parameters identification also requires extensive experiments, and therefore, it is usually not desirable in practice.

To address this issue, much attention has recently been focused on the nonparametric-based methods in the degradation prognostic of rotating machinery. For instance, artificial neural network (ANN) [17–20], fuzzy logic [21], and deep learning network (DPN) [22,23], etc., could be considered as successful nonparametric-based approaches in the PHM field. The benefit of using nonparametric-based methods lies in their ability to model the evolution of complex multi-dimensional degradation data, which can effectively extract latent features such as the spatio-temporal correlations (STC) among historical data. In addition, for the nonparametric-based methods, it is not necessary to establish an accurate linkage between the reliability health indicators and physical degradation such as crack propagation in an individual element. However, the common drawbacks associated with these nonparametric-based techniques are time-consuming and a large number of training samples are required in advance.

Due to time-variant and non-stationary characteristics of the health indicators, and the disadvantages of the model, which is sometimes case-specific, it is difficult to determine the best prognostic model. Therefore, it is intuitive to develop a fusion idea via combining parametric-based methods and nonparametric-based methods to blend their merits and enhance the prediction performance of degradation trajectories. However, it is not clear how to fuse parametric-based and nonparametric-based models to improve the prediction accuracy, especially when health indicators become abundant and redundant.

It is generally thought that the health indicators time series (HITS) is commonly composed of different sub-components which correspond to different time-variant characteristics of the systems. In this work, we assume that the HITS is mainly composed of high frequency component (HFC) and low frequency component (LFC). The HFC part corresponds to the stable change around the zero line of the HITS which frequently occurs; in contrast, the LFC part is essentially related to the evolutionary trend of the HITS which rarely occurs in practice, because the stable running trend occupies most of the life time, only the drastic trend can be clearly emerged when the failures are serious. This can be explained using degradation trajectories of several rotating machines (e.g., rolling bearings) as shown in Figure 1, where the degradation process of the bearings generally consists of two phases, i.e., phase I: the normal operation phase and phase II: the failure phase. The degradation health indicator in phase I is stable, which means that the bearing is normal. The degradation process changes from phase I to phase II once a fault occurs, where the HITS generally increases as the fault gets worse, the curve that rises steadily could be assumed as the LFC; in contrast, the volatile components with large and small amplitude could be treated as the HFC. To ensure better prediction accuracy, it is necessary to isolate the LFC associated with bearing defects and HRC associated with structural vibration from the degradation trajectories, and then develop a degradation algorithm based on both components.

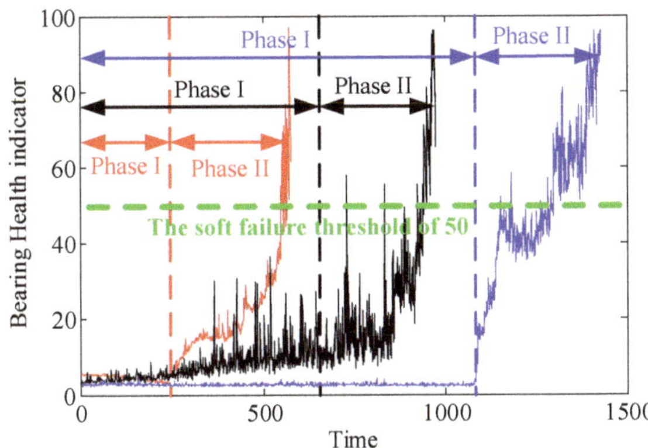

Figure 1. Typical bearing degradation trajectories described by the bearing health indicators.

To overcome this problem, a sparse low-rank matrix decomposition (SLMD) method is introduced in this research. In recent years, many researchers have laid their emphasis on fault diagnosis and sparse filtering using SLMD algorithm. For instance, in ref. [24], Selesnick et al. developed a convex low-rank matrix decomposition approach based on non-convex bivariate penalty function for sparse one-dimensional deconvolution. In ref. [25], He et al. proposed convex sparsity-based regularization scheme to extract multiple faults of motor bearing and locomotive bearing. In refs. [26,27], Li et al. proposed a bicomponent sparse low-rank matrix separation and group sparsity total variation de-noising approach to extract transient fault impulses from the noisy vibration signals, using the rolling bearing and gearbox as example. In ref. [28], Wang et al. utilized a sparse low-rank matrix decomposition method based on generalized minimax-concave (GMC) penalty to explore and diagnosis the localized faults in rolling bearings. In refs. [29,30], Parekh et al. proposed an enhanced low-rank matrix approximation and its effectiveness was verified by the synthetic data and nonlocal self-similarity based image. In ref. [31], Ding et al. proposed the sparse low-rank matrix decomposition (SLMD) algorithm based on split augmented Lagrangian shrinkage (SALS) algorithm with majorization-minimization (MM) to detect and extract the periodical oscillatory features of gearbox of an oil refinery.

Unfortunately, despite the effectiveness of the above approaches, the SLMD approaches still suffer from the following several challenges and their application development has been restricted:

(1) Generally, the penalty functions that are established in a low-rank matrix approximation model are symmetric function, e.g., absolute value function; the common drawback is that this penalty function is non-differentiable at zero point, which can lead to some numerical issues such as local optimum and early termination of algorithm.

(2) In conventional SLMD methods, the convex regularizer (e.g., L1-norm) which usually underestimates the sparse signal when the absolute value function is used as a sparsity regularizer. Additionally, both the convex and nonconvex methods shrink all the coefficients equally and remove too much energy of useful information, resulting in the separation of the LFC and HFC becoming more challenging.

To overcome these limitations and to robustly separate the LFC and HFC from degradation trajectories of health indicators, a novel asymmetric penalty sparse decomposition (APSD) algorithm with non-convex sparsity constraint is proposed in this work. The resulting non-convex regularization problem will be efficiently handled using the majorization-minimization (MM) method. Having the LFC and HFC separated, the LFC and HFC components (e.g., last 100 points) will be predicted using the wavelet neural network (WNN) and ARMA combined with recursive least squares algorithm (ARMA-RLS) methods, respectively. In an effort to establish a comprehensive assessment of degradation processes, the proposed fusion prognostics framework combines the WNN (a nonparametric-based approach) and the ARMA-RLS (a parametric-based approach) in an attempt to improve the prediction performance, in which the physical information of the degenerative process and massive sample data are not considered and accommodated. The final predicted data could be generated by combining the predicted LFC and HFC components. Finally, we demonstrate the proposed approach by four case studies of accelerated aging tests of rolling bearing.

The main contributions of this paper are summarized as follows:

(1) To address the issue of models merging and improve the prediction accuracy, the health indicators time series (HITS) is decomposed into different sub-components, i.e., low frequency component (LFC) and high frequency component (HFC), using the APSD algorithm.

(2) To address the drawback that penalty function is non-differentiable at zero point, a new asymmetric penalty function is proposed.

(3) To solve the proposed non-convex regularization problem based on asymmetric penalty function, the majorization-minimization (MM) algorithm is introduced.

(4) The decomposed LFC and HFC components can be predicted using the wavelet neural network (WNN) and ARMA combined with recursive least squares algorithm (ARMA-RLS) methods, respectively.

(5) The prediction accuracy is greatly improved compared with some state-of-the-art models, and the presented approach has powerful application potentials.

The frame of this paper is structured as follows. Section 2 contains the description of the asymmetric nonconvex sparse decomposition (APSD) algorithm. Section 3 provides the algorithms of degradation modeling, i.e., wavelet neural network and ARMA combined with recursive least squares algorithm (ARMA-RLS) methods. Experimental results and discussion are given in Section 4. Finally, Section 5 concludes with a discussion of future research.

2. Asymmetric Penalty Sparse Decomposition (APSD) Algorithm

2.1. Sparse Representation and Filter Banks

Generally, the fault vibration signal (observation signal) y of rotating machinery such as rolling bearing, which contains three parts below: the fault transient impulses x, the systematic natural signal f and the additive noise w, i.e.,

$$y = x + f + w \tag{1}$$

The core work of the fault diagnosis is to extract the fault transient impulses x from the noisy observation y. Assume that fault transient impulses and the systematic natural signal are estimated, we have,

$$\hat{x} \approx x, \ \hat{f} \approx f \tag{2}$$

Given an estimation $\hat{x}$ of x, we can estimate f as follows,

$$\hat{f} = LPF(y - \hat{x}) \tag{3}$$

where LPF is a specified low-pass filter. Substituting $\hat{f} = LPF(y - \hat{x})$ into $\hat{f} \approx f$, we have,

$$LPF(y - \hat{x}) \approx f \tag{4}$$

Substituting $y = x + f + w$ into $LPF(y - \hat{x}) \approx f$, we have,

$$LPF(y - \hat{x}) \approx y - x - w \tag{5}$$

Substituting $\hat{x} \approx x$ into $LPF(y - \hat{x}) \approx y - x - w$, we have,

$$LPF(y - \hat{x}) \approx y - \hat{x} - w \Leftrightarrow (I - LPF)(y - \hat{x}) \approx w \tag{6}$$

Defining $HPF = I - LPF = H$, which is a high-pass filter, thus we have,

$$HPF(y - \hat{x}) \approx w \tag{7}$$

On the other hand, it should be noted that Equation (1) is a highly underdetermined equation, i.e., ill-posed or $N - P$ hard problem [32,33], and has an infinite set of solutions because the number of unknowns is greater than the number of equations. Usually, convex optimization techniques are commonly used to estimate transient components from the observation signal, based on the aforementioned work [24–31], the estimation of x can be formulated as the optimization problem, i.e.,

$$\hat{x} = \underset{x}{\operatorname{argmin}} \left\{ \frac{1}{2} \|H(y - x)\|_2^2 + \lambda \|Dx\|_1 \right\} \tag{8}$$

where H is a specified high-pass filter, i.e., the HPF in Equation (7), λ represents regularization term parameter, D is a matrix defined as $D = \begin{bmatrix} -1, & 1 & & \\ & \ddots & \ddots & \\ & & -1, & 1 \end{bmatrix}$, which determines the sparsity degree of the approximating value of x. Commonly, if x is a sparse component, i.e., most of the values in x tend to zero, the correspondingly optimization problem in Equation (8) can be estimated by L1-norm fused lasso optimization (LFLO) model [24–31], i.e.,

$$\hat{x} = \underset{x}{\operatorname{argmin}} \left\{ \frac{1}{2} \|H(y-x)\|_2^2 + \lambda_0 \|x\|_1 + \lambda_1 \|Dx\|_1 \right\} \tag{9}$$

where λ_0 and λ_1 are regularization parameters. The solution of LFLO model can be obtained by the soft-threshold method [34] and total variation de-noising (TVD) algorithm [35–37], we have,

$$x = \operatorname{soft}(\operatorname{tvd}(y, \lambda_2), \lambda_1) \tag{10}$$

where function $\operatorname{tvd}(\cdot, \cdot)$ is the TVD algorithm and the soft-threshold model is given as follows,

$$\operatorname{soft}(x, \lambda) = \begin{cases} x - \lambda \frac{x}{|x|}, & |x| > \lambda \\ 0, & |x| \le \lambda \end{cases} \tag{11}$$

In addition, the high-pass filter H described above could be formulated as follows [36–38],

$$H = A^{-1}B \tag{12}$$

where A and B are Toeplitz matrices.

2.2. Asymmetric Penalty Regularization Model

Compare with optimization algorithm Equations (8) and (9), to estimate the fault transient impulses x precisely, this work introduces a novel penalty regularization method, i.e., asymmetric and symmetric nonconvex penalty regularization model,

$$\hat{x} = \underset{x}{\operatorname{argmin}} \left\{ F(x) = \frac{1}{2} \|H(y-x)\|_2^2 + \lambda_0 \sum_{n=0}^{N-1} \theta_\varepsilon(x_n; r) + \sum_{i=1}^{M} \lambda_i \sum_{n=0}^{N-1} \phi([D_i x]_n) \right\} \tag{13}$$

where $F(x)$ is the proposed objective cost function (OCF), penalty function $\theta_\varepsilon(x_n; r)$ is a asymmetric and differentiable function, and $\phi([D_i x]_n)$ is a symmetric and differentiable function, if term $M = 2$, matrix D_1 defined as $D_1 = \begin{bmatrix} -1, & 1 & & \\ & \ddots & & \\ & & \ddots & \\ & & -1, & 1 \end{bmatrix}$, and matrix D_2 defined as

$D_2 = \begin{bmatrix} -1, & 2, & -1 & \\ & -1, & 2, & -1, \\ & & \ddots & \\ & & -1, & 2, & -1 \end{bmatrix}$. The innovations of the novel compound regularizer model are as follows:

(I)   The M-term compound regularizers to estimate the fault transient impulses;
(II) The compound regularizer model consists of symmetric and asymmetric penalty functions, wherein the symmetric penalty function is a differentiable function compared with the nondifferentiable function $\|x_i\|$ at point $i = 0$.
(III) The MM algorithm is introduced for the solution of proposed compound regularization method, i.e., the OCF.

Based on this, the core issues of the proposed algorithm are (1) how to construct a symmetric and differentiable penalty function; (2) how to construct an asymmetric and differentiable penalty function and; (3) how to solve the proposed method based on the MM algorithm and make diagnosis results are more accurate than the traditional LFLO and nonconvex penalty regularization approaches.

For the first issue, traditional LFLO regularization approach uses the absolute value function $\phi_A(x) = \|x\|$ as the penalty function; however, the common drawback of function $\phi_A(x) = \|x\|$ is that this function is non-differentiable at zero point, which can cause some numerical problems. To address this problem, a non-linear approximation function of $\phi_B(x)$ or $\phi_c(x)$ is proposed, i.e.,

$$\phi_B(x) = \sqrt{|x|^2 + \varepsilon} \tag{14}$$

$$\phi_c(x) = |x| - \varepsilon \log(|x| + \varepsilon) \tag{15}$$

Note that when $\varepsilon = 0$, then the $\phi_B(x)$ and $\phi_c(x)$ degrade into the absolute value function $\phi_A(x)$, while the $\varepsilon > 0$, the $\phi_B(x)$ and $\phi_c(x)$ are differentiable at zero point. The function $\phi_A(x)$, $\phi_B(x)$, $\phi_c(x)$ and their first-order derivatives are listed in Table 1.

Table 1. Symmetric penalty functions and their derivatives.

Functions	$\phi(x)$	$\phi'(x)$
$\phi_A(x)$	$\|x\|$	Signal(x)
$\phi_B(x)$	$\sqrt{|x|^2 + \varepsilon}$	$x/\sqrt{|x|^2 + \varepsilon}$
$\phi_c(x)$	$|x| - \varepsilon \log(|x| + \varepsilon)$	$x/(|x| + \varepsilon)$

In order for the non-linear approximation functions to maintain the reliable sparsity-inducting behavior of the original LFLO algorithm, the parameter ε should be presented to an adequately small positive value. Fortunately, for example, the parameter $\varepsilon = 10^{-5}$ and $\varepsilon = 10^{-6}$ which are small enough so that the numerical issues can be avoided.

For the second issue, inspired by the absolute value function $\phi_A(x) = \|x\|$, and also in contrast to the symmetric and differentiable penalty function $\phi_B(x)$ and $\phi_c(x)$, here, a segmented function is proposed as follows,

$$\theta_\varepsilon(x) = \begin{cases} x, x > \varepsilon \\ f(x), \|x\| \le \varepsilon \\ -rx, x < \varepsilon \end{cases} \tag{16}$$

where $r > 0$ is a positive constant. It should be noted that if we exclude the intermediate function $f(x)$, the $\theta_\varepsilon(x)$ is also degrade into the absolute value function $\phi_A(x)$ when $r = 1$. Therefore, the main problem of Equation (16) will transform into a task of how to construct the intermediate function $f(x)$, $-\varepsilon \le x \le \varepsilon$. To address this issue, we seek a majorizer (here the Majorization-minimization algorithm is utilized) as the approximation function of $f(x)$, $-\varepsilon \le x \le \varepsilon$. In order to eliminate the issue that penalty function is non-differentiable at zero point, a simple quadratic equation (QE) is introduced accordingly,

$$g(x, v) = ax^2 + bx + c \tag{17}$$

According to the theory of majorization-minimization [39,40], we have,

$$\begin{aligned} g(v, v) = \theta(v, r), g'(v, v) = \theta'(v, r) \\ g(s, v) = \theta(s, r), g'(s, v) = \theta'(s, r) \end{aligned} \tag{18}$$

The parameter a, b, c and s are all functions of v, we have,

$$a = \frac{1+r}{4|v|}, \; b = \frac{1-r}{2}, \; c = \frac{(1+r)|v|}{4}, \; s = -v \tag{19}$$

Substituting Equation (19) into $g(x, v) = ax^2 + bx + c$, we have,

$$g(x, v) = \frac{1+r}{4|v|}x^2 + \frac{1-r}{2}x + \frac{(1+r)|v|}{4} \tag{20}$$

Similarly, the numerical issue of Equation (20) will appear if the parameter v approaches zero. To address this problem, the sufficiently small positive value ε is used instead of $|v|$, thus, the segmented function Equation (16) can be rewritten as,

$$\theta_\varepsilon(x) = \begin{cases} x, & x > \varepsilon \\ \frac{1+r}{4\varepsilon}x^2 + \frac{1-r}{2}x + \frac{(1+r)\varepsilon}{4}, & \|x\| \le \varepsilon \\ -rx, & x < \varepsilon \end{cases} \tag{21}$$

Hence, the new function $\theta_\varepsilon(x)$ is a continuously differentiable function. The plot of the continuously differentiable asymmetric penalty function $\theta_\varepsilon(x; r)$ is shown in Figure 2, and the function $\theta_\varepsilon(x; r)$ is a second-order polynomial on $[-\varepsilon, \varepsilon]$.

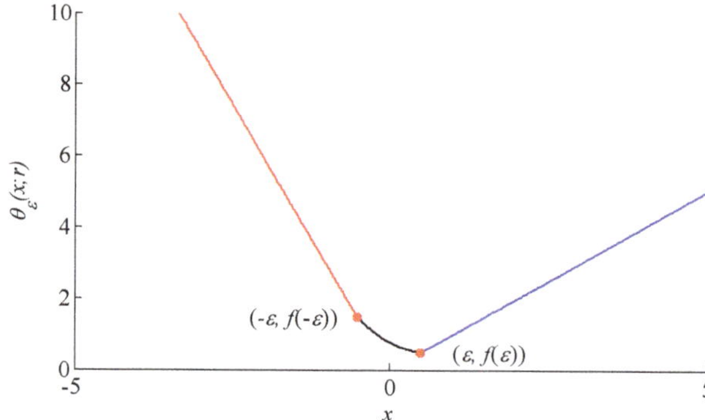

Figure 2. The plot of the asymmetric penalty function $\theta_\varepsilon(x; r)$

The third issue, will be solved and derived in Section 2.3 using the majorization-minimization algorithm.

2.3. The Solution of Proposed Model Based on Majorization-Minimization Algorithm

In this paper, the majorization-minimization (MM) algorithm is implemented to derive an iterative solution procedure for the proposed approach [38]. The $G(x,v)$ is chosen as the majorizer of $F(x)$. Specifically, the iterative solution procedure could be divided into three phases:

(a) Majorizer of symmetric and differentiable function $\phi([D_i x]_n)$ based on MM algorithm.
(b) Majorizer of asymmetric and differentiable function $\theta_\varepsilon(x_n; r)$ based on MM algorithm.
(c) Majorizer of objective cost $F(x)$ based on MM algorithm.

For problem (a), we first seek a majorizer $g(x, v)$ for $\phi(x)$, i.e.,

$$\begin{aligned} g(v, v) &= \phi(v), \text{for all } x, v \\ g(x, v) &\ge \phi(x) \end{aligned} \tag{22}$$

Since $\phi(x)$ is symmetric function, we set $g(x, v)$ to be an even second-order polynomial, i.e.,

$$g(x, v) = mx^2 + b \tag{23}$$

Thus, according to Equation (22) and $g(v, v) = \phi(v)$ and $g'(v, v) = \phi'(v)$, we have,

$$mx^2 + b = \phi(v) \text{ and } 2mv = \phi'(v) \tag{24}$$

The parameter m and b can be computed as,

$$m = \frac{\phi'(v)}{2v} \text{ and } b = \phi(v) - \frac{v}{2}\phi'(v) \tag{25}$$

Substituting Equation (25) in Equation (23), we have

$$g(x,v) = \frac{\phi'(v)}{2v}x^2 + \phi(v) - \frac{v}{2}\phi'(v) \tag{26}$$

Summing, we obtain,

$$\begin{aligned}
\sum_n g(x_n, v_n) &= \sum_n \left[\frac{\phi'(v_n)}{2v_n}x_n^2 + \phi(v_n) - \frac{v_n}{2}\phi'(v_n) \right] \\
&= \tfrac{1}{2}x^T \frac{\phi'(v_n)}{v_n}x + \sum_n \left[\phi(v_n) - \frac{v_n}{2}\phi'(v_n) \right] \\
&= \tfrac{1}{2}x^T [\Lambda(v)]_n x + c(v) \\
&\geq \sum_{n=0}^{N-1} \phi(x_n)
\end{aligned} \tag{27}$$

where $[\Lambda(v)]_n = \frac{\phi'(v_n)}{v_n}$ is diagonal matrix and $c(v) = \sum_n \left[\phi(v_n) - \frac{v_n}{2}\phi'(v_n) \right]$.

Therefore, based on Equation (27), we obtain,

$$\begin{aligned}
\sum_{i=0}^{M} \lambda_i \sum_{n=0}^{N-1} g([D_i x]_n, [D_i v]_n) &= \sum_{i=0}^{M} \left[\tfrac{\lambda_i}{2}(D_i x)^T \left[\frac{\phi'(D_i v)}{D_i v} \right](D_i x) + \sum_n \left[\phi([D_i v]_n) - \frac{[D_i v]_n}{2}\phi'([D_i v]_n) \right] \right] \\
&= \sum_{i=0}^{M} \left[\tfrac{\lambda_i}{2}(D_i x)^T [\Lambda(D_i v)](D_i x) + c_i(D_i v) \right] \\
&\geq \sum_{i=0}^{M} \lambda_i \sum_{n=0}^{N-1} \phi([D_i x]_n)
\end{aligned} \tag{28}$$

where $[\Lambda(D_i v)]_n = \frac{\phi'([D_i v]_n)}{[D_i v]_n}$ is diagonal matrix and $c_i(D_i v) = \sum_n \left[\phi([D_i v]_n) - \frac{[D_i v]_n}{2}\phi'([D_i v]_n) \right]$.

For problem (b), assume that $g_0(x,v)$ is the majorizer of the asymmetric and differentiable function $\theta_\varepsilon(x_n; r)$, since the $f(x) = \frac{1+r}{4\varepsilon}x^2 + \frac{1-r}{2}x + \frac{(1+r)\varepsilon}{4}$, $\|x\| \leq \varepsilon$, we have,

$$\begin{cases} g_0(x,v) = \frac{1+r}{4v}x^2 + \frac{1-r}{2}x + \frac{(1+r)v}{4} \geq f(x), v > \varepsilon \\ g_0(x,v) = -\frac{1+r}{4v}x^2 + \frac{1-r}{2}x - \frac{(1+r)v}{4} \geq f(x), v < -\varepsilon \end{cases} \tag{29}$$

when $v > \varepsilon$, then,

$$\begin{aligned}
g_0(x,v) - f(x) &= \left(\frac{1+r}{4v} - \frac{1+r}{4\varepsilon} \right)x^2 + (v - \varepsilon)\frac{1+r}{4} \\
&= \frac{(1+r)(v-\varepsilon)(v\varepsilon - x^2)}{4v\varepsilon} > 0
\end{aligned} \tag{30}$$

when $v < -\varepsilon$, then,

$$\begin{aligned}
g_0(x,v) - f(x) &= \left(-\frac{1+r}{4v} - \frac{1+r}{4\varepsilon} \right)x^2 - (v + \varepsilon)\frac{1+r}{4} \\
&= -\frac{(1+r)(v+\varepsilon)(v\varepsilon + x^2)}{4v\varepsilon} > 0
\end{aligned} \tag{31}$$

Therefore, the majorizer of the asymmetric and differentiable function $\theta_\varepsilon(x_n; r)$ is obtained,

$$\begin{cases} g_0(x,v) = \frac{1+r}{4|v|}x^2 + \frac{1+r}{2}x + |v|\frac{1+r}{4}, & |v| > \varepsilon \\ g_0(x,v) = \frac{1+r}{4\varepsilon}x^2 + \frac{1+r}{2}x + \varepsilon\frac{1+r}{4}, & |v| \leq \varepsilon \end{cases} \tag{32}$$

Summing, we obtain,

$$\sum_{n=0}^{N-1} g_0(x_n, v_n) = x^T [\mathbf{\Gamma}(v)]x + b^T x + c(v)$$
$$\geq \sum_{n=0}^{N-1} \theta_\varepsilon(x_n, r) \tag{33}$$

where $\mathbf{\Gamma}(v)$ is a diagonal matrix, i.e., $[\mathbf{\Gamma}(v)]_n = (1+r)/4|v_n|, |v_n| \geq \varepsilon$ and $[\mathbf{\Gamma}(v)]_n = (1+r)/4\varepsilon$, $|v_n| \leq \varepsilon$, and $[b]_n = (1-r)/2$.

For problem (c), based on Equations (28) and (33), the majorizer of $F(x)$ based on MM algorithm is given by,

$$G(x, v) = \tfrac{1}{2}\|H(y - x)\|_2^2 + \lambda_0 x^T [\mathbf{\Gamma}(v)]x + \lambda_0 b^T x$$
$$+ \sum_{i=1}^{M} \left[\tfrac{\lambda_i}{2}(D_i x)^T [\Lambda(D_i v)](D_i x) \right] + c(v) \tag{34}$$

Minimizing $G(x,v)$ with respect to x yields,

$$x = H^T H + 2\lambda_0 \mathbf{\Gamma}(v) + \sum_{i=1}^{M} \lambda_i D_i^T [\Lambda(D_i v) D_i]^{-1} \left(H^T H y - \lambda_0 b \right) \tag{35}$$

Substituting $H = BA^{-1}$ in Equation (35), we have,

$$x = A \left\{ B^T B + A^T \left(2\lambda_0 \mathbf{\Gamma}(v) + \sum_{i=1}^{M} \lambda_i D_i^T [\Lambda(D_i v)] D_i \right) A \right\}^{-1} \times (B^T B A^{-1} y - \lambda_0 A^T b)$$
$$= A(B^T B + A^T M A)^{-1}(B^T B A^{-1} y - \lambda_0 A^T b) \tag{36}$$
$$= A Q^{-1}(B^T B A^{-1} y - \lambda_0 A^T b)$$

where matrix $M = 2\lambda_0 \mathbf{\Gamma}(v) + \sum_{i=1}^{M} \lambda_i D_i^T [\Lambda(D_i v)] D_i$ and matrix $Q = B^T B + A^T M A$.

Finally, by using the above formulas, the fault transient impulses x can be obtained by the following iterations,

$$M^{(k)} = 2\lambda_0 \mathbf{\Gamma}\left(x^{(k)}\right) + \sum_{i=1}^{M} \lambda_i D_i^T \left[\Lambda\left(D_i x^{(k)}\right) \right] D_i \tag{37}$$

$$Q^{(k)} = B^T B + A^T M^{(k)} A \tag{38}$$

$$x^{(k+1)} = A \left[Q^{(k)} \right]^{-1} \left(B^T B A^{-1} y - \lambda_0 A^T b \right) \tag{39}$$

In conclusion, the complete steps of the proposed algorithm are summarized as follows,

(1) **Inputs**: signal y, $r \geq 1$, matrix A, matrix B, $\lambda_i, i = 0, 1, \ldots, M$, $k = 0$;
(2) $E = B^T B A^{-1} y - \lambda_0 A^T b$
(3) Initialize $x = y$;
(4) Repeat the following iterations:

$$[\mathbf{\Gamma}(v)]_n = (1+r)/4|v_n|, |v_n| \geq \varepsilon;$$

$$[\mathbf{\Gamma}(v)]_n = (1+r)/4\varepsilon, |v_n| \leq \varepsilon;$$

$$[\Lambda(D_i v)]_n = \frac{\phi'([D_i v]_n)}{[D_i v]_n}, i = 0, 1, 2, \ldots, M;$$

$$M^{(k)} = 2\lambda_0 \mathbf{\Gamma}\left(x^{(k)}\right) + \sum_{i=1}^{M} \lambda_i D_i^T \left[\Lambda\left(D_i x^{(k)}\right) \right] D_i;$$

$$Q^{(k)} = B^T B + A^T M^{(k)} A;$$

$$x^{(k+1)} = A\left[Q^{(k)}\right]^{-1} E;$$

(5) If the stopping criterion is satisfied, then output signal x, otherwise, $k = k + 1$, and go to step (4).

(6) **Output**: signal x.

2.4. Parameters Selection

In this section, the regularization parameters λ_i are set as [26]:

$$\begin{aligned}
\lambda_0 &= \beta_0 \sigma \\
\lambda_1 &= \beta_1 \sigma = \gamma(1 - \beta_0)\sigma \\
\lambda_2 &= \beta_0 \sigma
\end{aligned} \tag{40}$$

where β_0 and β_1 are the constants so as to maximize the signal-to-noise-rate (SNR), β_0 and γ are typically set up to be constants, i.e., $\beta_0 = [0.5, 1]$, $\gamma = [7.5, 8]$, and σ is the standard deviation (SD) of the external noise. In practical engineering applications, the SD of the external noise in Equation (40) could be computed using the healthy data and fault data under the same operating environment. Moreover, when the healthy data is not available or unknown, the SD of the external noise can still be estimated by the following formula,

$$\hat{\sigma} = MAD(y)/0.6745 \tag{41}$$

where the Equation (41) is a noise level estimator that used for traditional wavelet denoising in ref. [41], and $MAD(y)$ represents the median absolute deviation (MAD) of observation signal y, i.e.,

$$MAD(y) = \mathrm{median}(|y_i - \mathrm{median}(y)|), i = 1, 2, \ldots, N \tag{42}$$

To better understand the asymmetric penalty sparse decomposition algorithm, a chromatogram signal is analyzed by the proposed APSD algorithm [38]. For illustration, Figure 3a exhibits the raw chromatogram signal. Figure 3b,c respectively shows the LFC and HFC that generated by the proposed APSD. Note that the main evolutionary trend, i.e., low frequency component, is well estimated as illustrated in Figure 3b, meanwhile, the estimated peaks, i.e., high frequency component, illustrated in Figure 3c, are well delineated. Therefore, the LFC and HFC components could be captured by the asymmetric penalty sparse decomposition algorithm.

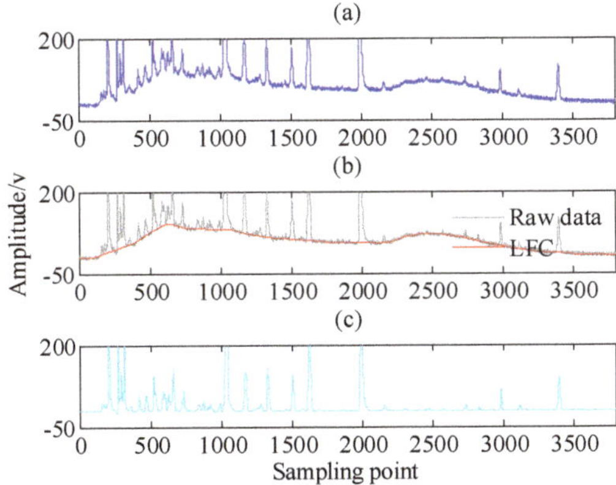

Figure 3. The decomposition results of noisy chromatogram data using proposed APSD algorithm [38]. (a) The raw data with additive noise; (b) The estimated LFC signal; (c) The estimated HFC signal.

3. Wavelet Neural Network and ARMA-RLS Algorithm

3.1. Wavelet Neural Network Algorithm

Wavelet neural network (WNN) is a recently developed neural network with a topology and structure that are similar to that of the back-propagation neural network (BPNN) [42–45]. The network structure is composed of the input layer, hidden layer and output layer; the topology structure of the WNN is shown in Figure 4, wherein the customary sigmoid function (CSF) is replaced by the wavelet basis functions $h(j)$ as activation function for the neurons in hidden layer. Hence, the combination of wavelet transform and BPNN has the advantage of both wavelet analysis and neural network; the data is transmitted forward and the prediction error is propagated backward, so as to achieve a more accurate predictive data.

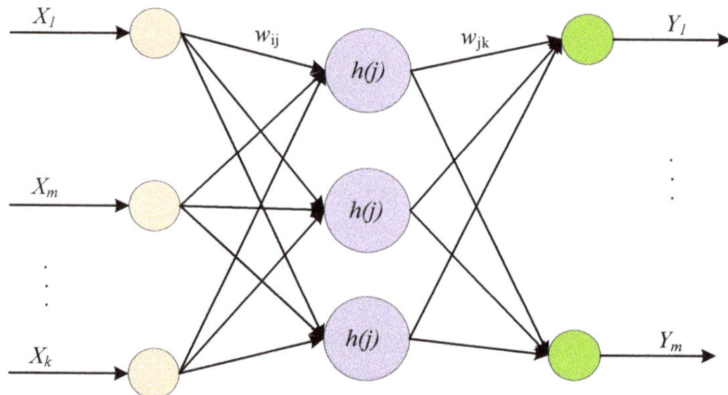

Figure 4. The structural diagram of wavelet neural network.

Generally, prediction accuracy and generalization ability will be affected by the choice of wavelet base function; compared to orthogonal wavelet, Gauss spline wavelet and Mexico hat wavelet, the Morlet wavelet has the smallest error and the reliable computational stability [46,47], thus this study employed Morlet wavelet as the activation function of hidden layer nodes, the formula is given below,

$$y = \cos(1.75x)e^{-x^2/2} \tag{43}$$

In Figure 4, the input data are represented by $X_1, X_2, \ldots, X_k$, and the predicted outputs are denoted by $Y_1, Y_2, \ldots, Y_m$, w_{ij} is the link-weight between the input layer and hidden layer. When the input data is x_i ($i = 1, 2, \ldots, k$), the output of hidden layer can be calculated as,

$$h(j) = h_j\left(\frac{\sum\limits_{i=1}^{k} w_{ij}x_i - b_j}{a_j}\right), j = 1, 2, \ldots, l \tag{44}$$

where $h(j)$ is the output of the j-th hidden layer and also represents the wavelet basis function, a_j is the scaling factor of function $h(j)$, b_j is the translation factor of function $h(j)$, The output of the WNN can be expressed as,

$$y(k) = \sum\limits_{i=1}^{l} w_{ik}h(i), k = 1, 2, \ldots, m \tag{45}$$

where w_{ik} denotes the link-weight between the hidden layer and output layer, $h(i)$ is the output of the i-th hidden layer, l represents the number of hidden layer and m represents the number of output layer. According to the fundamental principle of BPNN and gradient descent learning algorithm, the corresponding adjustment process of network weights and wavelet basis function is as follows,

(1) Calculate network prediction error, i.e.,

$$e = \sum_{k=1}^{m} y_n(k) - y(k) \tag{46}$$

where $y_n(k)$ is desired outputs and $y(k)$ predicted outputs generated by the WNN method.

(2) network weights and wavelet basis function can be adjusted by network prediction error, i.e.,

$$w_{n,k}^{(i+1)} = w_{n,k}^i + \Delta w_{n,k}^{(i+1)} \tag{47}$$

$$a_k^{(i+1)} = a_k^i + \Delta a_k^{(i+1)} \tag{48}$$

$$b_k^{(i+1)} = b_k^i + \Delta b_k^{(i+1)} \tag{49}$$

where the parameters $\Delta w_{n,k}^{(i+1)}$, $\Delta a_k^{(i+1)}$ and $\Delta b_k^{(i+1)}$ could be calculated as follows,

$$\Delta w_{n,k}^{(i+1)} = -\eta \frac{\partial e}{\partial w_{n,k}^{(i)}} \tag{50}$$

$$\Delta a_k^{(i+1)} = -\eta \frac{\partial e}{\partial a_k^{(i)}} \tag{51}$$

$$\Delta b_k^{(i+1)} = -\eta \frac{\partial e}{\partial b_k^{(i)}} \tag{52}$$

where η is learning rate. The training and correcting algorithm steps of the network parameters are as follows:

Step 1: Network initialization. The expansion parameters a_k and translation parameters b_k of the wavelet basis function as well as the network learning rate η, the link-weights are w_{ij} and w_{jk}, respectively, the error threshold ε, and maximum iterations T.

Step 2: Samples classification. Samples can be divided into training samples and test samples, wherein the training samples are applied for network training and the test samples are applied for testing prediction accuracy.

Step 3: Predicted output. The training samples are feed into the WNN network, and the predicted outputs are calculated. Then the error and gradient vectors of the output of WNN and the desired output are obtained.

Step 4: Weights adjustment. The parameters of the wavelet basis function and the back propagation of error corrects the weights of the WNN.

Step 5: End conditions. The training algorithm will judge whether the targeted error is less than the predetermined threshold ε ($\varepsilon > 0$) or exceeds the maximum iterations. If yes, the network training is stopped, otherwise, the algorithm returns to step 3. The training and correcting steps of the WNN are shown in Figure 5.

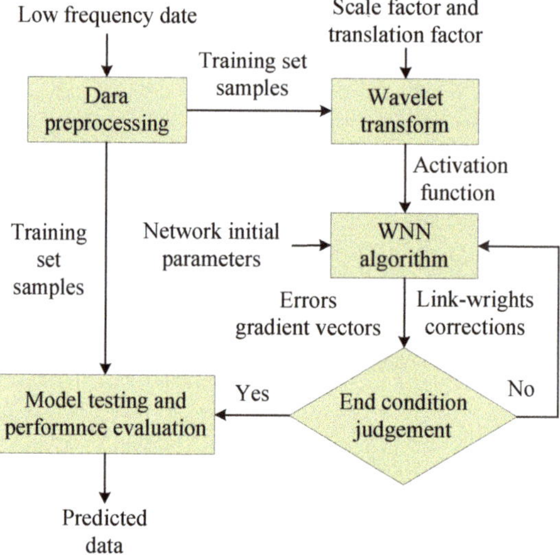

Figure 5. Flow chart of wavelet neural network (WNN) for prediction of low frequency data.

3.2. ARMA Combined With Recursive Least Squares (RLS) Algorithm

3.2.1. ARMA Review of ARMA Model

A stochastic process X_t is called an autoregressive moving-average process (ARMA) with order p and q, namely ARMA (p, q) [48,49], if the process is stationary and satisfies a linear stochastic difference equation of the form,

$$X_t = \phi_1 X_{t-1} + \ldots + \phi_p X_{t-p} + e_t + \theta_1 e_{t-1} + \ldots + \theta_q e_{t-q} \tag{53}$$

where e_t is white gaussian noise (WGN), i.e., $e_t \sim WN(0, \sigma^2)$, parameters $\phi_1, \phi_2, \ldots, \phi_p$ and $\theta_1, \theta_2, \ldots, \theta_q$ are coefficients of AR(p) and MA(q) models, and the polynomials are as follows,

$$\phi(z) = 1 - \phi_1 z - \phi_2 z - \ldots - \phi_p z^p \tag{54}$$

$$\theta(z) = 1 + \theta_1 z + \theta_2 z + \ldots + \theta_p z^p \tag{55}$$

3.2.2. Recursive Least Squares (RLS) Algorithm

Given an input samples set $\{u(1), u(2), \ldots, u(N)\}$ and a desired response set $\{d(1), d(2), \ldots, d(N)\}$, the output $y(n)$ could be computed by linear filter method as follows,

$$y(n) = \sum_{k=0}^{M} w_k u(n - k), n = 0, 1, 2, \ldots \tag{56}$$

Minimizing the sum of the error squares, we have,

$$\begin{aligned}
\varepsilon(n) &= \varepsilon(\omega_1(n), \omega_2(n), \ldots, \omega_{M-1}(n)) \\
&= \sum_{i=i_1}^{n} \beta(n, i) \times \left[e(i)^2\right] \\
&= \sum_{i=i_1}^{n} \beta(n, i) \times \left[d(i) - \sum_{k=0}^{M-1} \omega_k(n) u(i - k)\right]^2
\end{aligned} \tag{57}$$

where the error $e(i)$ is $e(i) = d(i) - y(i) = d(i) - \sum\limits_{k=0}^{M-1} \omega_k(n)u(i-k)$, and the $\beta(n,i)$ is the forgetting factor and $0 < \beta(n,i) \le 1, i = 1,2,\ldots,n$. To simplify the writing format, taking the form as $\beta(n,i) = \lambda^{n-i}, i = 1,2,\ldots,n$. Thus, the sum of error squares can be rewritten as,

$$
\begin{aligned}
\varepsilon(n) &= \varepsilon(\omega_1(n), \omega_2(n), \ldots, \omega_{M-1}(n)) \\
&= \sum_{i=i_1}^{n} \lambda^{n-i} \times \left[d(i) - \sum_{k=0}^{M-1} \omega_k(n)u(i-k) \right]^2
\end{aligned}
\tag{58}
$$

Defining the following formulas,

$$
u'(i) = \sqrt{\lambda^{n-i}}u(i); \quad d'(i) = \sqrt{\lambda^{n-i}}d(i)
\tag{59}
$$

Then the Equation (58) can be rewritten as,

$$
\begin{aligned}
\varepsilon(n) &= \sum_{i=i_1}^{n} \lambda^{n-i} \times \left[d(i) - \sum_{k=0}^{M-1} \omega_k(n)u(i-k) \right]^2 \\
&= \sum_{i=i_1}^{n} \left[d'(i) - \sum_{k=0}^{M-1} \omega_k(n)u'(i-k) \right]^2
\end{aligned}
\tag{60}
$$

which is the standard least squares (LS) criterion, the solution of the LS can be obtained as,

$$
\begin{aligned}
\omega(n) &= \left(\sum_{i=i_1}^{n} u'(i)u'(i)^T \right)^{-1} \sum_{i=i_1}^{n} u'(i)d'(i) \\
&= \left(\sum_{i=i_1}^{n} \lambda^{n-i}u(i)u(i)^T \right)^{-1} \sum_{i=i_1}^{n} \lambda^{n-i}u(i)d(i) \\
&= [\Phi(n)]^{-1}\psi(n)
\end{aligned}
\tag{61}
$$

where parameters $\Phi(n) = \sum\limits_{i=1}^{n} \lambda^{n-i}u(i)u(i)^T$ and $\psi(n) = \sum\limits_{i=1}^{n} \lambda^{n-i}u(i)d(i)$. Based on $\omega(n) = [\Phi(n)]^{-1}\psi(n)$, thus the $\omega(n)$ at time $n-1$ could be obtained, i.e.,

$$
\omega(n-1) = [\Phi(n-1)]^{-1}\psi(n-1)
\tag{62}
$$

Therefore, the variables $\Phi(n)$ and $\psi(n)$ can be rewritten using $\Phi(n-1)$ and $\psi(n-1)$, i.e.,

$$
\begin{aligned}
\Phi(n) &= \sum_{i=1}^{n} \lambda^{n-i}u(i)u(i)^T = \lambda \sum_{i=1}^{n-1} \lambda^{n-1-i}u(i)u(i)^T + u(n)u(n)^T \\
&= \lambda\Phi(n-1) + u(n)u(n)^T
\end{aligned}
\tag{63}
$$

and

$$
\begin{aligned}
\psi(n) &= \sum_{i=1}^{n} \lambda^{n-i}u(i)d(i) = \lambda \sum_{i=1}^{n-1} \lambda^{n-1-i}u(i)d(i) + u(n)d(n) \\
&= \lambda\psi(n-1) + u(n)d(n)
\end{aligned}
\tag{64}
$$

The matrix inversion formula of $\Phi^{-1}(n)$ is as follows,

$$
\Phi^{-1}(n) = \lambda^{-1}\Phi^{-1}(n-1) - \frac{\lambda^{-2}\Phi^{-1}(n-1)u(n)u^T(n)\Phi^T(n-1)}{1 + \lambda^{-1}u^T(n)\Phi^{-1}(n-1)u(n)}
\tag{65}
$$

Denoting

$$
P(n) = \Phi^{-1}(n)
\tag{66}
$$

and

$$k(n) = \frac{\lambda^{-1}P(n-1)u(n)}{1+\lambda^{-1}u^T(n)P(n-1)u(n)} = P(n)u(n) \tag{67}$$

Therefore,

$$P(n) = \lambda^{-1}P(n-1) - \lambda^{-1}k(n)u^T(n)P(n-1) \tag{68}$$

The main time-update equation $w(n)$ can be derived as,

$$
\begin{aligned}
w(n) &= [\Phi(n)]^{-1}\psi(n) = P(n)\psi(n) \\
&= P(n)(\lambda\psi(n-1) + u(n)d(n)) \\
&= P(n)(\lambda\Phi(n-1)w(n-1) + u(n)d(n)) \\
&= P(n)\{[\Phi(n) - u(n)u^T(n)]w(n-1) + u(n)d(n)\} \\
&= w(n-1) - P(n)u(n)u^T(n)w(n-1) + P(n)u(n)d(n) \\
&= w(n-1) + P(n)u(n)[d(n) - u^T(n)w(n-1)] \\
&= w(n-1) + P(n)u(n)\alpha(n) \\
&= w(n-1) + k(n)\alpha(n)
\end{aligned} \tag{69}
$$

where $\alpha(n) = d(n) - u^T(n)w(n-1)$ is the innovation process. In conclusion, the prediction processes of the ARMA combined with RLS algorithm are summarized as follows,

Step 1: Algorithm initialization. The forgetting factor $0 < \lambda \leq 1$, the predicted steps n.

Step 2: Initial parameter calculation. Calculate the initial order p and q of ARMA (p, q) model using Akaike information criterion (AIC) [50] based on historical data.

Step 3: Coefficients extraction. Extract the coefficients $\phi_1, \phi_2, \ldots, \phi_p$ and $\theta_1, \theta_2, \ldots, \theta_q$ as the input $u(i)$ of RLS algorithm, i.e., $u(i) = [1, \phi_1, \phi_2, \ldots, \phi_p, \theta_1, \theta_2, \ldots, \theta_q], i = 1, 2, \ldots, p + q$.

Step 4: Prediction iteration based on RLS algorithm. Repeat the following iterations,

$$k(i) = \frac{\lambda^{-1}P(i-1)u(i)}{1+\lambda^{-1}u^T(i)P(i-1)u(i)};$$

$$\alpha(i) = d(i) - u^T(i)w(i-1);$$

$$w(i) = w(i-1) + k(i)\alpha(i);$$

$$P(i) = \lambda^{-1}P(i-1) - \lambda^{-1}k(i)u^T(i)P(i-1);$$

$$y(i) = y(i-1) + k \times \alpha(i);$$

Step 5: Judgment of the end conditions. If the iterative step is satisfied, then output predicted signal $y(n)$, otherwise, $i = i + 1$, and go to step (4).

4. Experimental Validations

The vibration data collected from accelerated life tests (ALT) of rolling bearings were employed to validate the effectiveness of the proposed approach, the experimental setup is shown in Figure 6a. The experimental data were acquired and published by the IEEE-PHM Association [51,52]. The experimental platform included National Instruments (NI) data acquisition card, pressure regulator, cylinder pressure, force sensors, motor, speed sensor, torque-meter, accelerometers and tested bearing, etc. During the process of the experiment, the rotating speeds of the bearing were set to 1650 r/min and 1800 r/min, and 17 bearings were chosen during all those experiments. The sampling frequency was 25.6 kHz, and 2560 sample points (i.e., 0.1 s) were recorded each 10 s. The experimental tests were stopped if the amplitude of time-domain data exceeded 20 g. As shown in Figure 6b, the severe wear in bearing elements and the severe spalling failure in inner race were observed after dismantling.

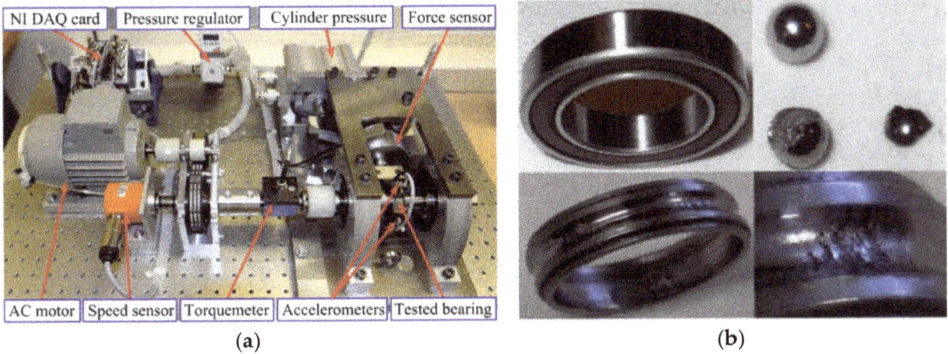

Figure 6. Experimental setup. (**a**) Overview of experimental setup; (**b**) Normal and degraded bearings [51,52].

Figure 7a shows the whole lifetime data of a tested bearing in the time-domain, and the time-domain waveform of the tested bearing in the normal stage and the time-domain waveform of the tested bearing in the catastrophic failure stage are shown in Figure 7b,c, respectively. In the normal stage, the failure impulses cannot be observed in the time-domain, the range of the amplitude is [−2 g, 2 g]. However, in the failure stage, the range of the amplitude is [−50 g, 50 g], and there exist obvious impulses with time interval of $\Delta T = 0.0059$ s, which is corresponding to the fault frequency of 169 Hz.

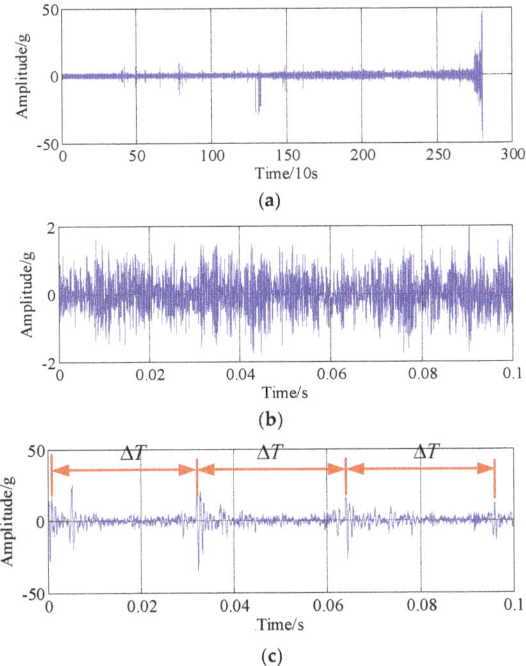

Figure 7. Raw signal of a tested bearing. (**a**) The historical signals of the whole lifetime; (**b**) The time-domain waveform of the bearing in the normal stage; (**c**) The time-domain waveform of the bearing in the failure stage.

In this experiment, four bearings (bearing 1, bearing 2, bearing 3 and bearing 4) are randomly selected and employed as testing targets to evaluate the prediction performance. Each bearing is degraded during the accelerated life tests without implanting any artificial fault in advance.

(1)   bearing 1: operating conditions: speed 1800 rpm and load 4000 N; whole test life: 28,030 s;
(1)   bearing 2: operating conditions: speed 1800 rpm and load 4000 N; whole test life: 18,020 s;
(1)   bearing 3: operating conditions: speed 1650 rpm and load 4200 N; whole test life: 7970 s;
(1)   bearing 4: operating conditions: speed 1800 rpm and load 4000 N; whole test life: 14,280 s.

Depending on the diversity of operating conditions and the manufacturing accuracy of tested bearings, the effective running time may be different for different bearings. Therefore, the degradation trajectories of the health indicators are different for each tested bearing. Figure 8 shows the peak-to-peak values of the whole lifetime of bearing 1. Accordingly, the health indicators, i.e., equivalent vibration intensity (EVI) [9,11], Kurtosis and EVI of bearing 2, bearing 3 and bearing 4 are illustrated in Figure 9a–c, respectively. It is seen that the amplitudes of bearings 1, 3 and 4 have gradual increasing trends, in addition, the whole test life of bearing 3 is the shortest due to harsh operating conditions, which indicates that the extremely failures are occurred before the experiment stops, thus, representing abrupt degradation processes, whereas the EVI amplitudes of bearings 2 show gradual increases; it might be concluded that the design/manufacturing quality and fatigue resistance strength are much higher than others under the same operating conditions.

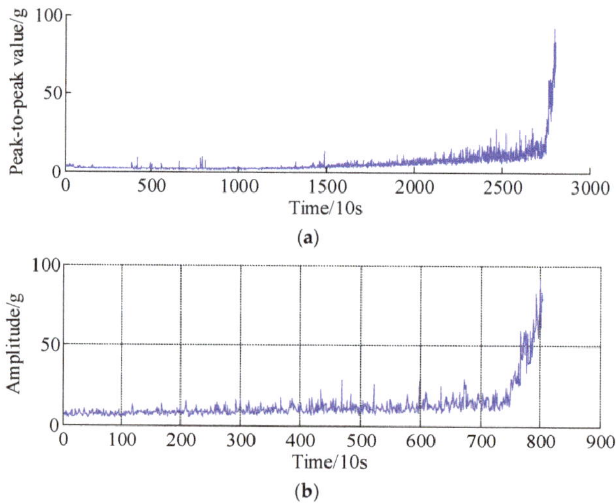

Figure 8. The health indicator curve of peak-to-peak value of bearing 1. (**a**) The peak-to-peak value of the whole lifetime of bearing 1; (**b**) The peak-to-peak value of the point 2001 from to point 2803.

Taking bearing 1 as an example, as shown in Figure 8b, because of the abrupt degradation time series is the most interesting and difficult part; thus, datasets 2001 to 2803 are selected as survey regions, where the dataset from 2001 to 2703 are selected as a historical curve, and the remaining 100 datasets, i.e., dataset 2704 to dataset 2803, are a predicted region. Specifically, the proposed APSD method is adopted to process the peak-to-peak curve of bearing 1, since the standard deviation (SD) of the peak-to-peak is unknown, it can be calculated by the following equation, i.e., $\hat{\sigma} = MAD(y)/0.6745 = 2.1750$. Therefore, the related parameter specification of the APSD approach are summarized in Table 2. The decomposition results are presented in Figure 10, Figure 10a is the low frequency component and Figure 10b is the high frequency component, respectively. It should be noted that the low frequency component almost coincides with the actual peak-to-peak trends, especially in the abrupt degradation regions, which means the degradation processes and trend could be reflected by the low frequency component.

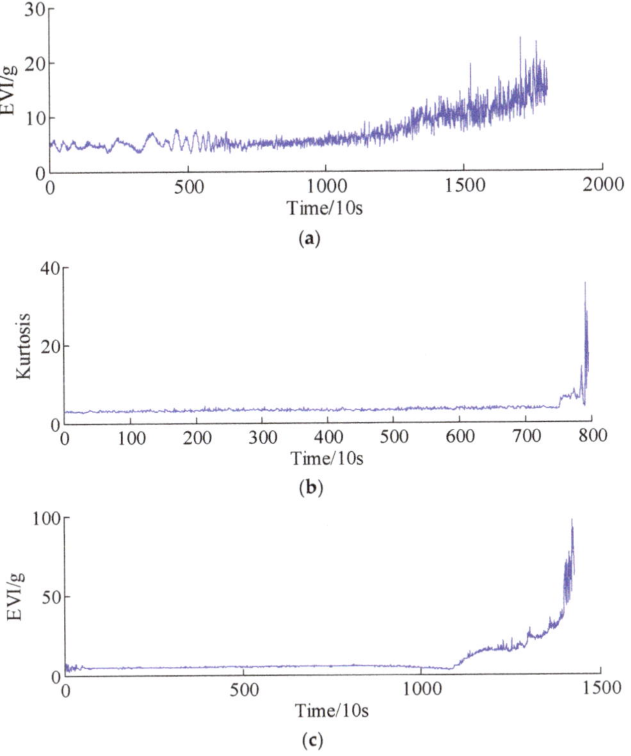

Figure 9. The health indicator curves of different bearings. (**a**) The equivalent vibration intensity curve of bearing 2; (**b**) The Kurtosis curve of bearing 3; (**c**) The equivalent vibration intensity curve of bearing 4.

Furthermore, the WNN algorithm and ARMA-RLS model are introduced for prediction. For the prediction of a low frequency component, the number of input neurons, hidden neurons, output neurons, training iterations, learning rate for WNN are set at 7, 10, 1, 300 and 0.01, respectively. For the prediction of high frequency components, the modelling orders, i.e., p and q, are calculated by Akaike information criterion (AIC) criterion, and then all the parameters of ARMA are fed into the RLS learning algorithm; to recursively estimate and update the related parameters, the forgetting factor is set to 0.99. The prediction curves of the LFC and HFC for rolling bearing are illustrated in Figure 11a,b, respectively. As can be seen in Figure 11, the prediction data generated by the ANN and ARMA-RLS models are able to reasonably trace the variation trend of original peak-to-peak time series. The final predicted result can be correspondingly obtained by integrating the predicted LFC and HFC components, as shown in Figure 12.

Table 2. The parameters setting of APSD method for bearing 1.

Parameters β_0	Parameters γ	Regularization Parameter λ_0	Regularization Parameter λ_1	Regularization Parameter λ_2	M-Term	Iteration Times
$\beta_0 = 0.8$	$\gamma = 7.5$	$\lambda_0 = 1.7400$	$\lambda_1 = 3.2625$	$\lambda_2 = 1.7400$	2	50

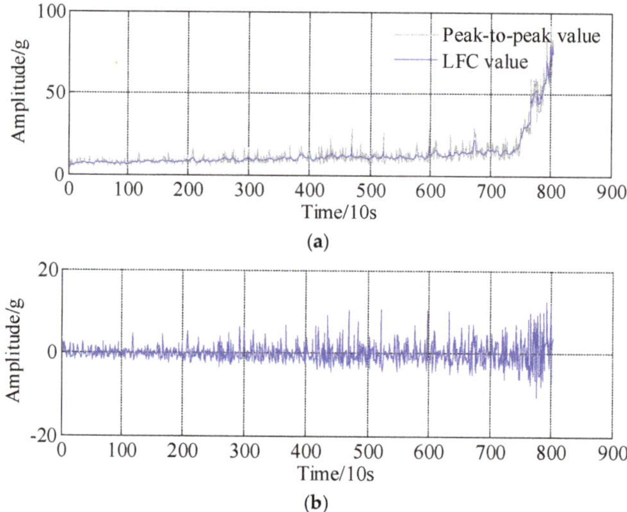

Figure 10. Decomposition results of low frequency component (LFC) and high frequency component (HFC) based on asymmetric penalty sparse decomposition. (**a**) The low frequency component (LFC); (**b**) The high frequency component (HFC).

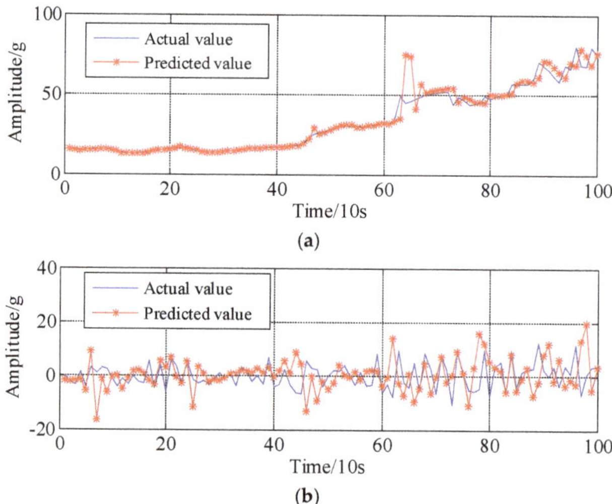

Figure 11. Prediction results of LFC and HFC based on WNN and ARMA-RLS method, respectively. (**a**) Predicted result of LFC based on WNN method; (**b**) Predicted results of HFC based on ARMA-RLS method.

As benchmarking approaches for the prognosis of bearing degradation trajectories, the ARMA, FARIMA and WNN, largest Lyapunov (LLyap) exponent models are employed for comparison. All the predicted results with the actual values are illustrated in Figure 12. As shown in Figure 12, all the benchmarking approaches could generally track the peak-to-peak fluctuation trend well except FARIMA because the memory function might be eliminated due to difference and inverse difference operating. Importantly, compared with the predicted results generated by benchmark approaches, the predictions created by the proposed method are closer to the actual values, and the specific quantitative comparison of the prediction performance are presented in following section.

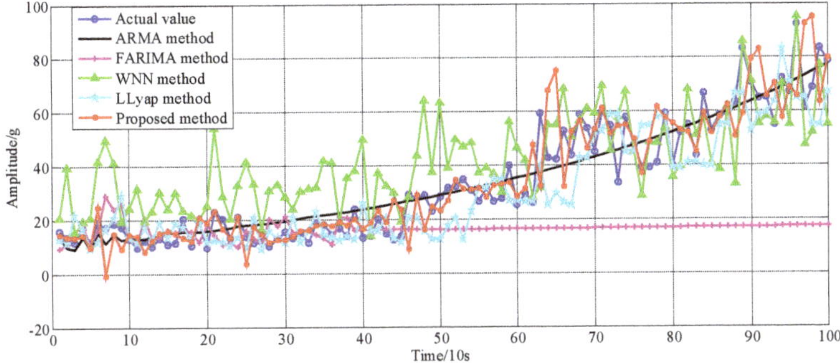

Figure 12. Comparison of the forecasting for peak-to-peak series using benchmarking methods and proposed approach.

Five assessment criteria of prediction results are introduced and calculated including mean absolute error (*MAE*), average relative error (*ARE*), root-mean-square error (*RMSE*), normalized mean square error (*NMSE*) and maximum of absolute error (*Max-AE*), which are denoted by,

$$MAE = \frac{1}{N}\sum_{k=1}^{N}\left|x(k) - \hat{x}(k)\right| \tag{70}$$

$$ARE = \frac{1}{N}\sum_{k=1}^{N}\frac{\left|x(k) - \hat{x}(k)\right|}{x(k)} \tag{71}$$

$$RMSE = \sqrt{\sum_{k=1}^{N}\frac{1}{N-1}\left[x(k) - \hat{x}(k)\right]^2} \tag{72}$$

$$NMSE = \frac{\sum_{k=1}^{N}\left[x(k) - \hat{x})(k)\right]^2}{\sum_{k=1}^{N}\left[x(k) - \overline{x(k)}\right]^2} \tag{73}$$

$$Max - AE = \max\left(\left|x(k) - \hat{x}(k)\right|\right) \tag{74}$$

where N is the number of time points, $x(k)$ and $\hat{x}(k)$ are actual data and predicted data, respectively. Generally speaking, obviously, the smaller *MAE*, *ARE*, *RMSE*, *NMSE* and *Max-AE* values, which means lower prediction errors and higher prediction accuracies. The performance of the proposed methodology is evaluated by computed prediction errors. The computed prediction errors are summarized in Table 3. From the results in Table 3, it is noticed that the proposed method has relatively higher accuracy than FARIMA, WNN and L-Lyap methods. In addition, it is observed that the ARMA model has smaller errors among these models, but the tracking trend of LFC and fluctuation trend of HFC cannot be reflected at all; see black line in Figure 12. This indicates that the presented prediction approach has certain application potentials.

Meanwhile, to better evaluate the performances of all the methods from a statistical perspective, box plots were introduced. The box plots of all the errors based on benchmark approaches and proposed method are shown in Figure 13. The results show that the median values of the prediction errors converge to 0 based on proposed method, which demonstrate that the predicted data generated by the proposed approach are close to their true values.

Table 3. The computed prediction errors of proposed method and benchmark methods.

Index	ARMA	FARIMA	WNN	L-Lyap	Proposed
Mean Absolute Error-*MAE*	0.5880	17.2700	8.2364	3.2453	0.9770
Average relative error-*ARE*	0.1872	0.4666	0.1583	0.2691	0.2599
Root-Mean-Square Error-*RMSE*	5.9094	173.5702	82.7793	32.6164	9.8190
Normalized Mean Square Error-*NMSE*	0.1657	2.4144	0.7173	0.3008	0.2244
Maximum of Absolute Error-*MaxAE*	21.9865	75.0981	36.4112	28.3548	32.9561

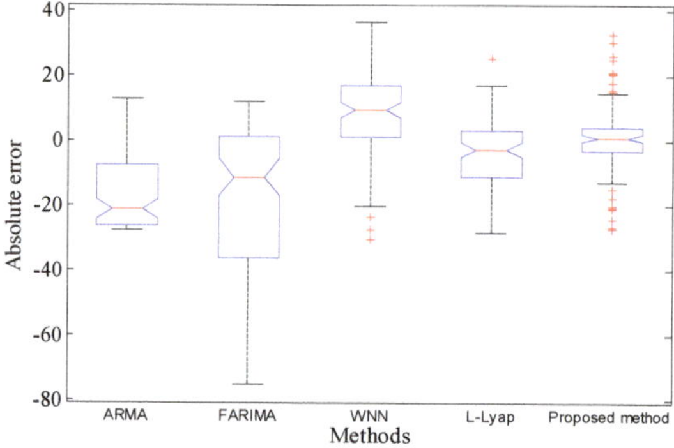

Figure 13. Quantitative performance evaluation based on absolute error boxplot.

Furthermore, the prediction performances of the remaining bearings are illustrated in this section. The model parameter settings of the APSD, WNN, ARMA-RLS are similar to those in the aforementioned steps, the specific parameters are omitted here for the sake of simplification. The decomposition results, i.e., the LFC and HFC, of the bearing 2, bearing 3 and bearing 4 are respectively presented in Figure 14a–f; overall, it can be found that the low frequency component almost coincides with the actual health indicators trends, especially in the abrupt degradation regions.

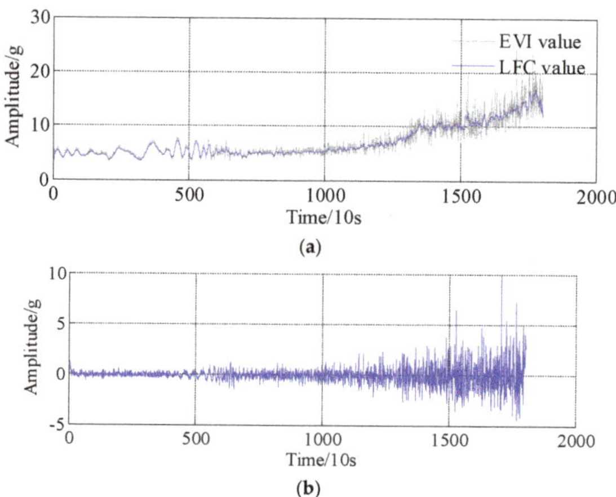

Figure 14. *Cont.*

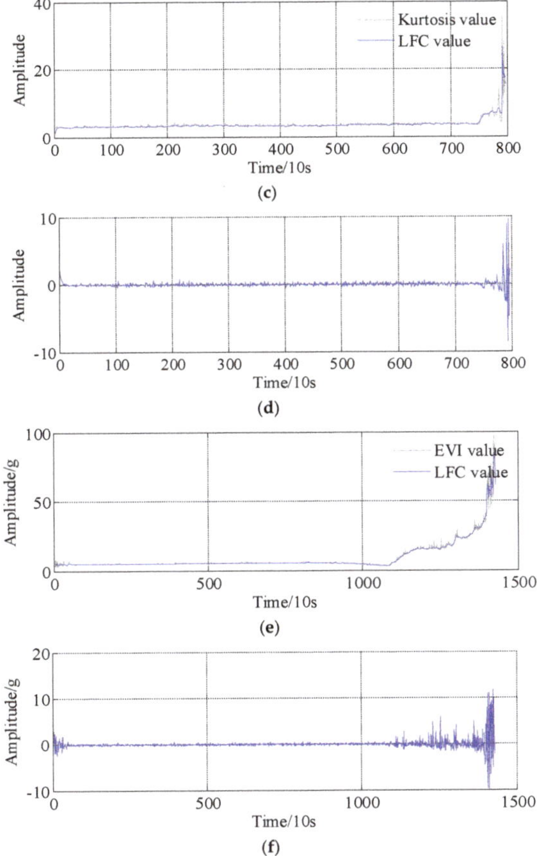

Figure 14. Decomposition results of LFC and HFC based on asymmetric penalty sparse decomposition. (**a**) The low frequency component of bearing 2; (**b**) The High frequency component of bearing 2; (**c**) The low frequency component of bearing 3; (**d**) The High frequency component of bearing 3; (**e**) The low frequency component of bearing 4; (**f**) The High frequency component of bearing 4.

The final predicted results of bearing 2, bearing 3 and bearing 4 are shown in Figure 15a–c, respectively. It is clear that all the predicted data converge to the actual data as time goes on. Interestingly, one can find that, the abrupt degradation points cannot be followed starting point, for example, the point 89 and point 90 in Figure 15b and point 72 in Figure 15c, nevertheless, subsequent points after those abrupt points could be restored, the reason is that the modeling parameters are updated by the artificial neural network and recursive least squares learning algorithm step by step, and the error is eliminated to the minimal range. The proposed two-step prediction method is more robust to the data collection errors. In conclusion, the proposed method performs best in the degradation trend prediction of the rolling bearings.

Additionally, Figure 16 summarizes the box plots of the absolute error (AE) of the remaining three bearings. Analogously, the box plots show that the median values of the absolute error of those three bearings converge to 0, which further reflects the effectiveness of the proposed method. Therefore, the proposed decomposition algorithm and prediction models have accurate prediction results.

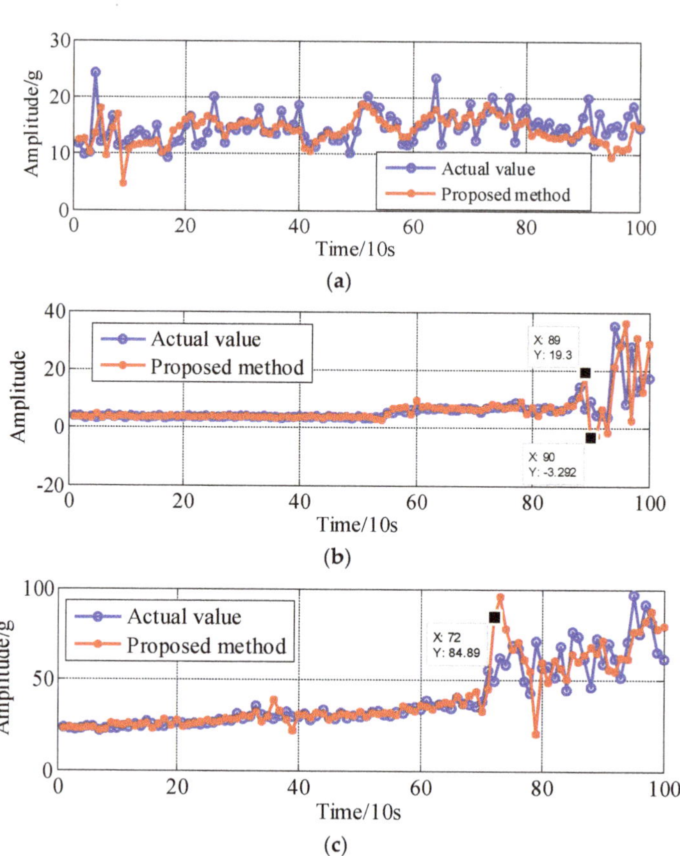

Figure 15. The predicted results of remaining three bearings. (**a**) The predicted results of bearing 2; (**b**) the predicted results of bearing 3; (**c**) the predicted results of bearing 4.

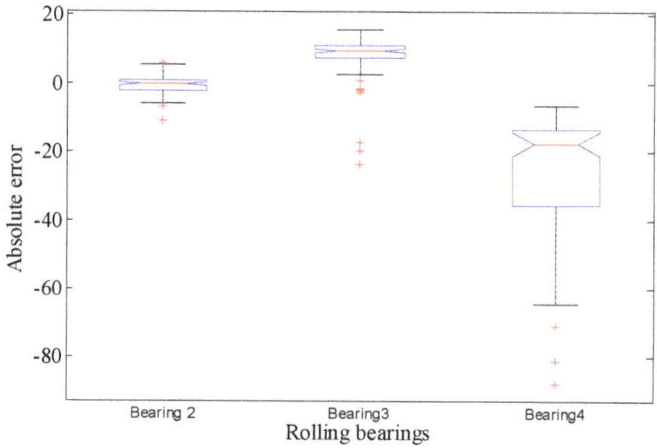

Figure 16. Quantitative performance evaluation based on absolute error boxplot of remaining three bearings.

5. Conclusions

Accurate prognosis of the degradation trend of rotating machinery plays an important role in industrial applications. Currently, most developments in the mechanical fault prognostics area have been targeted towards directly utilizing degradation-based data to trace the degradation trajectories, very few studies have used the idea of sparse decomposition. In this work, a novel intelligent prediction approach based on asymmetric penalty sparse decomposition (APSD) combined with WNN and ARMA-RLS models for health indicator degradation trajectories of four rolling bearings is proposed. The original health indicators od degradation trajectory are rearranged as two components, i.e., LFC and HFC. In particular, the HFC corresponds to the stable change around the zero line of health indicators, whereas the LFC is essentially related to the evolutionary trend of health indicators which rarely occurs in practice. The LFC and HFC are, respectively, predicted using the WNN and ARMA-RLS models. The final degradation regions (e.g., last 100 points) is correspondingly obtained by combining the predicted LFC and predicted HFC. Experimental results of four rolling bearings have demonstrated the superiority of the proposed method in terms of quantitative and qualitative evaluation compared to three commonly-used parametric-based and nonparametric-based methods, i.e., ARMA, FARIMA, WNN, L-Lyap methods.

This observation motivates the study of an integrated method for bearing prediction, which combines the strength of both parametric-based and nonparametric-based techniques. This paper focuses on the development of an intelligent degradation prognosis model that the error reverse transmission is employed to optimize the initial parameters of WNN, and the initial modelling parameters of ARMA are also optimized by the RLS algorithm step by step. The proposed method is more robust to different operating conditions and outperforms other one-step prediction methods by taking the LFC and HFC properties of health indicators into account. Therefore, the proposed method has huge potential in the field of PHM of mechanical equipment.

For future research, it would be interesting to investigate more advanced a sparse low-rank matrix decomposition (SLMD) algorithm to robustly separate the health indicator time series (HITS). Both LFC and HFC could be predicted using more sophisticated adaptive algorithms under more complex and harsh environments.

Author Contributions: Algorithms improvement, programming, experimental analysis and paper writing were done by Q.L. Review and suggestions were provided by S.Y.L. All authors have read and approved the final manuscript.

Acknowledgments: This research is supported the Fundamental Research Funds for the Central Universities (Grant Nos. CUSF-DH-D-2017059 and BCZD2018013) and the Research Funds of Worldtech Transmission Technology (Grant No. 12966EM).

Conflicts of Interest: The authors declare no conflict of interest.

References

1. Chen, Z.Y.; Li, W.H. Multisensor feature fusion for bearing fault diagnosis using sparse autoencoder and deep belief network. *IEEE Trans. Instrum. Meas.* **2017**, *66*, 1693–1702. [CrossRef]
2. Singleton, R.K.; Strangas, E.G.; Aviyente, S. The use of bearing currents and vibrations in lifetime estimation of bearings. *IEEE Trans. Ind. Inform.* **2017**, *13*, 1301–1309. [CrossRef]
3. Li, Q.; Ji, X.; Liang, S.Y. Incipient fault feature extraction for rotating machinery based on improved AR-minimum entropy deconvolution combined with variational mode decomposition approach. *Entropy* **2017**, *19*, 317. [CrossRef]
4. Lei, Y.G.; Li, N.P.; Guo, L.; Li, N.B.; Yan, T.; Lin, J. Machinery health prognostics: A systematic review from data acquisition to RUL prediction. *Mech. Syst. Signal Process.* **2018**, *104*, 799–834. [CrossRef]
5. Li, Q.; Liang, S.Y.; Song, W.Q. Revision of bearing fault characteristic spectrum using LMD and interpolation correction algorithm. *Procedia CIRP* **2016**, *56*, 182–187. [CrossRef]

6. Li, Q.; Hu, W.; Peng, E.F.; Liang, S.Y. Multichannel signals reconstruction based on tunable Q-factor wavelet transform-morphological component analysis and sparse Bayesian iteration for rotating machines. *Entropy* **2018**, *20*, 263. [CrossRef]

7. Hong, S.; Zhou, Z.; Zio, E.; Wang, W.B. An adaptive method for health trend prediction of rotating bearings. *Digit. Signal Process.* **2014**, *35*, 17–123. [CrossRef]

8. Rojas, I.; Valenzuela, O.; Rojasa, F.; Guillen, A.; Herrera, L.J.; Pomares, H.; Marquez, L.; Pasadas, M. Soft-computing techniques and ARMA model for time series prediction. *Neurocomputing* **2008**, *71*, 519–537. [CrossRef]

9. Li, Q.; Liang, S.Y.; Yang, J.G.; Li, B.Z. Long range dependence prognostics for bearing vibration intensity chaotic time series. *Entropy* **2016**, *18*, 23. [CrossRef]

10. Li, M.; Zhang, P.D.; Leng, J.X. Improving autocorrelation regression for the Hurst parameter estimation of long-range dependent time series based on golden section search. *Phys. A* **2016**, *445C*, 189–199. [CrossRef]

11. Li, Q.; Liang, S.Y. Degradation trend prognostics for rolling bearing using improved R/S statistic model and fractional Brownian motion approach. *IEEE Access* **2018**, *6*, 21103–21114. [CrossRef]

12. Song, W.; Li, M.; Liang, J.K. Prediction of bearing fault using fractional brownian motion and minimum entropy deconvolution. *Entropy* **2016**, *18*, 418. [CrossRef]

13. Peng, Y.; Dong, M. A prognosis method using age-dependent hidden semi-Markov model for equipment health prediction. *Mech. Syst. Signal Process.* **2011**, *25*, 237–252. [CrossRef]

14. Yan, J.H.; Guo, C.Z.; Wang, X. A dynamic multi-scale Markov model based methodology for remaining life prediction. *Mech. Syst. Signal Process.* **2011**, *25*, 1364–1376. [CrossRef]

15. Tien, T.L. A research on the prediction of machining accuracy by the deterministic grey dynamic model DGDM (1,1,1). *Appl. Math. Comput.* **2005**, *161*, 923–945. [CrossRef]

16. Abdulshahed, A.M.; Longstaff, A.P.; Fletcher, S.; Potdar, A. Thermal error modelling of a gantry-type 5-axis machine tool using a grey neural network model. *J. Manuf. Syst.* **2016**, *41*, 130–142. [CrossRef]

17. Santhosh, T.V.; Gopika, V.; Ghosh, A.K.; Fernandes, B.G. An approach for reliability prediction of instrumentation & control cables by artificial neural networks and Weibull theory for probabilistic safety assessment of NPPs. *Reliab. Eng. Syst. Saf.* **2018**, *170*, 31–44.

18. Torkashvand, A.M.; Ahmadi, A.; Nikravesh, N.L. Prediction of kiwifruit firmness using fruit mineral nutrient concentration by artificial neural network (ANN) and multiple linear regressions (MLR). *J. Integr. Agric.* **2017**, *16*, 1634–1644. [CrossRef]

19. Lazakis, I.; Raptodimos, Y.; Varelas, T. Predicting ship machinery system condition through analytical reliability tools and artificial neural networks. *Ocean Eng.* **2018**, *152*, 404–415. [CrossRef]

20. Dhamande, L.S.; Chaudhari, M.B. Detection of combined gear-bearing fault in single stage spur gear box using artificial neural network. *Procedia Eng.* **2016**, *144*, 759–766. [CrossRef]

21. Chen, C.C.; Vachtsevanos, G. Bearing condition prediction considering uncertainty: An interval type-2 fuzzy neural network approach. *Robot. Comput. Int. Manuf.* **2012**, *28*, 509–516. [CrossRef]

22. Ren, L.; Cui, J.; Sun, Y.Q.; Cheng, X.J. Multi-bearing remaining useful life collaborative prediction: A deep learning approach. *J. Manuf. Syst.* **2017**, *43*, 248–256. [CrossRef]

23. Deutsch, J.; He, D. Using deep learning-based approach to predict remaining useful life of rotating components. *IEEE Trans. Syst. Man Cybern. Syst.* **2018**, *48*, 11–20. [CrossRef]

24. Selesnick, I.W.; Bayram, I. Enhanced sparsity by non-separable regularization. *IEEE Trans. Signal Process.* **2016**, *64*, 2298–2313. [CrossRef]

25. He, W.P.; Ding, Y.; Zi, Y.Y.; Selesnick, I.W. Repetitive transients extraction algorithm for detecting bearing faults. *Mech. Syst. Signal Process.* **2017**, *84*, 227–244. [CrossRef]

26. Li, Q.; Liang, S.Y. Multiple faults detection for rotating machinery based on Bi-component sparse low-rank matrix separation approach. *IEEE Access* **2018**, *6*, 20242–20254. [CrossRef]

27. Li, Q.; Liang, S.Y. Bearing incipient fault diagnosis based upon maximal spectral kurtosis TQWT and group sparsity total variation de-noising approach. *J. Vibroeng.* **2018**, *20*, 1409–1425. [CrossRef]

28. Wang, S.B.; Selesnick, I.W.; Cai, G.G.; Feng, Y.N.; Sui, X.; Chen, X.F. Nonconvex sparse regularization and convex optimization for bearing fault diagnosis. *IEEE Trans. Ind. Electron.* **2018**, *65*, 7332–7342. [CrossRef]

29. Parekh, A.; Selesnick, I.W. Enhanced low-rank matrix approximation. *IEEE Signal Process. Lett.* **2016**, *23*, 493–497. [CrossRef]

30. Parekh, A.; Selesnick, I.W. Convex fused lasso denoising with non-convex regularization and its use for pulse detection. In Proceedings of the IEEE Signal Processing in Medicine and Biology Symposium, Philadelphia, PA, USA, 12 December 2015; pp. 1–6.

31. Ding, Y.; He, W.P.; Chen, B.Q.; Zi, Y.Y.; Selesnick, I.W. Detection of faults in rotating machinery using periodic time-frequency sparsity. *J. Sound Vib.* **2016**, *382*, 357–378. [CrossRef]

32. Donoho, D.L. Compressed sensing. *IEEE Trans. Inf. Theory* **2006**, *52*, 1289–1306. [CrossRef]

33. Donoho, D.L.; Tsaig, Y. Fast solution of ℓ1-norm minimization problems when the solution may be sparse. *IEEE Trans. Inf. Theory* **2008**, *54*, 4789–4812. [CrossRef]

34. Donoho, D.L. De-noising by soft-thresholding. *IEEE Trans. Inf. Theory* **1995**, *41*, 613–627. [CrossRef]

35. Selesnick, I.W. Total variation denoising via the Moreau envelope. *IEEE Signal Process. Lett.* **2017**, *24*, 216–220. [CrossRef]

36. Selesnick, I.W.; Parekh, A.; Bayram, I. Convex 1-D total variation denoising with non-convex regularization. *IEEE Signal Process. Lett.* **2015**, *22*, 141–144. [CrossRef]

37. Selesnick, I.W.; Graber, H.L.; Pfeil, D.S.; Barbour, R.L. Simultaneous low-pass filtering and total variation denoising. *IEEE Trans. Signal Process.* **2014**, *62*, 1109–1124. [CrossRef]

38. Ning, X.R.; Selesnick, I.W.; Duval, L. Chromatogram baseline estimation and denoising using sparsity (BEADS). *Chemom. Intell. Lab. Syst.* **2014**, *139*, 156–167. [CrossRef]

39. Mourad, N.; Reilly, J.P.; Kirubarajan, T. Majorization-minimization for blind source separation of sparse sources. *Signal Process.* **2017**, *131*, 120–133. [CrossRef]

40. Hunter, D.R.; Lange, K. Quantile Regression via an MM Algorithm. *J. Comput. Graph. Stat.* **2000**, *9*, 60–77.

41. Donoho, D.; Johnstone, I.; Johnstone, I.M. Ideal spatial adaptation by wavelet shrinkage. *Biometrika* **1993**, *81*, 425–455. [CrossRef]

42. Singh, K.R.; Chaudhury, S. Efficient technique for rice grain classification using back-propagation neural network and wavelet decomposition. *IET Comput. Vis.* **2016**, *10*, 780–787. [CrossRef]

43. Heermann, P.D.; Khazenie, N. Classification of multispectral remote sensing data using a back-propagation neural network. *IEEE Trans. Geosci. Remote* **1992**, *30*, 81–88. [CrossRef]

44. Chen, F.C. Back-propagation neural networks for nonlinear self-tuning adaptive control. *IEEE Control. Syst. Mag.* **1990**, *10*, 44–48. [CrossRef]

45. Zeng, Y.R.; Zeng, Y.; Choi, B.; Wang, L. Multifactor-influenced energy consumption forecasting using enhanced back-propagation neural network. *Energy* **2017**, *127*, 381–396. [CrossRef]

46. Osofsky, S.S. Calculation of transient sinusoidal signal amplitudes using the Morlet wavelet. *IEEE Trans. Signal Process.* **1999**, *47*, 3426–3428. [CrossRef]

47. Büssow, R. An algorithm for the continuous Morlet wavelet transform. *Mech. Syst. Signal Process.* **2007**, *21*, 2970–2979. [CrossRef]

48. Ji, W.; Chee, K.C. Prediction of hourly solar radiation using a novel hybrid model of ARMA and TDNN. *Sol. Energy* **2011**, *85*, 808–817. [CrossRef]

49. Brockwell, P.J.; Davis, R.A. *Time Series: Theory and Methods*, 2nd ed.; Springer: New York, NY, USA, 2009; ISBN 9781441903198.

50. Ogasawara, H. Bias correction of the Akaike information criterion in factor analysis. *J. Multivar. Anal.* **2016**, *149*, 144–159. [CrossRef]

51. Nectoux, P.; Gouriveau, R.; Medjaher, K.; Ramasso, E.; Morello, B.; Zerhouni, N.; Varnier, C. Pronostia: An experimental platform for bearings accelerated life test. In Proceedings of the IEEE International Conference on Prognostics and Health Management, Denver, CO, USA, 12 June 2012.

52. Benkedjouh, T.; Medjaher, K.; Zerhouni, N.; Rechak, S. Remaining useful life estimation based on nonlinear feature reduction and support vector regression. *Eng. Appl. Artif. Intell.* **2013**, *26*, 1751–1760. [CrossRef]

Article

Laplacian Spectra for Categorical Product Networks and Its Applications

Shin Min Kang [1,2,*], **Muhammad Kamran Siddiqui** [3,4], **Najma Abdul Rehman** [3], **Muhammad Imran** [4,5] **and Mehwish Hussain Muhammad** [6]

1   Department of Mathematics and RINS, Gyeongsang National University, Jinju 52828, Korea
2   Center for General Education, China Medical University, Taichung 40402, Taiwan
3   Department of Mathematics, COMSATS University Islamabad, Sahiwal Campus 57000, Pakistan;
    kamransiddiqui75@gmail.com (M.K.S.); najma_ar@hotmail.com (N.A.R.)
4   Department of Mathematical Sciences, United Arab Emirates University, P. O. Box 15551, Al Ain,
    United Arab Emirates; imrandhab@gmail.com
5   Department of Mathematics, School of Natural Sciences (SNS), National University of Sciences and
    Technology (NUST), Sector H-12, Islamabad 44000, Pakistan
6   College of Chemistry and Molecular Engineering, Zhengzhou University, Zhengzhou 450001, China;
    mehwish@foxmail.com
*   Correspondence: smkang@gnu.ac.kr

Received: 26 May 2018; Accepted: 6 June 2018; Published: 7 June 2018

Abstract: The Kirchhoff index, global mean-first passage time, average path length and number of spanning trees are of great importance in the field of networking. The "Kirchhoff index" is known as a structure descriptor index. The "global mean-first passage time" is known as a measure for nodes that are quickly reachable from the whole network. The "average path length" is a measure of the efficiency of information or mass transport on a network, and the "number of spanning trees" is used to minimize the cost of power networks, wiring connections, etc. In this paper, we have selected a complex network based on a categorical product and have used the spectrum approach to find the Kirchhoff index, global mean-first passage time, average path length and number of spanning trees. We find the expressions for the product and sum of reciprocals of all nonzero eigenvalues of a categorical product network with the help of the eigenvalues of the path and cycles.

Keywords: Laplacian spectra; categorical product; Kirchhoff index; global mean-first passage time; spanning tree

MSC: 05C12, 05C90

---

1. Introduction

The impact of the study of Laplacian spectra for graphs has increased due to its applications in different fields. Laplacian spectra have miscellaneous applications in graph theory, combinatorial optimization, mathematical biology, computer science, machine learning and in differential geometry, as well. Due to the wide range of applications, the Laplacian spectra of networks is a very interesting and attractive field of research. Computations of the Laplacian spectra of networks are involved in many results related to topological structures and dynamical processes.

Arenas et al. [1] developed a method for understanding synchronization phenomena in networks using Laplacian spectra. Synchronization processes in populations of locally-interacting elements are the focus of intense research in physical, biological, chemical, technological and complex network topologies.

Boccaletti et al. [2] focused on coupled biological and chemical systems, neural networks, socially-interacting species, the Internet and the World Wide Web, which are only a few examples of systems composed of a large number of highly interconnected dynamical units. The first approach to capture the global properties of such systems is to model them as graphs whose nodes represent the dynamical units and whose links stand for the interactions between them.

Liu et al. [3] discussed and investigated the properties of the Laplacian matrices for n-prism networks. They calculated the Laplacian spectra of n-prism graphs, which are both planar and polyhedral. In particular, they derived the analytical expressions for the product and the sum of the reciprocals of all nonzero Laplacian eigenvalues. Moreover, these results were used to handle various problems that often arise in the study of networks including the Kirchhoff index, global mean-first passage time, average path length and the number of spanning trees.

Ding et al. [4] discussed the Laplacian spectra of a three-prism graph and applied them. This graph is both planar and polyhedral and belongs to the generalized Petersen graph. Using the regular structures of this graph, they obtained the recurrent relationships for the Laplacian matrix between this graph and its initial state of a triangle and further derived the corresponding relationships for Laplacian eigenvalues between them. By these relationships, they obtained the analytical expressions for the product and the sum of the reciprocals of all nonzero Laplacian eigenvalues. Finally, they applied these expressions to calculate the number of spanning trees and mean first-passage time (MFPT) and saw the scaling of MFPT with the network size n, which is larger than those performed on some uniformly recursive trees.

Therefore, it is of great interest to compute the Laplacian spectra of different networks. In the last decades, networks and applications of Laplacian spectra have been studied by many scientists, i.e., [5].

Researchers have not paid much attention to applications of Laplacian spectra for networks based on different types of graph operations. Since graph products have a very significant contribution to describing very useful complex networks, we have considered networks based on a categorical product. In this paper, we study the Laplacian spectra of the complex network as a categorical product network. Categorical graph products are used to study complex networks in computer science, to understand structures in structural mechanics and to describe multilayer networks and have many applications in network topologies. Considering the structure of categorical product networks, we derive the expressions for the product and sum of reciprocals of nonzero eigenvalues of the categorical product with the help of the eigenvalues of the path and cycle. Furthermore, we compute the Kirchhoff index, global mean first passage time, average path length and number of spanning trees using the relation of nonzero eigenvalues to these applications.

The Kirchhoff index Kf, also simply called the resistance and denoted by R [6], of a connected graph G on n nodes is defined by:

$$Kf = \frac{1}{2} \sum_{i=1}^{n} \sum_{j=1}^{n} (\Omega)_{ij} \tag{1}$$

where $(\Omega)_{ij}$ is the resistance distance matrix. This formula for the Kirchhoff index reduces to [7]:

$$Kf(G) = N \sum_{i=2}^{N} \frac{1}{v_i} \tag{2}$$

where v_i represents the eigenvalue of the Laplacian matrix of the graph.

The global mean-first passage time (MFPT) F_{ij} has been studied with respect to transport and networks. The mean first passage time (MFPT) is very useful to estimate the speed of transport for random walks on complex networks [8,9]. In fact, MFPT denoted by $< F_N >$ measures the diffusion efficiency of random walks, which is obtained by averaging F_{ij} over $(N-1)$ possible destinations and N origins of particles:

$$< F_N > = \frac{1}{N(N-1)} \sum_{i \neq j} F_{ij}(N) \tag{3}$$

From ([10]), let commuting time C_{ij} between nodes i and j be exactly $2 \mid E \mid r_{ij}$, then we have:

$$C_{ij} = F_{ij} + F_{ji} = 2 \mid E \mid r_{ij} \tag{4}$$

where $\mid E \mid$ denotes the number of edges in G and r_{ij} is the effective resistance between two nodes i and j. By combining the two above relations, the MFPT can be computed by the following formula:

$$< F_N > = \frac{2E_t}{N_t(N_t - 1)} \sum_{i<j}^{n} r_{ij}(G) = \frac{2E_t}{N_t - 1} \sum_{k=2}^{N_t} \frac{1}{v_k} \tag{5}$$

where v_k are the eigenvalues of the Laplacian matrix of the graph after t iterations.

The average path length defines the average number of steps along the shortest path d_{ij} for all possible pairs of network nodes, which is the measure of the efficiency of information or mass transport on the network, then the average path length D_t, for G(t) (the graph after t iterations) is defined as:

$$D_t = \frac{2}{N_t - 1} \sum_{j>i} d_{ij} \tag{6}$$

Moreover, the shortest path d_{ij} and effective resistance r_{ij} are related by expression $r_{ij} = \frac{2d_{ij}}{|N|}$, where N represents the number of nodes in the complete graph. Then, by these two relations, we have:

$$D_t = \frac{2}{N_t(N_t - 1)} \frac{N_t}{2} \sum_{j>i} r_{ij} = \frac{N_t}{N_t - 1} \frac{N_t}{2} N_t \sum_{k=2}^{N_t} \frac{1}{v_k} \tag{7}$$

Spanning trees are very important in complex networks and play a key role in various networks. The exact number of spanning trees $N_{st}(G_t)$, for $G(t)$, $(t \geq 1)$, where t shows the iterations in constructing a graph, discussed in the next section, can be computed by Kirchhoff's matrix tree theorem [11].

More precisely, in this paper, we have computed Laplacian spectra for categorical product networks and have discussed their applications. After the Introduction, in the second section, the materials and methods are discussed; in the main section, the categorical product network is defined in an iterative way, and then, the Laplacian spectra are computed. Finally, the applications of the Laplacian spectra are computed.

2. Materials and Methods

In this section, we state the materials and methods that are used in the main section.

In graph theory, the direct product $G \times H$ of graphs G and H is defined as a graph such that the vertex set of $G \times H$ is the Cartesian product $V(G) \times V(H)$ and any two vertices (u, u_1) and (v, v_1) are adjacent in $G \times H$ if and only if u is adjacent with v and u_1 is adjacent with v_1. The direct product is also called the tensor product or categorical product.

Let G be a graph with vertices $1, 2, \ldots, n$. The Laplacian matrix of G is $L(G) = D(G) - A(G)$, where A(G) is $n \times n$ adjacency, and the matrix of G with $(i, j) - entry$ is equal to 1 if vertices i and j are adjacent and 0 otherwise. $D(G)$ is the diagonal matrix of vertices' degrees.

Definition 1. *Consider two matrices A and B. The Kronecker product $A \otimes B$ of two matrices A and B is the matrix that is obtained by taking (i, j)-th entries as $a_{ij}B$ for all i, j. The Kronecker product of the matrix $A \in M^{(p,q)}$ with the matrix $B \in M^{(r,s)}$ is defined as (see [12,13]):*

$$\begin{pmatrix} a_{11}B & \cdots & a_{1q}B \\ .. & . & . \\ a_{p1}B & \cdots & a_{pq}B \end{pmatrix}$$

The Kronecker product has the following main properties.

$$\begin{aligned}
(\alpha A) \otimes B &= A \otimes (\alpha B) = \alpha(A \otimes B), \quad \forall, A \in M^{(p,q)}, B \in M^{(r,s)} \quad \text{and scalar } \alpha \\
(A \otimes B)^T &= A^T \otimes B^T, \quad \forall A \in M^{p,q}, B \in M^{r,s} \\
(A \otimes B) \otimes C &= A \otimes (B \otimes C), \quad \forall A \in M^{(m,n)}, B \in M^{(p,q)}, C \in M^{(r,s)} \\
(A \otimes B)(C \otimes D) &= AC \otimes BD, \quad \forall A \in M^{(p,q)}, B \in M^{(r,s)}, C \in M^{(q,k)}, D \in M^{(s,l)} \\
trace(A \otimes B) &= trace(B \otimes A) = trace(A)trace(B), \quad \forall A \in M^m, B \in M^n.
\end{aligned}$$

We have used the Kronecker product of matrices to find the Laplacian spectra for the categorical product considering the spectra for the path and cycle. Let P_n denote the path with n vertices. Let C_n be a cycle of length n. Then, their spectra can be stated as [14]:

Lemma 1. *The Laplacian eigenvalues of a path P_n are $2 - 2\cos\frac{i\pi}{n}$, $(i = 0, 1, \ldots, n-1)$. The Laplacian eigenvalues of a cycle C_n are $2 - 2\cos\frac{2j\pi}{n}$, $j = 0, 1, \ldots, n-1$.*

With the help of the following flowchart in Figure 1, we will facilitate the understanding of the proposed approach in this paper clearly.

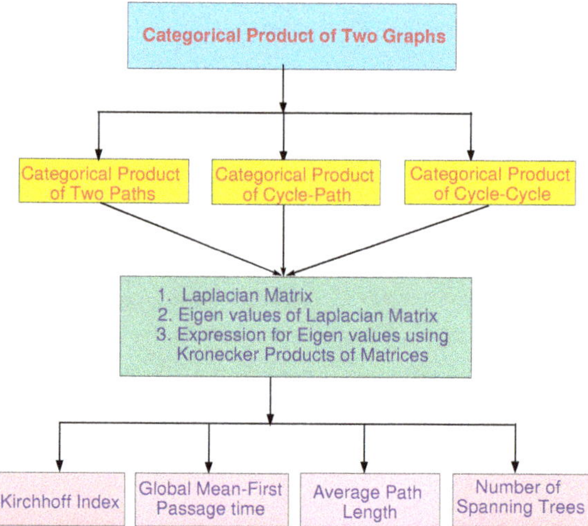

Figure 1. Flowchart for the methods.

3. Main Results

3.1. Categorical Product of Two Paths and Laplacian Spectra

Let a categorical product network of paths be constructed in an iterative way. We take categorical product network $G(t)$, $(t \geq 1)$ after $t-1$ iterations. Initially at $t = 1, G(1)$ is a path with n vertices. For $t \geq 2$, $G(t)$ is constructed from $G(t-1)$; from every existing vertex in $G(t-1)$, a new vertex is created so that a new path with n vertices is constructed; also, each new vertex in $G(t)$ is connected to the vertices in $G(t-1)$, shown by Figure 2. The number of vertices and edges in $G(t)$ is $N_t = nt$ and $E_t = (2t-2)n - (2t-2)$, $n \geq 2$.

Let λ_t and μ_t be the product of all nonzero eigenvalues of G_t and the sum of reciprocals of these eigenvalues, respectively, i.e., $\lambda_t = \prod_{i=2}^{N_t} \nu_i$ and $\mu_t = \sum_{i=2}^{N_t} \frac{1}{\nu_i}$, where $\nu_1 = 0$ and $\nu_i, i = 2, 3, \ldots, N_t$ denote the $N_t - 1$ nonzero eigenvalues of $L(G_t)$.

Theorem 1. *The product and sum of reciprocal nonzero eigenvalues of $L(G_t)$, the Laplacian matrix of $G(t)$, are:*

$$\lambda_t = \prod_{i=0}^{t-1}\prod_{j=0}^{n-1}\left((2-2\cos\frac{j\pi}{n})d_{i+1}+d_{j+1}(2-2\cos\frac{i\pi}{t})-(2-2\cos\frac{j\pi}{n})(2-2\cos\frac{i\pi}{t})\right)$$

$$\mu_t = \sum_{i=0}^{t-1}\sum_{j=0}^{n-1}\frac{1}{\left((2-2\cos\frac{j\pi}{n})d_{i+1}+d_{j+1}(2-2\cos\frac{i\pi}{t})-(2-2\cos\frac{j\pi}{n})(2-2\cos\frac{i\pi}{t})\right)}$$

d_i *shows the degree of vertex i.*

Proof. Consider a categorical product network G as is shown in the figure. By the properties of the Kronecker product of matrices, we can write the Laplacian matrix for G as [15],

$$L(G_t) = L(P_n)\otimes D(P_t) + D(P_n)\otimes L(P_t) - L(P_n)\otimes L(P_t)$$

where $D(P_n)$ is the diagonal matrix of order $n\times n$; with the diagonal elements' degree of vertices. By using the results from linear algebra, there exists invertible matrices P and Q such that:

$$(L(P_n))' = P^{-1}L(P_n)P, \quad (L(P_t))' = Q^{-1}L(P_t)Q$$

are the upper triangular matrices with diagonal elements, $2-2\cos\frac{\pi j}{n}$, $j = 0,1,\ldots,n-1$ and $2-2\cos\frac{\pi i}{t}$, $i = 0,1,\ldots,t-1$, respectively. Then, using the fact that:

$$(P\otimes Q)^{-1}\cdot(L(P_n)\otimes L(P_t))\cdot(P\otimes Q) = (P^{-1}\otimes Q^{-1})\cdot(L(P_n)P\otimes L(P_t)Q) = P^{-1}L(P_n)P\otimes Q^{-1}L(P_t)Q$$

is the upper triangular matrix with diagonal elements, the matrix:

$$(P\otimes Q)^{-1}\cdot(L(P_n)\otimes D(P_t) + D(P_n)\otimes L(P_t) - L(P_n)\otimes L(P_t))\cdot(P\otimes Q)$$

is upper triangular matrix with diagonal elements,

$$(2-2\cos\frac{j\pi}{n})d_{i+1}+d_{j+1}(2-2\cos\frac{i\pi}{t})-(2-2\cos\frac{j\pi}{n})(2-2\cos\frac{i\pi}{t}); \tag{8}$$
$$i = 0,1,\ldots,t-1, \quad j = 0,1,\ldots,n-1,$$
$$d_1 = d_t = d_n = d_{nt} = 1,$$
$$d_{k+1} = d_{nt-k} = 2, \quad k = 1,\ldots,n-2, \quad d_{1+nq} = d_{n+nq} = 2, \quad q = 1,\ldots,t-2,$$
$$\text{all other } d_i \text{ have a value of four.}$$

that are the eigenvalues for the categorical product network. Therefore:

$$\lambda_t = \prod_{i=0}^{t-1}\prod_{j=0}^{n-1}v_{i,j}$$

$$= \prod_{i=0}^{t-1}\prod_{j=0}^{n-1}\left((2-2\cos\frac{j\pi}{n})d_{i+1}+d_{j+1}(2-2\cos\frac{i\pi}{t})-(2-2\cos\frac{j\pi}{n})(2-2\cos\frac{i\pi}{t})\right) \tag{9}$$

$$\mu_t = \sum_{i=0}^{t-1}\sum_{j=0}^{n-1}\frac{1}{v_{i,j}}$$

$$= \sum_{i=0}^{t-1}\sum_{j=0}^{n-1}\frac{1}{\left((2-2\cos\frac{j\pi}{n})d_{i+1}+d_{j+1}(2-2\cos\frac{i\pi}{t})-(2-2\cos\frac{j\pi}{n})(2-2\cos\frac{i\pi}{t})\right)} \tag{10}$$

Corollary 1. *Let $L(G_2)$ be the Laplacian matrix of $P_n \times P_2$, the categorical product graph with n vertices after the first iteration, $t = 2$, then the product and sum of the reciprocal nonzero eigenvalues of $L(G_2)$ are:*

$$\lambda_2 = \prod_{i=0}^{1}\prod_{j=0}^{n-1} \nu_{i,j} = \prod_{i=0}^{1}\prod_{j=0}^{n-1}\left((2-2\cos\tfrac{i\pi}{n})d_{i+1} + d_{j+1}(2-2\cos\tfrac{i\pi}{t}) - (2-2\cos\tfrac{i\pi}{n})(2-2\cos\tfrac{i\pi}{t})\right)$$

$$\mu_2 = \sum_{i=0}^{1}\sum_{j=0}^{n-1}\frac{1}{\nu_{i,j}} = \sum_{i=0}^{1}\sum_{j=0}^{n-1}\frac{1}{\left((2-2\cos\tfrac{i\pi}{n})d_{i+1}+d_{j+1}(2-2\cos\tfrac{i\pi}{t})-(2-2\cos\tfrac{i\pi}{n})(2-2\cos\tfrac{i\pi}{t})\right)}$$

$\square$

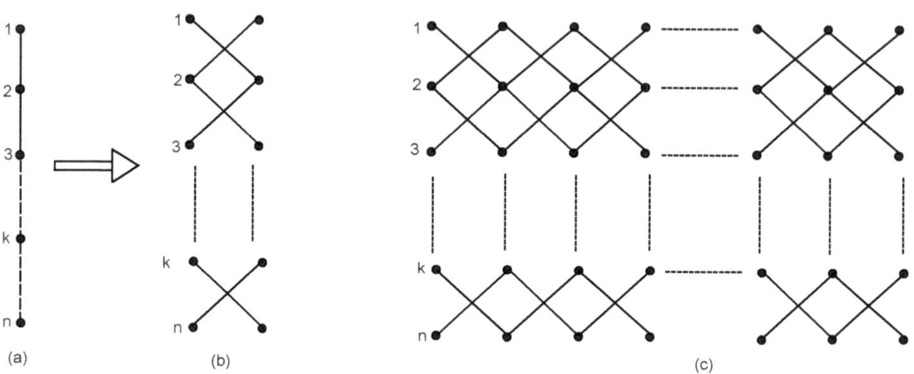

Figure 2. (a) Path P_n. (b) Grid $P_n \times P_2$ after the first iteration for $t = 2$. (c) General grid $P_n \times P_t$ structure.

3.2. Categorical Product of the Cycle-Path and Laplacian Spectra

Let a categorical product network of the cycle and path be constructed in an iterative way. We take the initial categorical product network $G(t)$, $(t \geq 1)$ after $t - 1$ iterations. Initially, at $t = 1, G(1)$ is a cycle with n vertices. For $t \geq 2$, $G(t)$ is constructed from $G(t - 1)$, from every existing vertex in $G(t - 1)$, a new vertex is created so that a new path with n vertices is constructed; also, each new vertex in $G(t)$ is connected to vertices in $G(t - 1)$, shown by Figure 3. The number of vertices and edges in $G(t)$ are $N_t = nt$ and $E_t = 2n(t - 1)$, $n \geq 2$.

Let λ_t and μ_t be the product of all nonzero eigenvalues of G_t and the sum of reciprocals of these eigenvalues, respectively, i.e., $\lambda_t = \prod_{i=2}^{N_t} \nu_i$ and $\mu_t = \sum_{i=2}^{N_t} \frac{1}{\nu_i}$, where $\nu_1 = 0$ and ν_i, $i = 2, 3, \ldots, N_t$ denote the $N_t - 1$ nonzero eigenvalues of $L(G_t)$.

Theorem 2. *The product and sum of reciprocal nonzero eigenvalues of $L(G_t)$, the Laplacian matrix of $G(t)$, are:*

$$\lambda_t = \prod_{i=0}^{t-1}\prod_{j=0}^{n-1}\left((2-2\cos\tfrac{2j\pi}{n})d_{i+1} + d_{j+1}(2-2\cos\tfrac{i\pi}{t}) - (2-2\cos\tfrac{2j\pi}{n})(2-2\cos\tfrac{i\pi}{t})\right)$$

$$\mu_t = \sum_{i=0}^{t-1}\sum_{j=0}^{n-1}\frac{1}{\left((2-2\cos\tfrac{2j\pi}{n})d_{i+1} + d_{j+1}(2-2\cos\tfrac{i\pi}{t}) - (2-2\cos\tfrac{2j\pi}{n})(2-2\cos\tfrac{i\pi}{t})\right)}$$

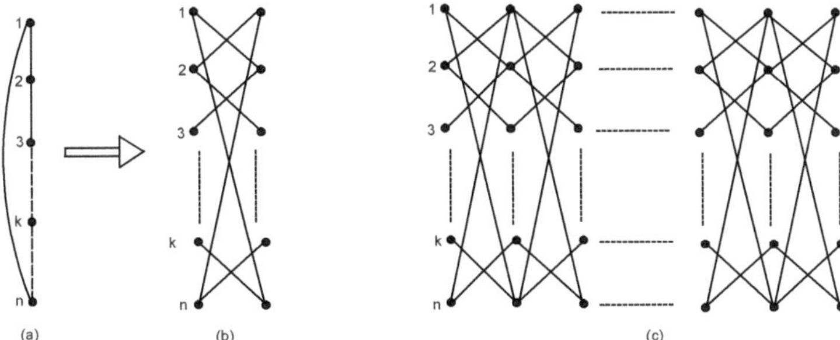

Figure 3. (a) Cycle C_n. (b) Grid $C_n \times P_2$ after the first iteration for $t = 2$. (c) General grid $C_n \times P_t$ structure.

Proof. Consider a categorical product network G as is shown in the figure. By the properties of the Kronecker product of matrices, we can write the Laplacian matrix for G as:

$$L(G_t) = L(C_n) \otimes D(P_t) + D(C_n) \otimes L(P_t) - L(C_n) \otimes L(P_t)$$

where $D(C_n)$ is the diagonal matrix of order $n \times n$; with the diagonal elements' degree of vertices. By using the results from linear algebra, there exists invertible matrices P and Q such that:

$$(L(C_n))' = P^{-1}L(C_n)P, \quad (L(P_t))' = Q^{-1}L(P_t)Q$$

are the upper triangular matrices with diagonal elements, $2 - 2\cos\frac{2\pi j}{n}$, $j = 0, 1, \ldots, n - 1$ and $2 - 2\cos\frac{\pi i}{t}$, $i = 0, 1, \ldots, t - 1$, respectively. Then, using the fact that:

$$(P \otimes Q)^{-1} \cdot (L(C_n) \otimes L(P_t)) \cdot (P \otimes Q) = (P^{-1} \otimes Q^{-1}) \cdot (L(C_n)P \otimes L(P_t)Q)$$

$$= P^{-1}L(C_n)P \otimes Q^{-1}L(P_t)Q$$

is the upper triangular matrix with diagonal elements, the matrix:

$$(P \otimes Q)^{-1} \cdot (L(C_n) \otimes D(P_t) + D(C_n) \otimes L(P_t) - L(C_n) \otimes L(P_t)) \cdot (P \otimes Q)$$

is the upper triangular matrix with diagonal elements,

$$(2 - 2\cos\frac{2j\pi}{n})d_{i+1} + d_{j+1}(2 - 2\cos\frac{i\pi}{t}) - (2 - 2\cos\frac{2j\pi}{n})(2 - 2\cos\frac{i\pi}{t}); \tag{11}$$
$$i = 0, 1, \ldots, t - 1, \quad j = 0, 1, \ldots, n - 1,$$
$$\text{all } d_i \text{ have a value of two.}$$

that are the eigenvalues for the categorical product network. Therefore,

$$\begin{aligned}
\lambda_t &= \prod_{i=0}^{t-1}\prod_{j=0}^{n-1} v_{i,j} \\
&= \prod_{i=0}^{t-1}\prod_{j=0}^{n-1}\left((2 - 2\cos\frac{2j\pi}{n})d_{i+1} + d_{j+1}(2 - 2\cos\frac{i\pi}{t}) - (2 - 2\cos\frac{2j\pi}{n})(2 - 2\cos\frac{i\pi}{t})\right)
\end{aligned} \tag{12}$$

$$\begin{aligned}
\mu_t &= \sum_{i=0}^{t-1}\sum_{j=0}^{n-1} \frac{1}{v_{i,j}} \\
&= \sum_{i=0}^{t-1}\sum_{j=0}^{n-1} \frac{1}{\left((2-2\cos\frac{2j\pi}{n})d_{i+1}+d_{j+1}(2-2\cos\frac{i\pi}{t})-(2-2\cos\frac{2j\pi}{n})(2-2\cos\frac{i\pi}{t})\right)}
\end{aligned} \tag{13}$$

$\square$

3.3. Categorical Product of the Cycle-Cycle and Laplacian Spectra

Let a categorical product network of cycles be constructed in an iterative way. We take initial categorical product network $G(t)$, $(t \geq 1)$ after $t - 1$ iterations. Initially, at $t = 1, G(1)$ is a categorical product network of cycle $C_n \times C_3$ with n vertices. For $t \geq 2$, $G(t)$ is constructed from $G(t - 1)$, and from every existing vertex in $G(t - 1)$, a new vertex is created so that n vertices are constructed; also, each new vertex in $G(t)$ is connected to the vertices in $G(t - 1)$, shown by Figure 4. The number of vertices and edges in $G(t)$ are $N_t = n(t + 2)$ and $E_t = 2n(t + 2) - 2$.

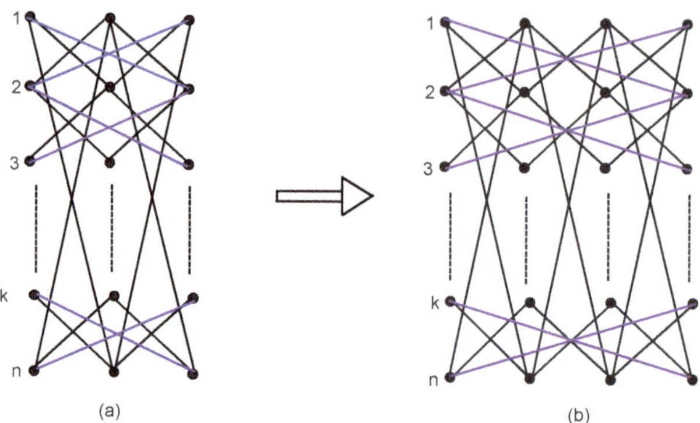

Figure 4. (a) Initial product $C_n \times C_3$ for $t = 1$. (b) Grid $C_n \times C_4$ after the first iteration for $t = 2$.

Let λ_t and μ_t be the product of all nonzero eigenvalues of G_t and the sum of reciprocals of these eigenvalues, respectively, i.e., $\lambda_t = \prod_{i=2}^{N_t} v_i$ and $\mu_t = \sum_{i=2}^{N_t} \frac{1}{v_i}$, where $v_1 = 0$ and v_i, $i = 2, 3, \ldots, N_t$ denote the $N_t - 1$ nonzero eigenvalues of $L(G_t)$.

Theorem 3. *The product and sum of the reciprocal nonzero eigenvalues of $L(G_t)$, the Laplacian matrix of $G(t)$, are:*

$$\lambda_t = \prod_{i=0}^{t+1} \prod_{j=0}^{n-1} \left((2 - 2\cos\frac{2j\pi}{n})d_{i+1} + d_{j+1}(2 - 2\cos\frac{2i\pi}{t+2}) - (2 - 2\cos\frac{2j\pi}{n})(2 - 2\cos\frac{2i\pi}{t+2}) \right)$$

$$\mu_t = \sum_{i=0}^{t+1} \sum_{j=0}^{n-1} \frac{1}{\left((2 - 2\cos\frac{2j\pi}{n})d_{i+1} + d_{j+1}(2 - 2\cos\frac{2i\pi}{t+2}) - (2 - 2\cos\frac{2j\pi}{n})(2 - 2\cos\frac{2i\pi}{t+2}) \right)}$$

Proof. Consider a categorical product network G as is shown in the figure. By the properties of the Kronecker product of matrices, we can write the Laplacian matrix for G as:

$$L(G_t) = L(C_n) \otimes D(C_t) + D(C_n) \otimes L(C_t) - L(C_n) \otimes L(C_t)$$

where $D(C_n)$ is the diagonal matrix of order $n \times n$; with the diagonal elements' degree of vertices. By using the results from linear algebra, there exists invertible matrices P and Q such that:

$$(L(C_n))' = P^{-1}L(C_n)P, \quad (L(C_t))' = Q^{-1}L(C_t)Q$$

are the upper triangular matrices with diagonal elements, $2 - 2\cos\frac{2\pi j}{n}$, $j = 0, 1, \ldots, n - 1$ and $2 - 2\cos\frac{2\pi i}{t+2}$, $i = 0, 1, \ldots, t + 1$, respectively. Then, using the fact that:

$$(P \otimes Q)^{-1} \cdot (L(C_n) \otimes L(C_t)) \cdot (P \otimes Q) =$$

$$(P^{-1} \otimes Q^{-1}) \cdot (L(C_n)P \otimes L(C_t)Q) = P^{-1}L(C_n)P \otimes Q^{-1}L(C_t)Q$$

is the upper triangular matrix with diagonal elements, the matrix:

$$(P \otimes Q)^{-1} \cdot (L(C_n) \otimes D(C_t) + D(C_n) \otimes L(C_t) - L(C_n) \otimes L(C_t)) \cdot (P \otimes Q)$$

is the upper triangular matrix with diagonal elements,

$$\begin{aligned}
&(2 - 2\cos\tfrac{2j\pi}{n})d_{i+1} + d_{j+1}(2 - 2\cos\tfrac{2i\pi}{t+2}) - (2 - 2\cos\tfrac{2j\pi}{n})(2 - 2\cos\tfrac{2i\pi}{t+2}); \\
&i = 0, 1, \ldots, t+1, \quad j = 0, 1, \ldots, n-1, \\
&d_1 = d_t = d_n = d_{nt} = 3, \quad \text{all other } d_i \text{ have a value of four.}
\end{aligned} \tag{14}$$

that are the eigenvalues for the categorical product network. Therefore:

$$\begin{aligned}
\lambda_t &= \prod_{i=0}^{t+1}\prod_{j=0}^{n-1} v_{i,j} \\
&= \prod_{i=0}^{t+1}\prod_{j=0}^{n-1}\left((2 - 2\cos\tfrac{2j\pi}{n})d_{i+1} + d_{j+1}(2 - 2\cos\tfrac{2i\pi}{t+2}) - (2 - 2\cos\tfrac{2j\pi}{n})(2 - 2\cos\tfrac{2i\pi}{t+2})\right)
\end{aligned} \tag{15}$$

$$\begin{aligned}
\mu_t &= \sum_{i=0}^{t+1}\sum_{j=0}^{n-1} \frac{1}{v_{i,j}} \\
&= \sum_{i=0}^{t+1}\sum_{j=0}^{n-1} \frac{1}{\left((2-2\cos\tfrac{2j\pi}{n})d_{i+1}+d_{j+1}(2-2\cos\tfrac{2i\pi}{t+2})-(2-2\cos\tfrac{2j\pi}{n})(2-2\cos\tfrac{2i\pi}{t+2})\right)}
\end{aligned} \tag{16}$$

$\square$

4. Applications of Laplacian Spectra

In this section, by the use of Expressions (15) and (16), we will find the Kirchhoff index, global mean-first passage time, average path length and the number of spanning trees.

4.1. Kirchhoff Index

The Kirchhoff index of a connected graph G is defined as the sum of resistance distances between all pairs of vertices, mathematically:

$$Kf(G(t)) = \sum_{i<j} r_{ij}(G)$$

where $r_{ij}(G)$ represents the resistance distance between a pair of vertices. In terms of eigenvalues for a connected network G of order N with all its nonzero eigenvalues represented by v_i; $i = 2, \ldots, N$, a well-known identity Kirchhoff index is defined as:

$$Kf(G(t)) = N_t \sum_{k=2}^{N_t} \frac{1}{v_k} = N_t \sum_{i=0}^{t-1}\sum_{j=0}^{n-1} \frac{1}{v_{i,j}}, \quad (i,j) \neq (0,0)$$

The Kirchhoff index for the categorical path-path product network:

$$Kf(G(t)) = nt \sum_{i=0}^{t-1}\sum_{j=0}^{n-1} \frac{1}{\left((2 - 2\cos\tfrac{j\pi}{n})d_{i+1} + d_{j+1}(2 - 2\cos\tfrac{i\pi}{t}) - (2 - 2\cos\tfrac{j\pi}{n})(2 - 2\cos\tfrac{i\pi}{t})\right)}$$

$$(i,j) \neq (0,0)$$

The Kirchhoff index for the categorical cycle-path product network:

$$Kf(G(t)) \;=\; nt \sum_{i=0}^{t-1} \sum_{j=0}^{n-1} \frac{1}{(2 - 2\cos\frac{2j\pi}{n})d_{i+1} + d_{j+1}(2 - 2\cos\frac{i\pi}{t}) - (2 - 2\cos\frac{2j\pi}{n})(2 - 2\cos\frac{i\pi}{t})}$$

The Kirchhoff index for the categorical cycle-cycle network:

$$Kf(G(t)) \;=\; n(t+2) \sum_{i=0}^{t+1} \sum_{j=0}^{n-1} \frac{1}{(2-2\cos\frac{2j\pi}{n})d_{i+1} + d_{j+1}(2-2\cos\frac{2i\pi}{t+2}) - (2-2\cos\frac{2j\pi}{n})(2-2\cos\frac{2i\pi}{t+2})}$$

4.2. Global Mean-First Passage Time

The global mean first-passage time (MFPT) is defined as the average of the first-passage time (FPT) between two nodes of a network. Mathematically, the global mean-first passage time (MFPT) for $G(t)$ is:

$$< F_N > \;=\; \frac{2E_t}{(N_t - 1)} \sum_{k=2}^{N_t} \frac{1}{v_k} = \frac{2E_t}{(N_t - 1)} \sum_{i=0}^{t-1} \sum_{j=0}^{n-1} \frac{1}{v_k}, \quad (i,j) \neq (0,0)$$

The global mean-first passage time for the categorical path-path product network:
Since $E_t = (2t - 2)n - (2t - 2)$ and $N_t = nt$:

$$< F_N > \;=\; \frac{2((2t - 2)n - (2t - 2))}{(nt - 1)}$$

$$\times \; \sum_{i=0}^{t-1} \sum_{j=0}^{n-1} \frac{1}{\left((2 - 2\cos\frac{j\pi}{n})d_{i+1} + d_{j+1}(2 - 2\cos\frac{i\pi}{t}) - (2 - 2\cos\frac{j\pi}{n})(2 - 2\cos\frac{i\pi}{t})\right)}$$

$$(i,j) \;\neq\; (0,0)$$

The global mean-first passage time for the categorical cycle-path product network:
Since $E_t = 2n(t - 1)$ and $N_t = nt$:

$$< F_N > \;=\; \frac{2(2n(t - 1))}{(nt - 1)}$$

$$\times \; \sum_{i=0}^{t-1} \sum_{j=0}^{n-1} \frac{1}{\left((2 - 2\cos\frac{2j\pi}{n})d_{i+1} + d_{j+1}(2 - 2\cos\frac{i\pi}{t}) - (2 - 2\cos\frac{2j\pi}{n})(2 - 2\cos\frac{i\pi}{t})\right)}$$

The global mean-first passage time for the categorical cycle-cycle product network:
Since $E_t = 2n(t + 2) - 2$ and $N_t = n(t + 2)$:

$$< F_N > \;=\; \frac{2(2n(t + 2) - 2)}{n(t + 2) - 1}$$

$$\times \; \sum_{i=0}^{t+1} \sum_{j=0}^{n-1} \frac{1}{\left((2 - 2\cos\frac{2j\pi}{n})d_{i+1} + d_{j+1}(2 - 2\cos\frac{2i\pi}{t+2}) - (2 - 2\cos\frac{2j\pi}{n})(2 - 2\cos\frac{2i\pi}{t+2})\right)}$$

4.3. Average Path Length

The average path length is an idea in network topology that can be defined as the average number of steps along with the shortest paths for all possible pairs of vertices of a network. The average path length in terms of the Laplacian eigenvalues is defined as:

$$D_t = \frac{N_t}{N_t - 1} \sum_{k=2}^{N_t} \frac{1}{v_k}, \quad (i,j) \neq (0,0)$$

The average path length for the categorical path-path product network:

$$D_t = \frac{nt}{nt-1} \sum_{i=0}^{t-1} \sum_{j=0}^{n-1} \frac{1}{\left((2-2\cos\frac{j\pi}{n})d_{i+1} + d_{j+1}(2-2\cos\frac{i\pi}{t}) - (2-2\cos\frac{j\pi}{n})(2-2\cos\frac{i\pi}{t}) \right)}$$

The average path length for the categorical cycle-path product network:

$$D_t = \frac{nt}{nt-1} \sum_{i=0}^{t-1} \sum_{j=0}^{n-1} \frac{1}{\left((2-2\cos\frac{2j\pi}{n})d_{i+1} + d_{j+1}(2-2\cos\frac{i\pi}{t}) - (2-2\cos\frac{2j\pi}{n})(2-2\cos\frac{i\pi}{t}) \right)}$$

The average path length for the categorical cycle-cycle product network:

$$D_t = \frac{n(t+2)}{n(t+2)-1}$$
$$\times \sum_{i=0}^{t+1} \sum_{j=0}^{n-1} \frac{1}{\left((2-2\cos\frac{2j\pi}{n})d_{i+1} + d_{j+1}(2-2\cos\frac{2i\pi}{t+2}) - (2-2\cos\frac{2j\pi}{n})(2-2\cos\frac{2i\pi}{t+2}) \right)}$$

4.4. The Number of Spanning Trees

The number of spanning trees has a very important role in various networks, and it can be computed using the product of nonzero eigenvalues of the Laplacian matrix [16];

$$N_{ST}(G_t) = \frac{\prod_{k=2}^{N_t} v_i}{N_t} = \frac{A_t}{N_t} = \frac{\prod_{i=0}^{t-1} \prod_{j=0}^{n-1} v_{i,j}}{N_t}, \quad (i,j) \neq (0,0)$$

The number of spanning trees for the categorical path-path product network:

$$N_{ST}(G_t) = \frac{\prod_{i=0}^{t-1} \prod_{j=0}^{n-1} \left((2-2\cos\frac{j\pi}{n})d_{i+1} + d_{j+1}(2-2\cos\frac{i\pi}{t}) - (2-2\cos\frac{j\pi}{n})(2-2\cos\frac{i\pi}{t}) \right)}{nt}$$

The number of spanning trees for the categorical cycle-path product network:

$$N_{ST}(G_t) = \frac{\prod_{i=0}^{t-1} \prod_{j=0}^{n-1} \left((2-2\cos\frac{2j\pi}{n})d_{i+1} + d_{j+1}(2-2\cos\frac{i\pi}{t}) - (2-2\cos\frac{2j\pi}{n})(2-2\cos\frac{i\pi}{t}) \right)}{nt}$$

The number of spanning trees for the categorical cycle-cycle product network:

$$N_{ST}(G_t) = \frac{\prod_{i=0}^{t+1} \prod_{j=0}^{n-1} \left((2-2\cos\frac{2j\pi}{n})d_{i+1} + d_{j+1}(2-2\cos\frac{2i\pi}{t+2}) - (2-2\cos\frac{2j\pi}{n})(2-2\cos\frac{2i\pi}{t+2}) \right)}{n(t+2)}$$

5. Discussion

Laplacian spectra and their applications for different networks have been studied for many years, but the Laplacian spectra and their applications for networks based on graph operations are very rare. Laplacian spectra are used as a tool to analyze the structure of a network, and the importance of graph products cannot be denied in multilayer networking. Therefore, we have considered all cases of categorical product networks to compute Laplacian spectra and further to find their applications. The Laplacian spectra of a family of recursive trees and their applications in network coherence were discussed in [17]. Furthermore, we have computed the Laplacian spectra of the categorical product network and have established the expressions for the product of the nonzero Laplacian eigenvalues and the sum of the reciprocals of all nonzero Laplacian eigenvalues. Using these expressions, we have computed the Kirchhoff index, also called the "network criticality", for the categorical product network. The enumeration of spanning trees on generalized pseudofractal networks is discussed in [18].

6. Conclusions

In this paper, we discuss a complex network based on the categorical product and have used the spectrum approach to find the Kirchhoff index. The global mean first-passage time (MFPT) is computed for a complex network based on the categorical product. Moreover, using the Laplacian spectra for the categorical product network, we compute the average path length, which is the basic idea in network topologies. It describes the measure of the efficiency of transport (mass or information) on a network. The last application of Laplacian spectra for the categorical product network that we have computed was for spanning trees, which is the direct application in designing a network. We can extend our work to different networks based on graph operations other than the categorical product.

These results are indirectly related to entropy. Using the same spectra, we have future plans to compute the global first-passage time for maximal-entropy random walks in categorical product networks, and the categorical product network entropy is also in our future plans using the idea given in [19,20].

Author Contributions: S.M.K. contribute for conceptualization, funding, and analyzed the data. M.K.S. contribute for supervision, methodology, software, validation, designing the experiments and formal analysing. N.A.R. and M.H.M. contribute for performed experiments, resources, some computations and wrote the initial draft of the paper which were investigated and approved by M.K.S. and M.I., and wrote the final draft. All authors read and approved the final version of the paper.

Acknowledgments: The authors are grateful to the anonymous referees for their valuable comments and suggestions that improved this paper. This research is supported by the Start-Up Research Grant 2016 of the United Arab Emirates University (UAEU), Al Ain, United Arab Emirates via Grant No. G00002233 and UPAR Grant of UAEU via Grant No. G00002590. Also This research is supported by The Higher Education Commission of Pakistan Under Research and Development Division, National Research Program for Universities via Grant No.: 5282/Federal/NRPU/R&D/HEC/2016.

Conflicts of Interest: The authors declare no conflict of interest.

References

1. Arenas, A.; Diaz-Guilera, A.; Kurths, J.; Moreno, Y.; Zhou, C.S. Synchronization in complex networks. *Phys. Rep.* **2008**, *469*, 93–153. [CrossRef]
2. Boccaletti, S.; Latora, V.; Moreno, Y.; Chavez, M.; Hwang, D.U. Complex networks: structure and dynamics. *Phys. Rep.* **2006**, *424*, 175–308. [CrossRef]
3. Liu, J.B.; Cao, J.; Alofi, A.; Al-Mazrooei, A.; Elaiw, A. Applications of Laplacian Spectra for n-prism networks. *Neurocomputing* **2016**, *198*, 69–73. [CrossRef]
4. Ding, Q.Y.; Sun, W.G.; Chen, F.Y. Applications of Laplacian spectra on 3-prism graph. *Mod. Phys. Lett. B* **2014**, *28*, 1450–1459. [CrossRef]
5. Hou, B.Y.; Zhang, H.I.; Liu, L. Applications of Laplacian spectra for extended Koch networks. *Eur. Phys. J. B* **2012**, *85*, 303–310. [CrossRef]
6. Klien, D.I.; Randic, M. Resistance distance. *J. Math. Chem.* **1993**, *2*, 81–95. [CrossRef]
7. Liu, I.B.; Pan, X.F.; Cao, I.; Hu, F.F. A note on some physical and chemical indices of clique-inserted lattices. *J. Stat. Mech. Theory Exp.* **2014**, *6*, 60–66.
8. Kaminska, A.; Srokowski, T. Mean first passage time for a Markovian jumping process. *Acta Phys. Pol. B* **2007**, *38*, 3119–3123.
9. Zhang, Z.Z.; Liu, H.X.; Wu, B.; Zhou, S.G. Enumeration of spanning trees in a pseudofractal scale-free web. *Europhys. Lett.* **2010**, *90*, 680–688. [CrossRef]
10. Chandra, A.K.; Raghavan, P.; Ruzzo, W.L.; Smolensky, T.; Tiwari, P. The electrical resistance of a graph captures its commute and cover times. *Comput. Complex.* **1996**, *6*, 312–340. [CrossRef]
11. Godsil, C.; Royle, G. *Algebraic Graph Theory, Graduate Texts in Mathematics*; Springer: New York, NY, USA, 2001.

12. Gao, X.; Luo, Y.; Liu, W. Resistance distances and the Kirchhoff index in Cayley graphs. *Discret. Appl. Math.* **2011**, *159*, 2050–2057. [CrossRef]
13. Horn, R.A.; Johnson, C.R. *Matrix Analysis*; Cambridge University Press: Cambridge, UK, 1985.
14. Brouwer, A.E.; Haemers, W.H. *Spectra of Graphs*; Springer: New York, NY, USA, 2011.
15. Sayama, H. Estimation of Laplacian spectra of direct and strong product graphs. *Discret. Appl. Math.* **2016**, *205*, 160–170. [CrossRef]
16. Zhang, Z.Z.; Wu, B.; Comellas, F. The number of spanning trees in Apollonian networks. *Discret. Appl. Math.* **2014**, *169*, 206–213. [CrossRef]
17. Sun, W.; Xuan, T.; Qin, S. Laplacian spectrum of a family of recursive trees and its applications in network coherence. *J. Stat. Mech.* **2016**, *6*, 063205. [CrossRef]
18. Xiao, J.; Zhang, J.; Sun, W. Enumeration of spanning trees on generalized pseudofractal networks. *Fractals* **2015**, *23*. [CrossRef]
19. Lin, Y.; Zhang, Z. Mean first-passage time for maximal-entropy random walks in complex networks. *Sci. Rep.* **2014**. [CrossRef] [PubMed]
20. Mowshowitz, A.; Dehmer, M. Entropy and the Complexity of Graphs Revisited. *Entropy* **2012**, *14*, 559–570. [CrossRef]

Article

Dynamics of Trapped Solitary Waves for the Forced KdV Equation

Sunmi Lee

Department of Applied Mathematics, Kyunghee University, Yongin-si, Korea; sunmilee@khu.ac.kr;
Tel.: +82-31-201-2409

Received: 27 March 2018; Accepted: 23 April 2018; Published: 24 April 2018

Abstract: The forced Korteweg-de Vries equation is considered to investigate the impact of bottom configurations on the free surface waves in a two-dimensional channel flow. In the study of shallow water waves, the bottom topography plays a critical role, which can determine the characteristics of wave motions significantly. The interplay between solitary waves and the bottom topography can exhibit more interesting dynamics of the free surface waves when the bottom configuration is more complex. In the presence of two bumps, there are multiple *trapped solitary wave solutions*, which remain stable between two bumps up to a finite time when they evolve in time. In this work, various stationary trapped wave solutions of the forced KdV equation are explored as the bump sizes and the distance between two bumps are varied. Moreover, the semi-implicit finite difference method is employed to study their time evolutions in the presence of two-bump configurations. Our numerical results show that the interplay between trapped solitary waves and two bumps is the key determinant which influences the time evolution of those wave solutions. The trapped solitary waves tend to remain between two bumps for a longer time period as the distance between two bumps increases. Interestingly, there exists a nontrivial relationship between the bump size and the time until trapped solitary waves remain stable between two bumps.

Keywords: forced Korteweg-de Vries equation; trapped solitary wave solutions; numerical stability; two bumps or holes; finite difference method

---

1. Introduction

Free surface waves of a two-dimensional channel flow for an inviscid and incompressible fluid have been studied when the rigid bottom of the channel has some obstacles [1,2]. Such free surface waves of shallow water with obstacles can be modeled by the forced Korteweg-de Vries (fKdV) equations [3,4]. There have been enormous applications of the Korteweg-de Vries (KdV) equation in various research areas such as mathematics, physics, fluid mechanics and hydrodynamics. The KdV equation with a forcing approximately describes the evolution of the free surface when a fluid flows over an obstacle. This forced KdV equation is related to the physical problems such as shallow-water waves over rocks, tsunami waves over obstacles, oceanic stratified flows encountering topographic obstacles, and acoustic waves on a crystal lattice. Moreover, it can be useful for other nonlinear differential equations including a nonlinear Schrodinger equation and a sine Gordon equation. These applications range from magnetohydrodynamic waves, geostrophic turbulence, atmosphere dynamics and the propagation of short laser pulses in optical fibers (see [5,6] and references therein).

The dynamics of the free surface wave motions is characterized by the Froude number F defined as $F = C/\sqrt{gh}$, where C is the upstream velocity and $\sqrt{gh}$ is the critical speed of a shallow water wave with the finite depth of the channel h and the gravity constant g. It can be expressed as $F = 1 + \epsilon\lambda$ with the perturbation measurement λ to the critical value 1 for a positive small parameter ϵ, then the forced Korteweg-de Vries (fKdV) equation can be obtained. Further, the flow can be classified as supercritical if $\lambda > 0$ and subcritical if $\lambda < 0$. For supercritical flows, there is a critical value, λ_c, such that solitary wave solutions exist when $\lambda > \lambda_c$ and no such solutions exist when $0 < \lambda < \lambda_c$. Solitary wave solutions present some special features such as remaining stable without any changes in shape or speed when they evolve in time. Moreover, this critical value of λ_c is dependent on the bump configuration.

The bottom topography is a key determinant for the characteristic of wave motions. Interesting surface wave phenomena can be observed when a bottom topography is rather complicated. For one bump, there has been extensive analytic and numerical work [1,3,7,8]. Cusped stationary solitary wave solutions are founded analytically and their numerical stability is shown for the Dirac delta function as forcing [9]. It is well known that there exist two branches of stationary solitary wave solutions when a semi-circular bump is given [10]. One is the near zero solitary wave with lower amplitude and the other one is a higher amplitude one (sech^2-shaped wave). Analytic stability of solitary wave solutions is discussed for the time-dependent forced KdV equation when a compactly supported symmetric bump is given [7]. It is proved that the near zero solution is stable with respect to time for a sufficiently large $\lambda > 0$. The other solitary wave solution (sech^2-shaped solution) is unstable when it evolves in time.

Recently, more researchers pay attention to the cases where bottom configurations are more complicated [11,12]. For subcritical flows, the generalized hydraulic falls have been obtained when two bumps are given [13]. These solutions are characterized by a train of waves 'trapped' between two bumps satisfying the radiation conditions at the infinity. The steady surface flows over a step and a rectangular obstacle have been investigated for supercritical, subcritical waves and hydraulic falls [14]. Various stationary solutions are explored when the bottom configurations are complex including one positive bump, one negative hole, or combinations of both [15]. Stationary solutions of the forced KdV with obstacles with a negative hole have been illustrated as the amplitude and the width of a negative hole are varied [16].

Analytic and numeric stationary solutions of the forced KdV equations are studied when one bump or two bumps are given as forcing, which are in the form of sech^2- or sech^4-functions [17]. They also explored the stability of solitary waves and table-top solutions when they evolve in time. Multiple stationary supercritical solutions of the fKdV equation are discussed when two-semi-circular bumps are given [18]. Their results show that there exists only one supercritical positive solitary wave, which is stable when it evolves in time. Solitary wave solutions and their time evolutions are investigated when the rigid bottom has one negative hole [19,20]. Interestingly, there are five stationary solutions when a negative hole is given and only the near zero wave remains stable when they evolve in time [19].

In our previous study, stationary *trapped solitary wave solutions* have been studied for the forced Korteweg-de Vries equation when two bumps or two holes are given as the bottom configuration [21]. Some of multiple stationary solitary waves have been found and then their numerical stability has been investigated in the presence of two bumps [22]. Interestingly, multiple *trapped supercritical wave solutions* remain stable between two bumps for a very long time when they evolve in time. It is worth noting that the interplay between trapped solitary waves and two bumps plays a key role in determining their time evolutions. In this work, we extend the previous work by exploring the impact of different two-bump scenarios on the trapped solitary waves and their numerical stability. First, various stationary trapped solitary wave solutions of the forced KdV equation have been obtained in the presence of two bumps or two holes. As the bump size or the distance between two bumps are varied, stationary solitary waves are obtained. Next, we employ the semi-implicit finite difference method which has been developed for the homogeneous KdV equation [23]. This numerical method has been used to obtain the time evolution of the trapped stationary solitary waves. Our numerical results show that multiple *trapped solitary wave solutions* stay stable between two bumps for a very long time under various two-bump configurations. This indicates that the interplay between trapped solitary waves and two bumps highlights the importance in determining the dynamics of trapped solitary waves.

2. The Forced KdV Equation

2.1. Model Equation with Two-Bump Configurations

The forced KdV equation is written as follows

$$
\begin{aligned}
&\eta_t(x,t) + 2\lambda\eta_x(x,t) - \frac{3}{2}(\eta^2)_x(x,t) - \frac{1}{3}\eta_{xxx}(x,t) = f_x(x), \quad -\infty < x < \infty \\
&\eta(x,0) = \eta_0(x) \\
&\eta(x,t) = \eta_x(x,t) = \eta_{xx}(x,t) = 0 \text{ at } x \pm \infty,
\end{aligned}
\tag{1}
$$

where $\eta(x,t)$ is the free surface elevation measured from an undisturbed water level, and λ is the measurement of the perturbation of the upstream uniform flow velocity from its critical value. The external forcing $f(x)$ is given by the topography of the rigid bottom. Our main interests in this work are numerical stability when stationary wave solutions evolve in time. Hence, the problem (1) defined in $(-\infty, \infty)$ can be restricted to a finite interval $[-b, b]$ for a positive constant b, where this interval is large enough to use a finite difference scheme. First, at these finite boundaries, we employ the artificial boundary condition, which has been proposed in [21]. Then, the time-independent nonlinear boundary value problem is efficiently solved using the Newton method with the artificial boundary condition which is described as

$$
\begin{aligned}
&2\lambda\eta(x) - \frac{3}{2}\eta^2(x) - \frac{1}{3}\eta_{xx}(x) = f(x), \quad -b \le x \le b \\
&\eta_x(x) = \mp\sqrt{6\lambda}\eta(x) \text{ at } x = \pm b
\end{aligned}
\tag{2}
$$

As shown in [21], the method provides a good alternative to find various stationary solutions under arbitrary two-bump configurations. Once the stationary wave solutions have been found, we take these stationary solutions as initial conditions in (1). Next, the finite difference method described in the next subsection is performed to solve the time-dependent forced KdV equation in a finite interval $x \in [-b, b]$ for $t > 0$.

Moreover, the bottom topography $f(x)$ is modeled by the following functions for two bumps

$$
h(x; a, P_1, P_2) = \begin{cases}
P_1 \cos^4(\frac{\pi}{2}(x + a)) & \text{if } |x + a| \le 1, \\
P_2 \cos^4(\frac{\pi}{2}(x - a)) & \text{if } |x - a| \le 1, \\
0 & \text{elsewhere.}
\end{cases}
$$

Note that $h(x; a, P_1, P_2)$ consists of two bumps (centered at a and $-a$) with the distance between the boundaries of two bumps, $d = 2(a - 1)$.

2.2. Semi-Implicit Finite Difference Method

The problem (1) defined in $(-\infty, \infty)$ can be restricted to a finite interval $\Omega = [-b, b]$ for a positive constant b. This Initial Boundary Value Problem has a third order derivative with respect to x and, due to the presence of a forcing, the direction of wave propagation is unknown. Therefore, it is not a trivial task to impose the correct boundary conditions at finite boundaries. This leads us to solve the alternative form of the forced KdV equation. This results in the following Initial Boundary Value problem

$$
\begin{aligned}
&u_{tx}(x,t) + 2\lambda u_{xx}(x,t) - \frac{3}{2}(u^2)_{xx}(x,t) - \frac{1}{3}u_{xxxx}(x,t) = f_{xx}(x), \\
&u(x,0) = u_0(x) \\
&u_x(x,t) = u_{xx}(x,t) = 0 \text{ at } x = \pm b,
\end{aligned}
\tag{3}
$$

The problem involves the fourth derivative for the spatial variable, therefore four boundary conditions are required. This is done by imposing the zero Neumann boundary conditions at each boundary. Now, we employ the finite difference scheme to find the numerical solutions of equation above. Let us denote $x_0 = -b, x_1 = -b + \Delta x, \cdots, x_i = -b + i\Delta x, \cdots, x_N = b$, $t_0 = 0, t_1 = \Delta t, \cdots, t_n = n\Delta t, \cdots, t_{N_t} = T$, $u(x_i, t_n) = u_i^n$ and $f_{xx}(x_i) = \bar{f}_i$ where $\Delta x = 2b/N$ and $\Delta t = T/N_t$ with positive integers N and N_t, respectively. First, the Crank-Nicolson scheme with standard difference schemes is applied to (3) and the resulting finite difference equations are written as

$$\frac{D^0(u_i^{n+1} - u_i^n)}{2\Delta x \Delta t} + 2\lambda \Delta_h \frac{(u_i^{n+1} + u_i^n)}{2(\Delta x)^2} - \frac{3}{2}\Delta_h \frac{(F_i^{n+1} + F_i^n)}{2(\Delta x)^2} - \frac{1}{3}\Delta_h^2 \frac{(u_i^{n+1} + u_i^n)}{2(\Delta x)^4} = \bar{f}_i, \tag{4}$$

$$u_i^0 = u(x_i, 0)$$

$$D^-(u_0) = 0 \text{ and } D^+D^-(u_{-1}) = 0$$

$$D^+(u_N) = 0 \text{ and } D^+D^-(u_{N+1}) = 0,$$

for $i = 0, 1, \cdots, N, n = 0, 1, \cdots, N_t$. The standard finite difference is used for forward $D^+u_i^n = u_{i+1}^n - u_i^n$, backward $D^-u_i^n = u_i^n - u_{i-1}^n$, central $D^0u_i^n = u_{i+1}^n - u_{i-1}^n$, $\Delta_h = D^+D^-u_i^n = u_{i+1}^n - 2u_i^n + u_{i-1}^n$, and $\Delta_h^2 = \Delta_h\Delta_h u_i^n = u_{i+2}^n - 4u_{i+1}^n + 6u_i^n - 4u_{i-1}^n + u_{i-2}^n$.

Note that the nonlinear term $F = u^2$ is linearized by taking a Taylor expansion for the nonlinear convection term F_i^{n+1} as

$$F_i^{n+1} = F_i^n + \Delta t \left(\frac{\partial F}{\partial t}\right)_i^n + O(\Delta t^2), \tag{5}$$

where $\frac{\partial F}{\partial t} = \frac{\partial F}{\partial u}\Delta u = 2u_i^n(u_i^{n+1} - u_i^n)$. Therefore, the nonlinear convection term can be linearized in the following form

$$F_i^{n+1} + F_i^n = 2u_i^n u_i^{n+1}. \tag{6}$$

Substituting (6) into (4) leads to the following system

$$a_i^n u_{i-2}^{n+1} + b_i^n u_{i-1}^{n+1} + e_i^n u_i^{n+1} + c_i^n u_{i+1}^{n+1} + a_i^n u_{i+2}^{n+1} = d_i^n, \tag{7}$$

where

$$a_i^n = r, \quad b_i^n = 2p\lambda - 3pu_{i-1}^n - 4r - 1, \quad c_i^n = 2p\lambda - 3p\eta_{i+1}^n - 4r + 1 \quad e_i^n = -4p\lambda + 6pu_i^n + 6r$$

and

$$d_i^n = -ru_{i-2}^n + (-1 - 2p\lambda + 4r)\eta_{i-1}^n + (4p\lambda - 6r)u_i^n + (1 - 2p\lambda + 4r)u_{i+1}^n - r, u_{i+2}^2 + 2\Delta x \Delta t \bar{f}_i,$$

where $p = \Delta t/\Delta x, r = -\Delta t/3(\Delta x)^3$. This system is solved at each time step. The computational domain is taken to be b= 50 with $\Delta x = 0.1$ and $\Delta t = 0.05$. Our scheme is based on the linearized implicit scheme as proposed in [23,24]. This scheme is unconditionally linearly stable with a local truncation error, $O(\Delta x^2, \Delta t^2)$. Therefore, there is no restriction on Δx and Δt (more details on this numerical scheme are found in [23,24]).

3. Numerical Simulations

In this section, we present numerical simulations under various two-bump configurations including two bumps (positive forcing) or two holes (negative forcing). Our main focus is on trapped solitary waves between two bumps or two holes and their stability as they evolve in time. In particular, the bottom configuration is confined as two identical or symmetric positive bumps or negative holes. First, stationary trapped solitary waves are found as varying the distance between two bumps and the bump size. Then, we investigate the impact of various bump scenarios on their numerical stability.

3.1. Trapped Solitary Waves between Two Positive Bumps

As well known, the time independent KdV equation without any forcing (i.e. $f(x) = 0$) has two exact solutions such as $\eta_1(x) \equiv 0$ and $\eta_2(x) = 2\lambda sech^2[(6\lambda)^{1/2}(x - x_0)/2]$ with a phase shift x_0. Also, in the presence of a single positive symmetric forcing which satisfies a compact support condition, there are two symmetric solitary wave solutions: one is near $\eta_1(x)$ and the other one is near $\eta_2(x)$ for all $\lambda > \lambda_c$ with a critical value $\lambda_c > 0$ [7]. Under two-bump configurations, with the similar argument, there exist at least two stationary solitary wave solutions, since the bump $h(x; a, P_1, P_2)$ is positive and symmetric with a compact support. Recall that $h(x; a, P_1, P_2)$ is defined as each bump is centered at $-a$ and a and the bump height at each center is $P_1 = P_2$.

Figure 1 displays eight stationary wave solutions using the two-bump configuration $h(x; 2, 1, 1)$ and $\lambda = 1.5$. The left panel shows two symmetric solitary wave solutions including the lower amplitude wave (blue dashed), which is the near zero solution denoted by $\eta_s(x)$. Two non-symmetric wave solutions (with lower and higher amplitude, respectively) are shown in the middle panel. The right panel represents four stationary wave solutions which consist of two symmetric solutions and two non-symmetric ones. The blue dashed wave illustrated in the right panel of Figure 1 depicts the solitary wave solution which is trapped between two bumps similar to $\eta_2(x)$ ($sech^2$-type wave is placed in the middle of two bumps). Let this be referred to as a *trapped solitary wave solution* and denoted by $\eta_T(x)$. In this study, we focus on the dynamics of these trapped solitary waves, $\eta_T(x)$ (a higher amplitude wave in the middle of two bumps) under various bump configurations.

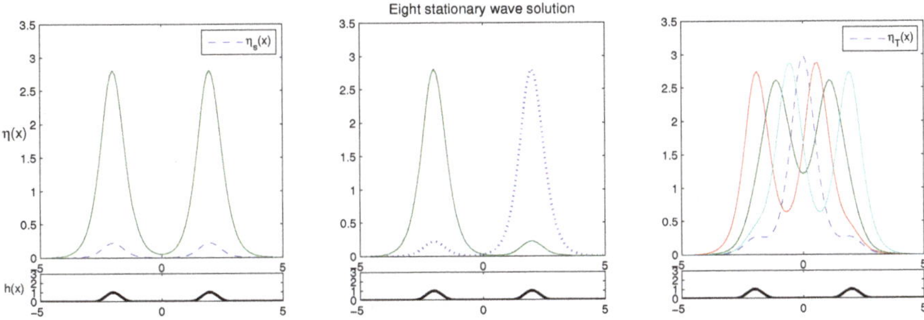

Figure 1. Stationary wave solutions and a two-bump configuration are shown with $\lambda = 1.5$ and $h(x; 2, 1, 1)$. The left panel shows the stable solitary wave $\eta_s(x)$ (the blue dashed wave) while the middle panel shows two non-symmetric waves. The right panel displays the trapped solitary wave solution $\eta_T(x)$, the blue dashed wave between two bumps.

It is well known that the near zero wave solution $\eta_s(x)$ is stable with respect to time when either single positive symmetric bump or two positive bumps is given. The solitary wave solutions with two higher amplitudes (the solid curves in the left panel of Figure 1) are moving out fast since they are placed right over the bumps and this makes them unstable in a very short time. However, the trapped solitary wave which is placed between two bumps ($\eta_T(x)$ in the right panel of Figure 1) remains stable for a very long time as observed [22]. Here, the time evolution of the trapped solitary wave, $\eta_T(x)$ is revisited in Figures 2 and 3. The unperturbed trapped solitary wave solution ($\eta_T(x)$) remains stable between two bumps until $t = 250$ in Figure 2. Figure 3 displays the time evolutions of perturbed trapped solitary wave solutions with $+5\%$ and -5% perturbation, respectively. Under $\pm5\%$ perturbations, trapped solitary waves start moving out of two bumps around $t = 70$. It is worth noting that the trapped solitary wave $\eta_T(x)$ consists of the unforced solitary wave $\eta_2(x)$ and the near zero solitary waves $\eta_s(x)$ with two bumps. Hence, we can conjecture that the trapped solitary wave $\eta_T(x)$ remains stable up to a certain time due to the stable interplay between $\eta_2(x)$ and $\eta_s(x)$.

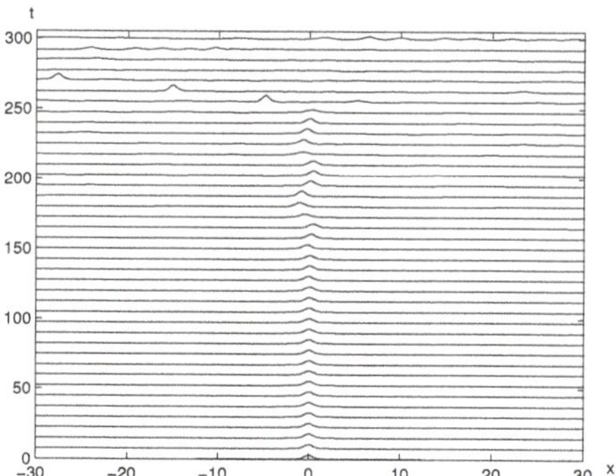

Figure 2. The time evolution of the trapped solitary wave solution is illustrated without any perturbation using $h(x, 2; 1, 1)$ and $\lambda = 1.5$. The trapped solitary wave remains stable up to a long time up to $t = 250$.

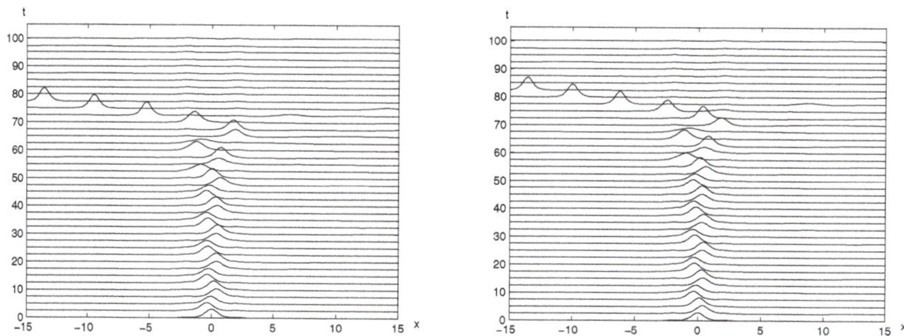

Figure 3. The time evolution of the trapped solitary wave solution is shown using $h(x, 2; 1, 1)$ and $\lambda = 1.5$ with $+5\%$ perturbation (**left**) and -5% perturbation (**right**), respectively. They start moving between two bumps and evolve out of two bumps around $t = 70$.

3.2. The Impact of the Bump Distance on the Stability of $\eta_T(x)$

We investigate the impact of the distance between two bumps on the trapped solitary waves. Figure 4 illustrates three stationary trapped solitary wave solutions using three different bump distances between two bumps. All of their shapes are similar to $\eta_2(x)$ around $x = 0$ with the near zero wave just over each bump. The distance between two bumps are varied as 2, 4 and 6 using $h(x, 2; 1, 1)$, $h(x, 3; 1, 1)$ and $h(x, 4; 1, 1)$, respectively. Their time evolutions are illustrated in Figures 5 and 6.

First, the time evolution of the unperturbed trapped solitary wave solution is displayed using $h(x, 3; 1, 1)$ (the distance between two bumps is 4) in Figure 5. It is interesting to observe that the trapped solitary wave solution is stable up to a very long time (simulated until $t = 500$). This can be compared with the unperturbed result with $h(x, 2; 1, 1)$ stable up to $t = 250$ in Figure 2. Next, the trapped solitary wave using $h(x, 3; 1, 1)$ are perturbed with $+5\%$ and -5% in Figure 6. They move back and forth between two bumps, however, they still remain between two bumps for a very long time (the results shown until $t = 100$).

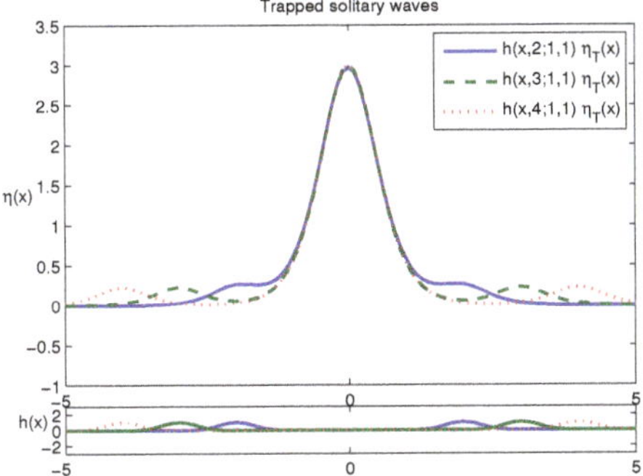

Figure 4. Stationary trapped solitary wave solutions and two-bump configurations are displayed using three bump distances $h(x, 2; 1, 1)$, $h(x, 3; 1, 1)$ and $h(x, 4; 1, 1)$, respectively.

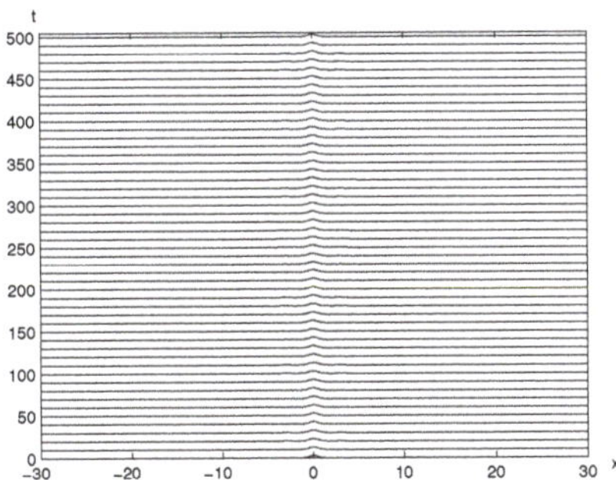

Figure 5. The time evolution of the trapped solitary wave solution is displayed with $h(x, 3; 1, 1)$ and $\lambda = 1.5$. The unperturbed trapped solitary wave remains stable between two bumps for a very long time (simulated up to $t = 500$).

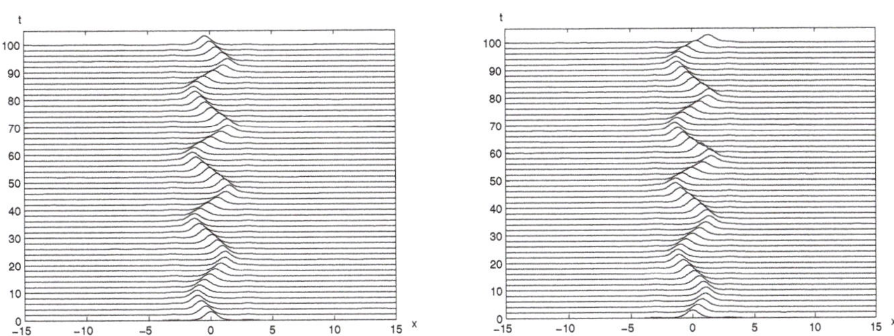

Figure 6. The time evolution of the trapped solitary wave solution is shown for $h(x, 3; 1, 1)$ and $\lambda = 1.5$ with $+5\%$ perturbation (**left**) and -5% perturbation (**right**), respectively. The perturbed trapped solitary waves bounce between two bumps for a long time (shown up to $t = 100$).

3.3. The Impact of the Bump Size on the Stability of $\eta_T(x)$

The impact of the size of two bumps on the time evolutions for trapped solitary waves has been explored. Let us consider our baseline bump size to be 1 (the bump size is defined as the maximum height of the bump at the center, which is $P_1 = P_2$). Now, the bump size is varied from 0.5 to 2. As shown in Figure 7, each stationary trapped solitary wave has the shape of $\eta_2(x)$ around $x = 0$ and the near zero wave over each bump. Also, the distance between two bumps is varied as well; the left panel shows the result with the bump distance 2 and the right panel with the bump distance 4. The amplitude of near zero wave gets higher as the bump size becomes larger in both panels.

The time evolutions of the perturbed trapped solitary wave solution using $h(x, 2; 0.5, 0.5)$ with $+5\%$ and -5% are displayed in Figure 8. Note that the perturbed trapped solitary wave solution stays much longer between two bumps when comparing with the results in Figure 4. That implies that the trapped solitary wave with the bump size 0.5 is more stable than the ones with the bump size 1. Now, using the larger bump size, Figure 9 illustrates that the perturbed trapped solitary wave solutions with the bump size 2 moves out of two bumps quickly than the results in Figure 4. This indicates that there is a critical bump size for trapped solitary waves to remain stable for a longer time. Therefore, we investigate the relationship between the bump size and the time until trapped solitary waves remain stable.

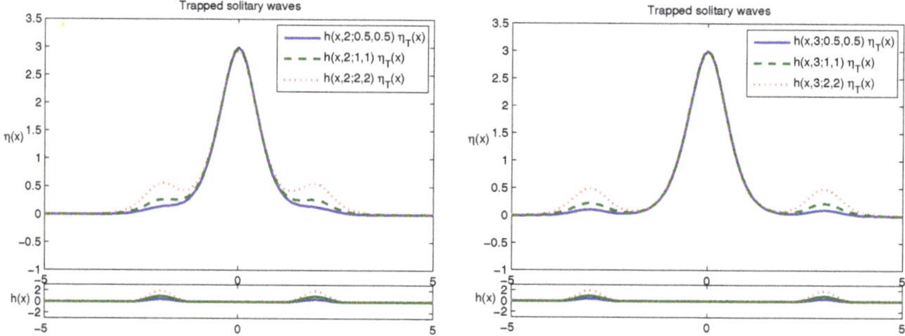

Figure 7. Stationary trapped solitary wave solutions are displayed using three different bump sizes, $P_1 = P_2 = 0.5$, $P_1 = P_2 = 1$ and $P_1 = P_2 = 2$ when the distance between two bumps is 2 (**left**) and 4 (**right**), respectively.

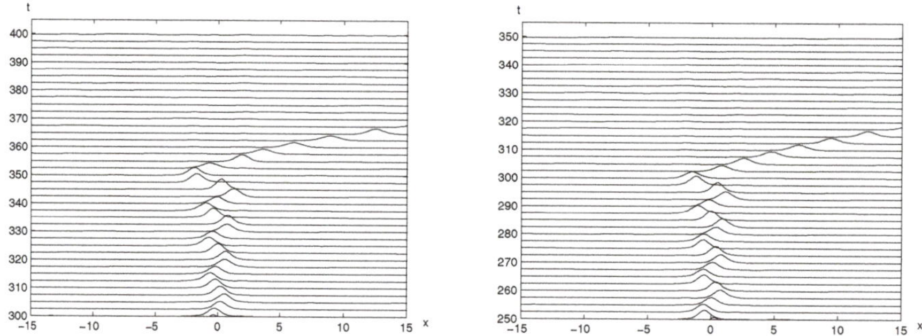

Figure 8. The time evolution of the trapped solitary wave solution using $h(x, 2; 0.5, 0.5)$ and $\lambda = 1.5$ with $+5\%$ perturbation (**left**) and -5% perturbation (**right**), respectively. They start moving between two bumps and evolve out of two bumps around $t = 340$ and $t = 300$, respectively.

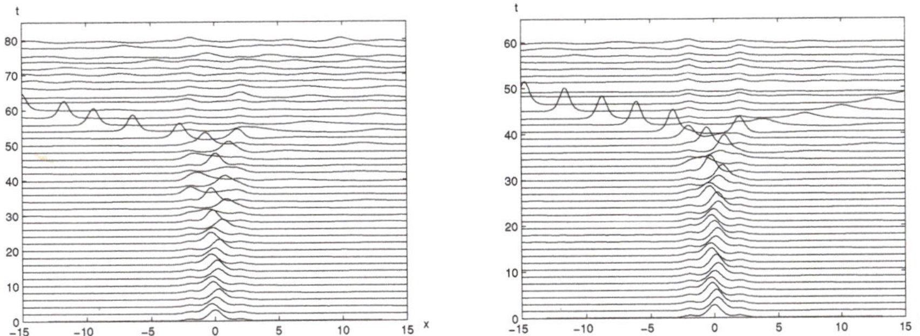

Figure 9. The time evolution of the trapped solitary wave solution using $h(x, 2; 2, 2)$ and $\lambda = 1.5$ with $+5\%$ perturbation (**left**) and -5% perturbation (**right**), respectively. They start moving between two bumps and evolve out of two bumps around $t = 50$ and $t = 40$, respectively.

The left panel of Figure 10 displays the time when the trapped solitary wave solutions move out of two bumps as the bump size is varied. The time is measured when trapped solitary waves under $+5\%$ (solid) or -5% (dotted) perturbation start moving out of two bumps. The time decreases when the bump size increases from 0.5 to 2 so that it takes the minimum around the bump size 2, then the time increases again significantly as the bump size increases. Next, we investigate the relationship between the bump distance and the time until trapped solitary waves remain stable. The right panel of Figure 10 illustrates the time when the trapped solitary wave solutions move out of two bumps as the bump distance is varied. Again, the time is measured when trapped solitary waves under $+5\%$ (solid) or -5% (dotted) perturbation start moving out of two bumps. As seen in the right panel, the time increases as the distance between two bumps increases linearly.

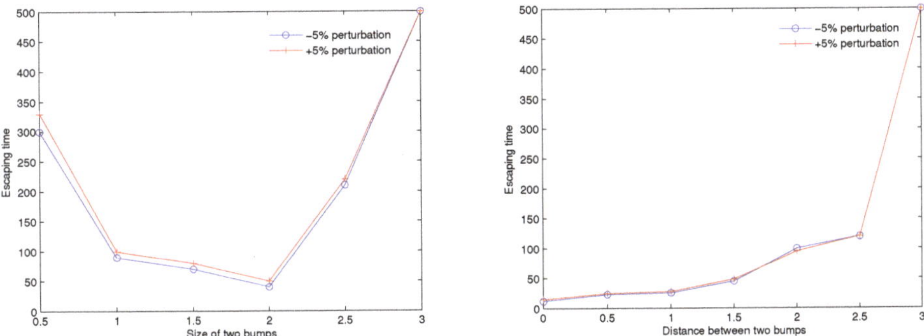

Figure 10. The impact of the bottom configurations is shown on the time when perturbed trapped solitary waves start evolving out of between two bumps (the left panel for the bump size and the right panel for the bump distance).

3.4. Trapped Solitary Waves between Two Negative Holes

In this subsection, we present the dynamics of trapped solitary waves when two negative holes are given as the bottom configuration. As studied in [19], there are five stationary wave solutions including the near zero wave (a negative solitary wave) and the near $\eta_2(x)$ wave in the presence of one negative bump or hole. Unlike the solitary wave of a positive bump, the near zero solitary wave exists for all $\lambda > 0$ when a negative bump is given. Moreover, the near zero wave is stable and the near $\eta_2(x)$ wave is unstable when they evolve in time [19]. In the presence of two symmetric negative holes, the number of stationary solitary wave solutions increase (the larger the distance between two holes, the more various trapped solitary waves [21]).

First, stationary trapped solitary wave solutions between two holes are obtained as the distance between two holes is varied. In Figure 11, three stationary trapped solitary wave solutions are displayed under three different bump distance using $h(x,2;-1,-1)$, $h(x,3;-1,-1)$ and $h(x,4;-1,-1)$, respectively. Each trapped solitary wave has the shape of $\eta_2(x)$ around $x=0$ and the near zero negative wave over each hole. Next, Figure 12 illustrates stationary trapped solitary waves as the hole depth is varied, -0.5, -1 and -2. Further, the left panel of Figure 12 shows the trapped solitary waves using the hole distance 2 and the right panel shows the trapped solitary waves using the hole distance 4.

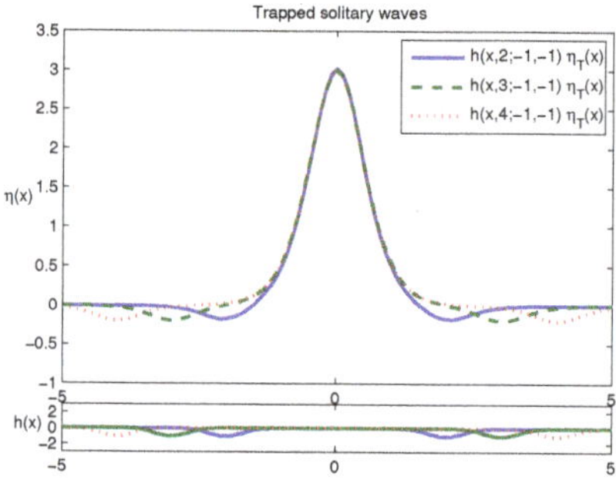

Figure 11. Stationary trapped solitary wave solutions are shown using three two-hole configurations with three different distances, $h(x,2;-1,-1)$, $h(x,3;-1,-1)$ and $h(x,4;-1,-1)$, respectively.

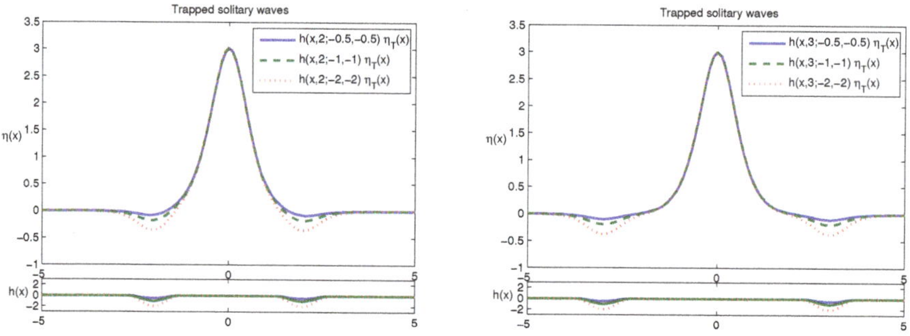

Figure 12. Stationary trapped solitary wave solutions are displayed using three different hole sizes, $P_1 = P_2 = -0.5$, $P_1 = P_2 = -1$ and $P_1 = P_2 = -2$ when the distance between two holes is 2 (**left**) and 4 (**right**), respectively.

Figures 13 and 14 illustrate the time evolutions of unperturbed trapped solitary wave solutions. As seen in both figures, the trapped solitary waves move out of two holes much faster than the cases of two symmetric positive bumps. In Figure 13, the trapped solitary wave stays between two holes up to $t = 100$ with $h(x,3;-1,-1)$ (right) while it stays only up to $t = 20$ with $h(x,2;-1,-1)$ (left). This is consistent with the results for two positive bumps; the larger the distance between two holes, the longer the trapped solitary wave solution remains stable between two holes. Similarly, as we vary the depth of two holes, the time evolutions of trapped solitary waves are changing as well in a complex way. As the depth of two holes is reduced half $h(x,3;-0.5,-0.5)$, the trapped solitary wave remains between two holes for a longer time (see the left panel of Figure 14). The depth of two holes is increased double $h(x, 3; -2, -2)$, the trapped solitary wave remains between two holes for a shorter time (see the right panel of Figure 14). Again, this is consistent with the results for two positive bumps.

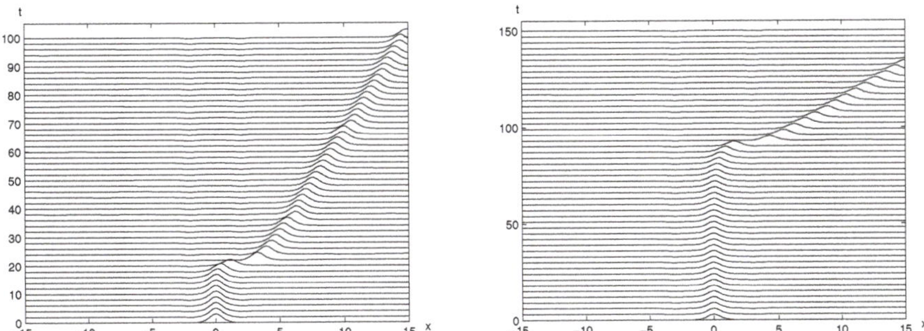

Figure 13. The time evolution of trapped solitary wave solution is displayed with $\lambda = 1.5$ and $h(x, 2; -1, -1)$ **(left)** and $h(x, 3; -1, -1)$ **(right)**, respectively. The unperturbed trapped solitary wave remains stable between two holes for a very short time $t = 20$ (see the **left** panel). It remains for a longer time when the distance increases (stable up to $t = 100$ in the **right** panel).

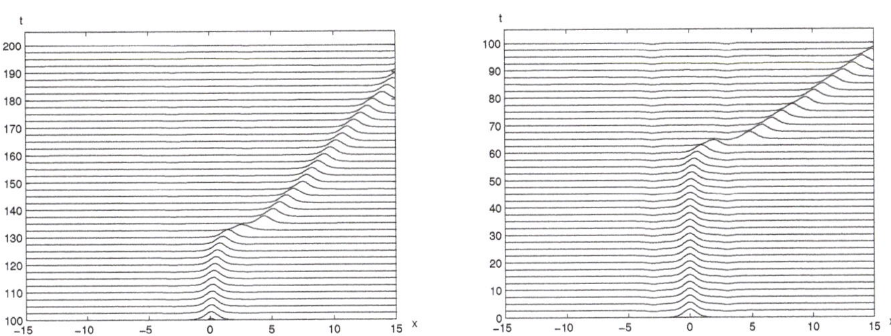

Figure 14. The time evolution of trapped solitary wave solution with $\lambda = 1.5$ and $h(x, 3; -0.5, -0.5)$ **(left)** and $h(x, 3; -2, -2)$ **(right)**, respectively. The unperturbed trapped solitary wave remains stable between two holes up to $t = 120$ (see the **left** panel). It remains for a shorter time when the depth of holes increases to -2 (stable up to $t = 60$ in the **right** panel).

4. Conclusions

In this work, the focus has been on the trapped solitary wave solutions of the forced KdV and their numerical stability in the presence of two bumps. We use the Newton method incorporating artificial boundary conditions to find various trapped solitary waves in the presence of two positive bumps and two negative holes. This method provides a good alternative to find various stationary wave solutions under arbitrary two-bump configurations when they are close enough (the distance between two bumps are not large) [21]. The semi-implicit finite difference method has been employed to investigate the numerical stability of these trapped solitary waves. Our previous results indicate that the number of stationary wave solutions increases as the bump distance increases and there exist multiple trapped solitary waves under different two-bump configurations [22]. In this study, we have extended the results under more various two bump configurations. Specifically, the impact of the bump distance and the bump size on numerical stability of trapped solitary waves have been investigated.

For both two positive bumps and two negative holes, the near zero solution is stable when it evolves in time. Moreover, it is worth to observe that a *trapped wave solution* remains stable for a very longer time between two positive bumps. As the distance between two bumps increases, trapped solitary waves stay longer in a straightforward fashion. On the other hand, the relationship for the bump size and the time trapped solitary waves to remain between the bumps is not trivial. There is a critical bump size for the trapped solitary to be stable for a longer time. Furthermore, for two negative holes, the trapped solitary waves do not remain for a long time as two positive bumps are given. Their time evolutions are characterized by the interplay between trapped solitary waves and two-bump configurations.

However, for non-symmetric bumps (two bumps or two holes with different sizes), trapped solitary waves become non-symmetric as well and they move out of two bumps in a much shorter time. Non-symmetry becomes even more significant in stationary solitary waves for the combined bottom configuration of a bump and a hole. They lose stability too fast to measure the length of time where solitary waves are stable. This highlights the importance of symmetry in the bumps and solitary waves for their stability. We find that the interplay between trapped solitary waves and two bumps becomes stronger so that they remain stable up to a certain time when two positive bumps are given. Hence, we can conjecture that two positive bumps provide a trap for some trapped wave solutions to be stable for a certain finite time.

Acknowledgments: This work was supported by the National Research Foundation of Korea (NRF) grant funded by the Korean government (MSIP) (NRF-2015R1C1A2A01054944).

Conflicts of Interest: The authors declare no conflict of interest.

Abbreviations

The following abbreviations are used in this manuscript:

MDPI	Multidisciplinary Digital Publishing Institute
DOAJ	Directory of open access journals
TLA	Three letter acronym
LD	linear dichroism

References

1. Camassa, R.; Wu, T. Stability of forced solitary waves. *Philos. Trans. R. Soc. Lond. A* **1991**, *337*, 429–466.
2. Zabuski, N.J.; Kruskal, M.D. Interaction of "solitons" in a collisionless plasma and the recurrence of initial states. *Phys. Rev. Lett.* **1965**, *15*, 240–243.
3. Dias, F.; Vanden-Broeck, J.M. Generalized critical free-surface flows. *J. Eng. Math.* **2002**, *42*, 291–301.
4. Shen, S.S. On the accuracy of the stationary forced Korteweg-De Vries equation as a model equation for flows over a bump. *Q. Appl. Math.* **1995**, *53*, 701–719.
5. Crighton, D.G. Applications of KdV. *Acta Appl. Math.* **1995**, *39*, 39–67.
6. Hereman, W. Shallow Water Waves and Solitary Waves. In *Mathematics of Complexity and Dynamical Systems*; Meyers, R., Ed.; Springer: New York, NY, USA, 2012.
7. Choi, J.W.; Sun, S.M.; Whang, S.I. Supercritical surface gravity waves generated by a positive forcing. *Eur. J. Mech. B Fluids* **2008**, *27*, 750–770.
8. Shen, S.S.; Shen, M.C.; Sun, S.M. A model equation for steady surface waves over a bump. *J. Eng. Math.* **1989**, *23*, 315–323.
9. Shen, S.S.; Manohar, R.P.; Gong, L. Stability of the lower cusped solitary waves. *Phys. Fluids* **1995**, *7*, 2507–2509.
10. Forbes, L.K. Critical free-surface flow over a semi-circular obstruction. *J. Eng. Math.* **1988**, *22*, 3–13.
11. Grimshaw, R.; Maleewong, M. Stability of steady gravity waves generated by a moving localised pressure disturbance in water of finite depth. *Phys. Fluids* **2013**, *25*, 006705. [CrossRef]
12. Pratt, L.J. On nonlinear flow with multiple obstructions. *Geophys. Astrphys. Fluid Dyn.* **1984**, *41*, 1214–1225.
13. Dias, F.; Vanden-Broeck, J.M. Trapped waves between submerged obstacles. *J. Fluid Mech.* **2004**, *509*, 93–102.

14. Binder, B.J.; Dias, F.; Vanden-Broeck, J.M. Influence of rapid changes in a channel bottom on free-surface flows. *IMA J. Appl. Math.* **2008**, *73*, 254–273.

15. Binder, B.J.; Blyth, M.G.; McCue, S.W. Free-surface flow past arbitrary topography and an inverse approach for wave-free solutions. *IMA J. Appl. Math.* **2013**, *78*, 685–696. [CrossRef]

16. Ee, B.K.; Grimshaw, R.H.J.; Zang, D.H.; Chow, K.W. Steady transcritical flow over a hole: Parametric map of solutions of the forced Korteweg-de Vries equation. *Phys. Fluids* **2010**, *22*, 056602. [CrossRef]

17. Chardard, F.; Dias, F.; Nguyen, H.Y.; Vanden-Broeck, J.M. Stability of some stationary solutions to the forced KdV equation with one or two bumps. *J. Eng. Math.* **2011**, *70*, 175–189.

18. Gong, L.; Shen, S.S. Multiple wave solutions of the staionary forced Korteweg-De Vries equation and their stability. *SIAM J. Appl. Math.* **1994**, *54*, 1268–1290.

19. Choi, J.W.; Lin, T.; Sun, S.M.; Whang, S.I. Supercritical surface waves generated by negative or oscillatory forcing. *Discret. Contin. Dyn. Syst. B* **2010**, *14*, 1313–1335.

20. Choi, J.W. Free surface waves over a depression. *Bull. Aust. Math. Soc.* **2002**, *65*, 329–335.

21. Whang, S.I.; Lee, S.M. Absorbing boundary conditions for the stationary forced KdV equation. *Appl. Math. Comput.* **2008**, *202*, 511–519.

22. Lee, S.M.; Whang, S.I. Trapped Supercritical wave for the forced KdV equation with two bumps. *Appl. Math. Model.* **2014**, *39*, 2649–2660. [CrossRef]

23. Feng, B.; Mitsui, T. A finite difference method for the Korteweg-de Vries and the Kadomtsev-Petviashvili equations. *J. Comput. Appl. Math.* **1998**, *90*, 95–116.

24. Djidjeli, K.; Price, W.; Twizell, E.; Wang, Y. Numerical methods for the solution of the third- and fifth-order dispersive Korteweg-de Vries equations. *J. Comput. Appl. Math.* **1995**, *58*, 307–336.

MDPI

St. Alban-Anlage 66

4052 Basel

Switzerland

Tel. +41 61 683 77 34

Fax +41 61 302 89 18

www.mdpi.com

Symmetry Editorial Office

E-mail: symmetry@mdpi.com

www.mdpi.com/journal/symmetry